A Practical Course in Differential Equations and Mathematical Modelling

Classical and New Methods
Nonlinear Mathematical Models
Symmetry and Invariance Principles

A Practical Course in Differential Equations and Mathematical Modelling

Classical and New Methods
Nonlinear Mathematical Models
Symmetry and Invariance Principles

Nail H. Ibragimov

Blekinge Institute of Technology, Sweden

Higher Education Press

World Scientific

NEW JERSEY · LONDON · SINGAPORE · BEIJING · SHANGHAI · HONG KONG · TAIPEI · CHENNAI

Nail H. Ibragimov
Department of Mathematics and Science
Blekinge Institute of Technology
S-371 79 Karlskrona, Sweden

Copyright © 2010 by
Higher Education Press
4 Dewai Dajie, 100011, Beijing, P.R. China and
World Scientific Publishing Co Pte Ltd
5 Toh Tuch Link, Singapore 596224

ISBN 13: 978-981-4291-94-1
ISBN 10: 981-4291-94-3

Preface

Modern mathematics has over 300 years of history. From the very beginning, it was focused on differential equations as a major tool for mathematical modelling. Most of mathematical models in physics, engineering sciences, biomathematics, etc. lead to nonlinear differential equations.

Today's engineering and science students and researchers routinely confront problems in mathematical modelling involving solution techniques for differential equations. Sometimes these solutions can be obtained analytically by numerous traditional ad hoc methods appropriate for integrating particular types of equations. More often, however, the solutions cannot be obtained by these methods, in spite of the fact that, e.g. over 400 types of integrable second-order ordinary differential equations were accumulated due to ad hoc approaches and summarized in voluminous catalogues.

On the other hand, the fundamental natural laws and technological problems formulated in terms of differential equations can be successfully treated and solved by Lie group methods. For example, Lie group analysis reduces the classical 400 types of equations to 4 types only! Development of group analysis furnished ample evidence that the theory provides a universal tool for tackling considerable numbers of differential equations even when other means of integration fail. In fact, group analysis is the only universal and effective method for solving nonlinear differential equations analytically. The old integration methods rely essentially on linearity as well as on constant coefficients. Group analysis deals equally easily with *linear and nonlinear* equations, as well as with constant and variable coefficients. For example, from the traditional point of view, the linear equation

$$\frac{\mathrm{d}^n y}{\mathrm{d} x^n} + a_1 \frac{\mathrm{d}^{n-1} y}{\mathrm{d} x^{n-1}} + \cdots + a_{n-1} \frac{\mathrm{d} y}{\mathrm{d} x} + a_n\, y = 0$$

with constant coefficients $a_1, \ldots, a_n$ is different from the equation

$$\bar{x}^n \frac{\mathrm{d}^n \bar{y}}{\mathrm{d} \bar{x}^n} + a_1\, \bar{x}^{n-1} \frac{\mathrm{d}^{n-1} \bar{y}}{\mathrm{d} \bar{x}^{n-1}} + \cdots + a_{n-1}\, \bar{x}\, \frac{\mathrm{d} \bar{y}}{\mathrm{d} \bar{x}} + a_n\, \bar{y} = 0$$

known as *Euler's equation*. From the group standpoint, however, these equations are merely two different representations of one and the same equation with two known commuting symmetries, namely,

$$X_1 = \frac{\partial}{\partial x}, \quad X_2 = y \frac{\partial}{\partial y} \quad \text{and} \quad \overline{X}_1 = \bar{x} \frac{\partial}{\partial \bar{x}}, \quad \overline{X}_2 = \bar{y} \frac{\partial}{\partial \bar{y}}$$

for the first and second equation, respectively. These symmetries span two similar Lie algebras and readily lead to the transformation $x = \ln |\bar{x}|$ converting Euler's equation to the equation with constant coefficients.

I believe that Lie groups are interesting first of all due to their utilization for solving differential equations. It was a mistake to isolate them from this natural application and treat as a branch of abstract mathematics. *"To isolate mathematics from the practical demands of the sciences is to invite the sterility of a cow shut away from the bulls"* (P.L. Chebyshev, 1821~1894).

Today group analysis is becoming part of curricula in differential equations and nonlinear mathematical modelling and attracts more and more students. For example, the course in *Partial Differential Equations* at Moscow Institute of Physics and Technology attracted more than 100 students when I used Lie group methods, instead of 10 students that we had in the traditional course. The same happened when I delivered similar lectures for science students in South Africa and Sweden.

The present text is based on these lectures and reflects, to a certain extent, my own taste and experience. Primarily, it has been designed for the course in differential equations delivered at the Blekinge Institute of Technology for engineering, mathematics and science students. Then the text has been revised, enlarged and is used now in the following courses:

Differential equations: The course covers both ordinary and partial differential equations; it combines basic classical methods, mainly for linear equations, with new methods for solving nonlinear equations analytically; designed for beginners; students learn how to find symmetries of differential equations by solving determining equations.

Analytical methods in mathematical modelling: The emphasis in this course is on nonlinear mathematical models in physics, biology and engineering sciences; the course covers such topics as nonlinear superposition, symmetry and conservation laws, group invariant solutions.

Group analysis of differential equations: The course introduces students of mathematics and engineering to those areas of the theory of transformations groups and Lie algebras which are most important in practical applications; during the course, students develop analytic skills in modern methods for solving nonlinear ordinary and partial differential equations.

Distributions and invariance principle in initial value problems: An easy to follow introduction to basic concepts of the distribution theory with emphasis on useful tools; Lie's infinitesimal technique is extended to the space of distributions and used, together with an invariance principle, for calculating fundamental solutions and solving initial value problems for equations with constant and variable coefficients.

In my presentation, I have striven to make the group analysis of differential equations more accessible for engineering and science students. Therefore, the emphasis in this book is on applications of known symmetries rather than on their computation. In order to formulate the essence of my experience in solving various types of differential equations, I rephrase the famous French aphorism *cherchez la femme* as follows:

If you cannot solve a nonlinear differential equation, cherchez le groupe.

My sincere thanks are due to my colleague Claes Jogréus for his lasting help. My wife Raisa read the manuscript at various stages of completion of the second edition, corrected misprints and contributed numerous valuable criticisms, for which I make grateful acknowledgement. It is also a pleasure to thank my daughters Sania and Alia for several helpful comments.

Karlskrona, 3 March 2009   Nail H. Ibragimov

Contents

Chapter 1

Selected topics from analysis

This preparatory chapter is designed to meet the needs of beginners and provides a background in elementary mathematics and mathematical analysis which is necessary in the succeeding parts of the book.

Additional reading: E. Goursat [10].

1.1 Elementary mathematics

1.1.1 Numbers, variables and elementary functions

Real numbers appear in our practical activities (e.g., while measuring distances, weights, etc.) as approximate decimal numbers. For example, the distance to the moon at perigee is S km, where the number S is approximately equal to 356630. A more accurate estimation of the distance is 356629 km and 744 m. Hence,

$$S \approx 356629.744 \equiv 356629 + \frac{744}{1000} = 356629 + \frac{7}{10} + \frac{4}{100} + \frac{4}{1000}.$$

If one will continue further, one will get even better approximations and obtain a representation of the number S as an infinite decimal. Thus, we use the following definition.

Definition 1.1.1. *Real numbers* are identified with infinite decimals

$$a = a_0.a_1a_2\ldots a_n\ldots, \tag{1.1.1}$$

where a_0 is an integer, and $a_1, a_2, \ldots, a_n \ldots$ are digits, i.e., they can assume any of ten Arabic number symbols, 0 through 9. Eq. (1.1.1) means that

$$a = a_0 + \frac{a_1}{10} + \frac{a_2}{100} + \cdots + \frac{a_n}{10^n} + \cdots. \tag{1.1.2}$$

Remark 1.1.1. If (1.1.1) is a periodical decimal, and only in this case, a is a *rational* number, i.e., $a = p/q$, where p and q are integers, $q \neq 0$. The real numbers determined by non-periodical infinite decimals are termed *irrational* numbers. The numbers 0. (9)= 0. 9999... and 1 are identified.

Example 1.1.1. Famous examples of irrational numbers are:

$$\sqrt{2} = 1.4142136\ldots \qquad \approx 1.41$$
$$\pi = 3.1415926535\ldots \qquad \approx 3.14$$
$$e = 2.718281828459045\ldots \approx 2.72$$
$$\gamma = \lim_{n\to\infty} \left(1 + \frac{1}{2} + \cdots + \frac{1}{n} - \ln n\right) \approx 0.58$$

where γ is known as *Euler's number.*

Remark 1.1.2. It is a historical accident that we represent real numbers in the decimal system. If Babylonian culture would last much longer we would probably use the Babylonian sexagesimal system and employ, instead of (1.1.2), the representation

$$a = a_0 + \frac{a_1}{60} + \frac{a_2}{60^2} + \cdots + \frac{a_n}{60^n} + \cdots . \tag{1.1.3}$$

Definition 1.1.2. A *variable* x is a quantity to which any numerical value can be assigned. A quantity with a fixed value is called a *constant*. One should distinguish *arbitrary* constants from *absolute* constants. An arbitrary constant retains any given value throughout the investigation, while an absolute constant retains the same value in all problems.

Example 1.1.2. In the equation of a circle, $x^2 + y^2 = R^2$, x and y are variables representing the coordinates of a point moving along the circle, while the radius R is an arbitrary constant. On the other hand, the formula $C = 2\pi R$ for the circumference of the circle contains, along with the arbitrary constant R, two absolute constants, 2 and $\pi \approx 3.14$.

Theorem 1.1.1. Any real number a is a limit of a sequence of rational numbers $r_n = p_n/q_n$, where p_n and $q_n \neq 0$ are integers:

$$a = \lim_{n\to\infty} r_n. \tag{1.1.4}$$

Proof. Let the real number a be given by Eq. (1.1.1). We take for r_n the finite sums of the corresponding infinite series (1.1.2):

$$r_1 = a_0 + \frac{a_1}{10}, \quad r_2 = a_0 + \frac{a_1}{10} + \frac{a_2}{100}, \quad \ldots, \quad r_n = a_0 + \frac{a_1}{10} + \frac{a_2}{100} + \cdots + \frac{a_n}{10^n}.$$

They provide a sequence of rational numbers $\{r_n\}$ satisfying Eq. (1.1.4).

The following definition is based on Theorem 1.1.1.

Definition 1.1.3. The exponential function $y = a^x$, where $a > 0$ is any real number, is defined by the following equations:

$$a^0 = 1, \quad a^1 = a, \quad a^n = \underbrace{a \cdots a}_{n}, \quad n = 2, 3, \dots, \qquad \text{(here } x = n\text{);}$$

$$a^{\frac{1}{n}} \equiv \sqrt[n]{a} = b \Leftrightarrow b^n = a; \quad a^{\frac{p}{q}} = \sqrt[q]{a^p}, \qquad \text{(here } x = p/q\text{);}$$

$$a^x = \lim_{n \to \infty} a^{x_n} \equiv \lim_{n \to \infty} \sqrt[q_n]{a^{p_n}}, \quad \text{(here } x = \lim_{n \to \infty} x_n, \ x_n = p_n/q_n\text{)} \cdot$$

Basic laws of exponents:

$$a^{-x} = \frac{1}{a^x}, \quad a^x a^y = a^{x+y}, \quad (ab)^x = a^x b^x, \quad (a^x)^y = a^{xy}.$$

Example 1.1.3. Consider the number $10^{\sqrt{2}} = 25.954\dots$. We have $\sqrt{2} = \lim_{n \to \infty} x_n$, where $x_0 = 1$, $x_1 = 1.4$, $x_2 = 1.41, \dots$. Accordingly, $10^{\sqrt{2}} = \lim_{n \to \infty} y_n$, where $y_0 = 10^{x_0} = 10$, $y_1 = 10^{x_1} \approx 25.12$, $y_2 = 10^{x_2} \approx 25.70, \dots$.

In solving differential equations, one often encounters the exponential function

$$y = e^x. \tag{1.1.5}$$

Here e is a real number determined by one of the most important limits in mathematical analysis:

$$e = \lim_{n \to \infty} \left(1 + \frac{1}{n}\right)^n. \tag{1.1.6}$$

Its value, accurate to fifteen decimal places, is given in Example 1.1.1.

Function (1.1.5) is a representative of so-called *elementary functions* defined as follows.

Definition 1.1.4. The *basic elementary functions* are:

$$y = C, \quad C = \text{const.};$$
$$y = x^\alpha, \quad \text{where } x > 0, \ \alpha \text{ is a real number;}$$
$$y = a^x, \quad \text{where } a > 0, \ a \neq 1;$$
$$y = \log_a x, \quad \text{where } a > 0, \ a \neq 1; \quad x > 0;$$
$$y = \sin x, \ y = \cos x, \ y = \operatorname{tg} x \ (\equiv \tan x), \ y = \operatorname{ctg} x;$$
$$y = \arcsin x, \ y = \arccos x, \ y = \operatorname{arctg} x, \ y = \operatorname{arcctg} x.$$

A function $y = f(x)$ is called an *elementary function* if it is obtained from the basic elementary functions by a finite number of operations involving *addition, subtraction, multiplication, division,* and *superposition.*

Remark 1.1.3. The logarithm $\log_a x$ with $a = e$ is called the *natural logarithm* and denoted by $\ln x$.

Remark 1.1.4. The basic trigonometric functions can be obtained from one of them, e.g., from $\sin x$, in combination with other basic elementary functions. Indeed,

$$\cos x = \sqrt{1 - \sin^2 x}, \quad \tan x = \frac{\sin x}{\sqrt{1 - \sin^2 x}}, \quad \cot x = \frac{\sqrt{1 - \sin^2 x}}{\sin x}.$$

Similar relations exist between inverse trigonometric functions as well, e.g.

$$\arcsin x = \arctan\frac{x}{\sqrt{1 - x^2}}. \tag{1.1.7}$$

Example 1.1.4. The following hyperbolic functions provide examples of non-basic elementary functions:

$$\sinh x = \frac{e^x - e^{-x}}{2}, \quad \cosh x = \frac{e^x + e^{-x}}{2}, \quad \tanh x = \frac{e^x - e^{-x}}{e^x + e^{-x}}. \tag{1.1.8}$$

Elementary functions of many variables are obtained in a similar way.

Example 1.1.5. The following function $\psi(t, x, z)$ is an elementary function of three variables, t, x, z :

$$\psi = -\frac{1}{4} \ln \left| M + \frac{1}{t}\left(2 \sin^2 x + l_1 e^{-z} \sin x + l_2 e^{-2z} \right) \right| \tag{1.1.9}$$

involving three arbitrary constants, l_1, l_2 and M.

Example 1.1.6. The following functions that often occur in applications are given by integrals and are not elementary:

$$\mathrm{Si}(x) = \int_0^x \frac{\sin t}{t}\, dt \quad \text{(the integral sine)}, \tag{1.1.10}$$

$$\mathrm{Ci}(x) = -\int_x^\infty \frac{\cos t}{t}\, dt \quad \text{(the integral cosine)}, \tag{1.1.11}$$

$$\mathrm{erf}(x) = \frac{2}{\sqrt{\pi}} \int_0^x e^{-t^2}\, dt \quad \text{(the error function)}, \tag{1.1.12}$$

$$\mathrm{Ei}(x) = -\int_{-\infty}^x \frac{e^t}{t}\, dt, \quad \mathrm{li}(x) = \int_0^x \frac{dt}{\ln t} \equiv \mathrm{Ei}(\ln x), \tag{1.1.13}$$

$$\Gamma(x) = \int_0^\infty e^{-t}\, t^{x-1} dt \quad \text{(the Gamma function)}. \tag{1.1.14}$$

The Gamma function plays an important part in analysis and differential equations. It has interesting general properties, e.g.

$$\Gamma(x + 1) = x\Gamma(x), \quad \Gamma(x)\Gamma(1 - x) = \frac{\pi}{\sin(\pi x)}, \tag{1.1.15}$$

and the remarkable numerical values (see, e.g. [34]):

$$\Gamma(1) = 1, \quad \Gamma\left(\frac{1}{2}\right) = \sqrt{\pi}, \quad \Gamma\left(\frac{n}{2}\right) = \frac{2\pi^{n/2}}{\omega_n}, \quad \Gamma(n + 1) = n!, \tag{1.1.16}$$

where ω_n is the surface area of the unit sphere in n dimensions.

1.1.2 Quadratic and cubic equations

Problems of elementary mathematics can often be solved by the method of transformations. Let us begin with elementary algebra.

Recall that the roots $x = x_1$ and $x = x_2$ of the general quadratic equation

$$ax^2 + bx + c = 0, \quad a \neq 0, \tag{1.1.17}$$

are given by

$$x_{1,2} = \frac{-b \pm \sqrt{b^2 - 4ac}}{2a}. \tag{1.1.18}$$

The expression

$$\Delta = b^2 - 4ac \tag{1.1.19}$$

is known as the *discriminant* of the quadratic equation (1.1.17). It is manifest from (1.1.18) that the vanishing of discriminant (1.1.19),

$$\Delta = b^2 - 4ac = 0, \tag{1.1.20}$$

is the condition for Eq. (1.1.17) to have two equal roots, $x_1 = x_2$.

In accordance with tradition, students learn from school to derive solution (1.1.18) by completing the square. Indeed, this method is simple but it is not suitable for tackling the general cubic as well as equations of higher degrees.

The idea of transformation of equations, unlike the method of completing the square appropriate only for the quadratic equation, furnishes a general method appropriate for solution of the quadratic equation as well as for a simplification of equations of higher degrees. The simplest transformation of equations is provided by a linear transformation of the variable x :

$$y = x + \varepsilon. \tag{1.1.21}$$

It converts any equation of degree n into an equation of the same degree. In particular, the quadratic equation (1.1.17) after the substitution $x = y - \varepsilon$ becomes $ay^2 + (b - 2a\varepsilon)y + a\varepsilon^2 - b\varepsilon + c = 0$. Hence, transformation (1.1.21) converts (1.1.17) into a new quadratic equation,

$$\bar{a}y^2 + \bar{b}y + \bar{c} = 0,$$

where

$$\bar{a} = a, \quad \bar{b} = b - 2a\varepsilon, \quad \bar{c} = c + a\varepsilon^2 - b\varepsilon. \tag{1.1.22}$$

Defining ε from $b - 2a\varepsilon = 0$, one obtains $\bar{b} = 0$ and $\bar{c} = c - b^2/(4a)$. Hence, the transformation

$$y = x + \frac{b}{2a} \tag{1.1.23}$$

converts (1.1.17) into the equation

$$ay^2 - \frac{b^2 - 4ac}{4a} = 0.$$

Substituting its roots

$$y_{1,2} = \pm \frac{\sqrt{b^2 - 4ac}}{2a}$$

in Eq. (1.1.23), one arrives at roots (1.1.18) of Eq. (1.1.17).

Consider now the general cubic equation written with binomial coefficients for convenience of calculations:

$$ax^3 + 3bx^2 + 3cx + d = 0, \quad a \neq 0. \tag{1.1.24}$$

After the linear transformation

$$y = ax + b \tag{1.1.25}$$

it takes the form

$$y^3 + 3py + 2q = 0, \tag{1.1.26}$$

where

$$p = ac - b^2, \quad 2q = a^2 d - 3abc + 2b^3. \tag{1.1.27}$$

The reduced equation (1.1.26) is readily solved by setting

$$y = \sqrt[3]{k} + \sqrt[3]{l}.$$

Then

$$y^3 - 3\sqrt[3]{kl}\, y - (k + l) = 0$$

and Eq. (1.1.26) yields

$$k + l = -2q, \quad \sqrt[3]{kl} = -p.$$

It follows that k and l are the roots of the quadratic equation

$$z^2 + 2qz - p^3 = 0.$$

Hence one of the roots, e.g. k, is given by

$$k = -q + \sqrt{q^2 + p^3}\,.$$

Let

$$u = \sqrt[3]{-q + \sqrt{q^2 + p^3}}$$

be any one of the three values of this cube root. Then all three values of $\sqrt[3]{k}$ are given by u, ϵu, $\epsilon^2 u$. Here ϵ is an imaginary cube root of unity, i.e. $\epsilon^3 = 1$, and has the following form (see Section 1.2.6, Example 1.2.1):

$$\epsilon = \frac{-1 + i\sqrt{3}}{2}, \quad \text{where} \quad i = \sqrt{-1}.$$

The square of ϵ is the complex conjugate cube root of unity:

$$\epsilon^2 = \frac{-1 - i\sqrt{3}}{2}.$$

Since $\sqrt[3]{kl} = -p$, the corresponding values of $\sqrt[3]{l}$ are

$$-\frac{p}{u}, \quad -\frac{p}{u}\epsilon^2, \quad -\frac{p}{u}\epsilon.$$

They can be rewritten in the form v, $v\epsilon^2$, $v\epsilon$, where

$$v = \sqrt[3]{-q - \sqrt{q^2 + p^3}}.$$

Summing up, one arrives at what is called *Cardan's solution* for the cubic equation. Namely, the roots of (1.1.26) are given by

$$y_1 = u + v, \quad y_2 = \epsilon u + \epsilon^2 v, \quad y_3 = \epsilon^2 u + \epsilon v, \tag{1.1.28}$$

where

$$u = \sqrt[3]{-q + \sqrt{q^2 + p^3}}, \quad v = \sqrt[3]{-q - \sqrt{q^2 + p^3}}, \tag{1.1.29}$$

The expression

$$\Delta = q^2 + p^3 \tag{1.1.30}$$

is termed the *discriminant* of cubic (1.1.26). It follows from (1.1.29) that the vanishing of the discriminant is the condition for (1.1.26) to have two equal roots. The roots of the general cubic equation (1.1.24) are obtained by substituting (1.1.28) in Eq. (1.1.25) and invoking (1.1.27).

Example 1.1.7. The equation $y^3 - 6y + 4 = 0$ has form (1.1.26) with $p = -2$ and $q = 2$. Here $q^2 + p^3 = -4$, and hence (1.1.29) is written

$$u = \sqrt[3]{2(-1 + i)}, \quad v = \sqrt[3]{2(-1 - i)}.$$

The reckoning shows that $u = 1 + i$, $v = 1 - i$, and formulae (1.1.28) provide three distinct real roots:

$$y_1 = 2, \quad y_2 = -(1 + \sqrt{3}), \quad y_3 = -1 + \sqrt{3}. \tag{1.1.31}$$

Remark 1.1.5. The discriminant of the general cubic equation (1.1.24) has the form

$$(ad)^2 - 6abcd + 4ac^3 - 3(bc)^2 + 4b^3d = -9 \begin{vmatrix} 3a & 2b & c & 0 \\ 3b & 2c & d & 0 \\ 0 & a & 2b & 3c \\ 0 & b & 2c & 3d \end{vmatrix}. \tag{1.1.32}$$

The vanishing of discriminant (1.1.32) is the condition for Eq. (1.1.24) to have two equal roots. Furthermore, the condition for Eq. (1.1.24) to have three equal roots is provided by the following two equations (cf. (1.1.20)):

$$b^2 - ac = 0, \quad b^3 - a^2d = 0. \tag{1.1.33}$$

An invariant description of discriminant (1.1.32) and of Eq. (1.1.33) is given in [21], Section 10.1.3.

1.1.3   Areas of similar figures. Ellipse as an example

Definition 1.1.5. The transformations

$$\bar{x} = x\cos\theta + y\sin\theta + a_1, \quad \bar{y} = y\cos\theta - x\sin\theta + a_2, \tag{1.1.34}$$

composed of rotations and translations, do not alter distances between points, and hence areas of geometric figures in the (x, y) plane. Therefore transformations (1.1.34) are termed *isometric* or *rigid* motions in the (x, y) plane. In geometry, two figures are said to be *equal* if one can be mapped to another by an appropriate isometric motion.

Consider a *scaling transformation*

$$\bar{x} = ax, \quad \bar{y} = by, \tag{1.1.35}$$

known also as a *similarity transformation* or *dilation*. The arbitrary constants $a \neq 0$ and $b \neq 0$ are called the parameters of the transformation. The scaling transformation determines a *uniform* expansion (contraction) from the origin if $a = b$, and to a *non-uniform* expansion (contraction) otherwise.

Definition 1.1.6. Two geometric figures obtained one from another by a scaling transformation (1.1.35) are said to be *similar*.

Example 1.1.8. Any rectangle is similar to the unit square

$$0 \leq x \leq 1, \quad 0 \leq y \leq 1.$$

Indeed, given a rectangle with sides a and b, we first move it by a proper translation and rotation to the "standard location" so that it will have the form $\{0 \leq x \leq a, 0 \leq y \leq b\}$. Then the stretching $\bar{x} = x/a$, $\bar{y} = y/b$ converts the rectangle to the unit square $\{0 \leq \bar{x} \leq 1, 0 \leq \bar{y} \leq 1\}$.

Theorem 1.1.2. Let two plain geometric figures, $\mathcal{M}$ and $\overline{\mathcal{M}}$, be similar and let the latter be obtained from the former by a scaling transformation $\bar{x} = ax, \bar{y} = by$. Then the areas S and $\overline{S}$ of $\mathcal{M}$ and $\overline{\mathcal{M}}$, respectively, are related by

$$\overline{S} = abS. \tag{1.1.36}$$

Proof. Let us first consider a rectangle with sides m and n in the x and y directions, respectively. After the dilation, one obtains a rectangle with sides $\overline{m} = am$ and $\overline{n} = bn$. Hence, the areas of the original and new rectangles, $S = mn$ and $\overline{S} = \overline{m}\,\overline{n}$, are related by (1.1.36). One can cover an arbitrary figure (provided that it is not too fancy) by a grid of rectangular areas and apply (1.1.36) to these rectangles. Imagine now the process repeated over and over again with finer and finer grids. Since the area S of the figure in question is the limit of the sums of the areas of the covering rectangles, this completes the proof.

A good example is provided by the ellipse. An *ellipse*, in the (x, y) plane, is the locus of points the sum of whose distances from two fixed points (called the *foci* of the ellipse) is constant. Let us use the standard equation of an ellipse in the rectangular Cartesian coordinates:

$$\frac{x^2}{a^2} + \frac{y^2}{b^2} = 1, \tag{1.1.37}$$

where $a \neq 0$ and $b \neq 0$ are arbitrary constants called the *major* and *minor* semi-axes of the ellipse, respectively.

Theorem 1.1.2 furnishes an elementary method for calculating the area of ellipses. We note that any ellipse (in particular, a circle) is similar to the unit circle

$$x^2 + y^2 = 1.$$

Indeed, the stretching

$$\overline{x} = \frac{x}{a}, \quad \overline{y} = \frac{y}{b}$$

converts ellipse (1.1.37) to the unit circle

$$\overline{x}^2 + \overline{y}^2 = 1,$$

the area of the latter being $\overline{S} = \pi$. Now formula (1.1.36) is written

$$\overline{S} = \frac{S}{ab},$$

where S is the area of the ellipse. Thus, the area of ellipse (1.1.37) is

$$S = \pi ab. \tag{1.1.38}$$

1.1.4 Algebraic curves of the second degree

Straight lines, ellipses (in particular, circles), hyperbolas and parabolas provide commonly known examples of algebraic curves on the plane.

Examples of non-algebraic curves are trigonometric curves, e.g. the sine, cosine, tangent curves:

$$y = \sin x, \quad y = \cos x, \quad y = \tan x,$$

logarithmic, exponential and probability curves:

$$y = \ln x, \quad y = e^x, \quad y = e^{-x^2},$$

and a variety of spirals such as the spiral of Archimedes, the logarithmic, hyperbolic and parabolic spirals:

$$r = a\theta, \quad \ln r = a\theta, \quad r\theta = a, \quad (r - c)^2 = a\theta,$$

where $r = \sqrt{x^2 + y^2}$, $\theta = \arctan(y/x)$ and $a, c = $ const.

In general, algebraic curves on the (x, y) plane are defined by equations $P(x, y) = 0$, where $P(x, y)$ is a polynomial of any degree in two variables, x and y. Let us consider equations of the second degree:

$$Ax^2 + 2Bxy + Cy^2 + ax + by + c = 0, \tag{1.1.39}$$

where $A, \ldots, c$ are arbitrary constants. If $A = B = C = 0$, Eq. (1.1.39) reduces to the linear equation $ax + by + c = 0$ defining straight lines. We will discuss the curves of the second degree when Eq. (1.1.39) contains at least one of the quadratic terms Ax^2, $2Bxy$ and Cy^2.

The general linear transformation

$$x = \alpha \overline{x} + \beta \overline{y} + \mu, \quad y = \gamma \overline{x} + \delta \overline{y} + \nu \tag{1.1.40}$$

on the plane maps any Eq. (1.1.39) into an equation of the same form. We consider here invertible transformations (1.1.40), i.e., such that

$$\Delta = \alpha\delta - \beta\gamma \neq 0. \tag{1.1.41}$$

Definition 1.1.7. Two equations of form (1.1.39) related by a linear transformation (1.1.40) are said to be *equivalent*. The curves determined by equivalent equations are also termed equivalent curves.

We will classify the algebraic curves according to their equivalence. Note, that the linear transformation (1.1.40) is composed of the transformation

$$x = \alpha \overline{x} + \beta \overline{y}, \quad y = \gamma \overline{x} + \delta \overline{y} \tag{1.1.42}$$

called the *homogeneous* linear transformation and of the translation

$$x = \overline{x} + \mu, \quad y = \overline{y} + \nu. \tag{1.1.43}$$

The term *homogeneous* is due to the fact that the transformation (1.1.42) comprises the terms of the first degree in $\overline{x}, \overline{y}$ and therefore maps the principal part of Eq. (1.1.39), i.e., the quadratic form

$$Ax^2 + 2Bxy + Cy^2 \tag{1.1.44}$$

into a quadratic form again, namely into the quadratic form

$$\widetilde{A}\overline{x}^2 + 2\widetilde{B}\overline{x}\overline{y} + \widetilde{C}\overline{y}^2,$$

where

$$
\begin{aligned}
\widetilde{A} &= \alpha^2 A + 2\alpha\gamma B + \gamma^2 C, \\
\widetilde{B} &= \alpha\beta A + (\alpha\delta + \beta\gamma)B + \gamma\delta C, \\
\widetilde{C} &= \beta^2 A + 2\beta\delta B + \delta^2 C.
\end{aligned}
\tag{1.1.45}
$$

Furthermore, the translation (1.1.43) does not alter the quadratic terms of Eq. (1.1.39). Therefore, we start the classification of algebraic curves according to the equivalence of the quadratic forms (1.1.44) with respect to the linear homogeneous transformations (1.1.42).

First we observe that one can try to annul simultaneously the coefficients $\widetilde{A}$ and $\widetilde{C}$ in (1.1.45). Namely, we write the equations $\widetilde{A} = 0$ and $\widetilde{C} = 0$, dividing them by γ^2 and δ^2, respectively, and denoting the quantities α/γ and β/δ by λ, as the following quadratic equation:

$$A\lambda^2 + 2B\lambda + C = 0. \tag{1.1.46}$$

If $B^2 - AC \neq 0$, Eq. (1.1.46) has two distinct roots, $\lambda_1 \neq \lambda_2$, and one can let $\alpha/\gamma = \lambda_1$ and $\beta/\delta = \lambda_2$ since the equations $\alpha = \lambda_1\gamma, \beta = \lambda_2\delta$ with $\gamma \neq 0, \delta \neq 0$ are compatible with condition (1.1.41) since $\lambda_1 \neq \lambda_2$. In consequence, one will annul $\widetilde{A}$ and $\widetilde{C}$ simultaneously. The expression $B^2 - AC$ is called the *discriminant* of the quadratic form (1.1.44). The following statement is derived from Eq. (1.1.45).

Lemma 1.1.1. The homogeneous linear transformation (1.1.42) changes the discriminant as follows:

$$\widetilde{B}^2 - \widetilde{A}\widetilde{C} = \Delta^2(B^2 - AC), \tag{1.1.47}$$

where Δ is defined by (1.1.41).

According to Lemma 1.1.1, each of the following three conditions:

$$B^2 - AC > 0, \quad B^2 - AC = 0, \quad B^2 - AC < 0 \tag{1.1.48}$$

remains unaltered under transformation (1.1.42), and hence under the general linear transformation (1.1.40). Therefore, we will consider each case (1.1.48) separately.

Let $B^2 - AC > 0$. Then Eq. (1.1.46) has two distinct real roots

$$\lambda_1 = \frac{-B + \sqrt{B^2 - AC}}{A}, \quad \lambda_2 = \frac{-B - \sqrt{B^2 - AC}}{A}.$$

One annuls $\widetilde{A}$ and $\widetilde{C}$ simultaneously by letting $\alpha/\gamma = \lambda_1$ and $\beta/\delta = \lambda_2$. Then, letting, e.g. $\gamma = \delta = 1$, substituting $\alpha = \lambda_1$, $\beta = \lambda_2$ in (1.1.45) and invoking that $\lambda_1\lambda_2 = C/A$, $\lambda_1 + \lambda_2 = -2B/A$, one obtains

$$\widetilde{B} = \frac{2(AC - B^2)}{A} \neq 0.$$

Thus, rewriting Eq. (1.1.39) in the variables $\overline{x}$, $\overline{y}$ defined by (1.1.42):

$$x = \lambda_1\overline{x} + \lambda_2\overline{y}, \quad y = \overline{x} + \overline{y} \tag{1.1.49}$$

and dividing by $\widetilde{B}$, one arrives at the following equation of a hyperbola:

$$\overline{x}\,\overline{y} + \tilde{a}\,\overline{x} + \tilde{b}\,\overline{y} + \tilde{c} = 0. \tag{1.1.50}$$

Consequently, we say that Eq. (1.1.39) with $B^2 - AC > 0$ represents algebraic curves of *hyperbolic type*.

One can further simplify Eq. (1.1.50) by translation (1.1.43),

$$\overline{x} = \tilde{x} + \mu, \quad \overline{y} = \tilde{y} + \nu.$$

Namely, letting $\mu = -\tilde{b}$ and $\nu = -\tilde{a}$ one reduces Eq. (1.1.50) to the standard form

$$\tilde{x}\,\tilde{y} = k, \quad k = \text{const.}$$

If $k = 0$ the hyperbola reduces to the pair of intersecting straight lines, $\tilde{x} = 0$ and $\tilde{y} = 0$. Note that setting $\tilde{x} = \xi + \eta$, $\tilde{y} = \xi - \eta$, one obtains the following second standard form of a hyperbola:

$$\xi^2 - \eta^2 = k, \quad k = \text{const.}$$

Remark 1.1.6. It is assumed above that $A \neq 0$. If $A = 0$ and $C \neq 0$, we exchange x and y and arrive again at the assumption $A \neq 0$. Finally, if $A = C = 0$, then $B \neq 0$ and Eq. (1.1.39) has already form (1.1.50).

Let $B^2 - AC = 0$. Then Eq. (1.1.46) has the repeated root $\lambda = -B/A$. We let $\beta/\delta = \lambda$ and obtain $\widetilde{C} = 0$. Moreover, the choice $\beta = \lambda\delta$ yields that $\widetilde{B} = 0$ and that $\widetilde{A} \neq 0$. One can set, e.g. $\gamma = \delta = 1$. Then condition (1.1.41) requires that $\alpha \neq \lambda$. Hence, rewriting Eq. (1.1.39) in the variables $\overline{x}$, $\overline{y}$ defined by (1.1.42):

$$x = \alpha\overline{x} + \lambda\overline{y}, \quad y = \overline{x} + \overline{y}, \quad (\alpha \neq \lambda), \tag{1.1.51}$$

and dividing by $\widetilde{A}$, one arrives at the following equation:

$$\overline{x}^2 + \tilde{a}\,\overline{x} + \tilde{b}\,\overline{y} + \tilde{c} = 0. \tag{1.1.52}$$

Now the translation $\overline{x} = \tilde{x} - (\tilde{a}/2)$, $\overline{y} = \tilde{y}$ reduces the latter equation to the standard equation for a parabola:

$$\tilde{x}^2 + \tilde{b}\,\tilde{y} + k = 0.$$

Therefore, we say that Eq. (1.1.39) with $B^2 - AC = 0$ represents algebraic curves of *parabolic type*.

Let $B^2 - AC < 0$. Then Eq. (1.1.46) has two complex roots, $\lambda_1 = p + iq$, $\lambda_2 = p - iq$, where

$$p = -\frac{B}{A}, \quad q = \frac{\sqrt{AC - B^2}}{A}.$$

One can annul the coefficients $\tilde{A}$ and $\tilde{C}$ simultaneously by letting $\alpha/\gamma = \lambda_1$ and $\beta/\delta = \lambda_2$ in (1.1.45). One can set, e.g. $\gamma = \delta = 1$ and obtain the complex change of variables

$$x = (p + iq)\,x' + (p - iq)\,y', \quad y = x' + y' \tag{1.1.53}$$

mapping Eq. (1.1.39) into an equation of a hyperbola (1.1.50). We want, however, to avoid complex change of variables and use only real variables. To this end, we solve Eqs. (1.1.53) with respect to x' and y' :

$$x' = \frac{y}{2} + i\frac{py - x}{2q}, \quad y' = \frac{y}{2} - i\frac{py - x}{2q},$$

and consider the real and imaginary parts of these complex conjugate variables (multiplied by 2 for the sake of simplicity) as new real variables:

$$\overline{x} = y, \quad \overline{y} = \frac{py - x}{q}.$$

Solving the latter equations with respect to x, y, we obtain the following real transformation of form (1.1.42):

$$x = p\overline{x} - q\overline{y}, \quad y = \overline{x}. \tag{1.1.54}$$

It maps Eq. (1.1.39) into an equation of the form

$$\overline{x}^2 + \overline{y}^2 + \tilde{a}\,\overline{x} + \tilde{b}\,\overline{y} + \tilde{c} = 0. \tag{1.1.55}$$

One can eliminate the terms of the first degree by the translation $\overline{x} = \tilde{x} - (\tilde{a}/2)$, $\overline{y} = \tilde{y} - (\tilde{b}/2)$ thus reducing (1.1.55) to the standard form

$$\tilde{x}^2 + \tilde{y}^2 = k.$$

This equation represents a circle if $k > 0$, a point if $k = 0$, and no real locus if $k < 0$. Thus, we say that Eq. (1.1.55) represents a circle with the understanding that the degenerate cases indicated above may occur. Since a circle is special case of an ellipse, we say that Eq. (1.1.39) with $B^2 - AC < 0$ represents algebraic curves of *elliptic type*.

We summarize the above results in the following statement.

Theorem 1.1.3. Any Eq. (1.1.39) of the second degree represents a curve of *hyperbolic type* and can be mapped by a homogeneous linear transformation to form (1.1.50):

$$\overline{x}\,\overline{y} + \tilde{a}\,\overline{x} + \tilde{b}\,\overline{y} + \tilde{c} = 0, \quad \text{if} \quad B^2 - AC > 0, \tag{1.1.56}$$

parabolic type and can be mapped to form (1.1.52):

$$\overline{x}^2 + \tilde{a}\,\overline{x} + \tilde{b}\,\overline{y} + \tilde{c} = 0, \quad \text{if} \quad B^2 - AC = 0, \tag{1.1.57}$$

and *elliptic type* and can be mapped to form (1.1.55):

$$\overline{x}^2 + \overline{y}^2 + \tilde{a}\,\overline{x} + \tilde{b}\,\overline{y} + \tilde{c} = 0, \quad \text{if} \quad B^2 - AC < 0. \tag{1.1.58}$$

Remark 1.1.7. It is manifest from transformation (1.1.53) that elliptic and hyperbolic curves are connected by complex linear transformations.

1.2   Differential and integral calculus

1.2.1   Rules for differentiation

Let $f(x)$ be a function of one variable. Its derivative $f'(x)$ at a point x is defined by

$$f'(x) = \lim_{\Delta x \to 0} \frac{f(x + \Delta x) - f(x)}{\Delta x}. \qquad (1.2.1)$$

It is customary to denote the derivative of $y = f(x)$ also by

$$\frac{df(x)}{dx}, \quad y', \quad \frac{dy}{dx}, \quad D(y) = D_x(y), \quad D(f(x)) = D_x(f(x)).$$

Let $u = f(x^1, x^2, \ldots, x^n)$ be a function of n variables. Its partial derivative with respect to one of the variables, e.g. x^i is defined by

$$\frac{\partial f}{\partial x^i} = \lim_{\Delta x^i \to 0} \frac{f(x^1, \ldots, x^i + \Delta x^i, \ldots, x^n) - f(x^1, x^2, \ldots, x^n)}{\Delta x^i}. \qquad (1.2.2)$$

It is often denoted also by

$$\frac{\partial u}{\partial x^i}, \quad u_i = u_{x^i}, \quad D_i(u) = D_{x^i}(u), \quad D_i(f(x)),$$

where $x = (x^1, x^2, \ldots, x^n)$.

The main rules for differentiation comprise:
   (i) the formulae

$$D(au + bv) = aD(u) + bD(v),$$
$$D(uv) = vD(u) + uD(v),$$
$$D(u^\alpha) = \alpha u^{\alpha-1} D(u), \qquad (1.2.3)$$
$$D\left(\frac{u}{v}\right) = \frac{vD(u) - uD(v)}{v^2},$$

where a, b and α are constants, u, v are functions and D is an ordinary or a partial derivation;
   (ii) the rule for the differentiation of the inverse function:

$$D_x(y) = \frac{1}{D_y(x)}, \qquad (1.2.4)$$

   (iii) the *chain rule* stating that if $y = y(u)$, $u = u(x)$, then

$$D_x(y) = D_u(y) \cdot D_x(u). \qquad (1.2.5)$$

The term *chain rule* is due to the possibility of iterating (1.2.5), e.g. if $y = y(u)$, $u = u(v)$, and $v = v(x)$, then

$$D_x(y) = D_u(y) \cdot D_v(u) \cdot D_x(v).$$

1.2.2 The mean value theorem

The name refers to the following statement.

Theorem 1.2.1. Let $f(x)$ be continuously differentiable in the interval $[a, b]$. Then, for any points x_1, x_2, where $a < x_1 < x_2 < b$, there is at least one point $\xi \in [x_1, x_2]$ such that

$$f(x_2) - f(x_1) = (x_2 - x_1) f'(\xi). \qquad (1.2.6)$$

The following consequence of the mean value theorem is crucial in the integral calculus and in the theory of differential equations as well.

Theorem 1.2.2. Let $y = f(x)$ be continuously differentiable in an interval $a \leq x \leq b$. Then $y' = 0$ within the interval $[a, b]$ if and only if $y = C = $ const.

Proof. If $y = C$, then it is obvious from definition (1.2.1) that $y' = 0$. Let us assume now that $y' = 0$. According to Theorem 1.2.1, we have

$$y(x_2) - y(x_1) = (x_2 - x_1)y'(\xi)$$

for any x_1, x_2 from the interval $[a, b]$, where $x_1 \leq \xi \leq x_2$. By hypothesis, $y' = 0$ in the whole interval $[a, b]$, and hence $y'(\xi) = 0$. Therefore, the above equation yields that $y(x_2) = y(x_1)$, i.e., $y = $ const. within the interval $[a, b]$.

Corollary 1.2.1. Two functions have the same derivative if and only if their difference is constant. In other words, $f'(x) = g'(x)$ if and only if

$$f(x) = g(x) + C, \quad C = \text{const.}$$

1.2.3 Invariance of the differential

The differential of a function $y = f(x)$ of one variable is

$$dy = y' dx.$$

Likewise, a differential of a function $u = f(x^1, x^2, \ldots, x^n)$ of several variables is defined by

$$du = \frac{\partial u}{\partial x^1} dx^1 + \cdots + \frac{\partial u}{\partial x^n} dx^n \equiv \sum_{i=1}^{n} \frac{\partial u}{\partial x^i} dx^i.$$

We will often use the usual *summation convention* to omit the sign of summation when the index of summation is repeated as subscript and superscript. For example, the above equation is written in the form

$$du = \frac{\partial u}{\partial x^i} dx^i.$$

The invariance of the differential refers to the following property of the differential of a function $f(u)$. If $u = u(x)$, then

$$df = \frac{df(u)}{du} du = \frac{df(u(x))}{dx} dx. \tag{1.2.7}$$

The general statement is as follows.

Theorem 1.2.3. Let $u = f(x)$ be a function of n variables $x = (x^1, \ldots, x^n)$ and let $x^i = x^i(t), i = 1, \ldots, n$, be functions of s variables $t = (t^1, \ldots, t^s)$. Then the differential of u regarded as a function $u = f(x)$ of x is identical with the differential of u regarded as a function $u = f(x(t))$ of t, i.e.,

$$du = \sum_{i=1}^{n} \frac{\partial f(x)}{\partial x^i} dx^i = \sum_{k=1}^{s} \frac{\partial f(x(t))}{\partial t^k} dt^k. \tag{1.2.8}$$

In particular, if $u = f(x, y)$ is a function of two variables, then an arbitrary change of variables

$$x = \varphi(\tilde{x}, \tilde{y}), \quad y = \psi(\tilde{x}, \tilde{y})$$

does not change the differential, i.e.,

$$\frac{\partial f(\varphi(\tilde{x}, \tilde{y}), \psi(\tilde{x}, \tilde{y}))}{\partial \tilde{x}} d\tilde{x} + \frac{\partial f(\varphi(\tilde{x}, \tilde{y}), \psi(\tilde{x}, \tilde{y}))}{\partial \tilde{y}} d\tilde{y} = \frac{\partial f(x, y)}{\partial x} dx + \frac{\partial f(x, y)}{\partial y} dy$$

or briefly:

$$du = \frac{\partial u}{\partial \tilde{x}} d\tilde{x} + \frac{\partial u}{\partial \tilde{y}} d\tilde{y} = \frac{\partial u}{\partial x} dx + \frac{\partial u}{\partial y} dy.$$

1.2.4   Rules for integration

Calculation of integrals is based on the following main rules.

1. *Integration is the inverse operation to differentiation:*

$$\int df(x) = f(x) + C \quad \text{or} \quad \int f'(x)dx = f(x) + C, \tag{1.2.9}$$

where C is an arbitrary constant termed the *constant of integration* (cf. Corollary 1.2.1).

2. *Integration is a linear operation:*

$$\int [au(x) + bv(x)]dx = a \int u(x)dx + b \int v(x)dx, \quad a, b = \text{const.} \tag{1.2.10}$$

3. *Change of variables in integrals (see Eq. (1.2.7)):*

$$\int f(x)dx = \int f(\varphi(t)) \cdot \varphi'(t)dt. \tag{1.2.11}$$

4. *Integration by parts:*

$$\int u\,dv = uv - \int v\,du. \tag{1.2.12}$$

5. *Differentiation of definite integrals:*

(i) $\dfrac{d}{dx} \displaystyle\int_a^x f(s)ds = f(x),$ \hfill (1.2.13)

(ii) $\dfrac{d}{dx} \displaystyle\int_{\varphi(x)}^{\psi(x)} g(s,x)ds = \int_{\varphi(x)}^{\psi(x)} \dfrac{\partial g(s,x)}{\partial x}\,ds + \psi'(x)g(\psi(x),x) - \varphi'(x)g(\varphi(x),x).$

1.2.5 The Taylor series

The *Taylor series* expansion of $f(x)$ at $x = a$ has the form:

$$f(x) = f(a) + f'(a)(x-a) + \cdots + \frac{f^{(n)}(a)}{n!}(x-a)^n + \cdots, \tag{1.2.14}$$

where $f^{(n)}(a)$ is the value of the nth derivative of $f(x)$ at $x = a$. If $a = 0$, the Taylor series (1.2.14) is known as the *Maclaurin series* and has the form

$$f(x) = f(0) + f'(0)x + \frac{f''(0)}{2!}x^2 + \cdots + \frac{f^{(n)}(0)}{n!}x^n + \cdots. \tag{1.2.15}$$

Let us compute the Maclaurin expansion of exponential function $f(x) = e^x$. Since $f'(x) = e^x$, we have $f(0) = f'(0) = \cdots = f^{(n)}(0) = 1$, and expansion (1.2.15) has the form

$$f(x) = 1 + x + \frac{x^2}{2!} + \cdots + \frac{x^n}{n!} + \cdots = \sum_{n=0}^{\infty} \frac{x^n}{n!}.$$

The following tables contain the Maclaurin and Taylor expansions of some frequently used functions.

(i) **The exponential function and logarithm**

$$e^x = 1 + x + \frac{x^2}{2!} + \frac{x^3}{3!} + \cdots; \qquad a^x = e^{x\ln a} = 1 + x\ln a + \cdots. \tag{1.2.16}$$

$$\ln x = (x-1) - \frac{(x-1)^2}{2} + \frac{(x-1)^3}{3} - \cdots \quad (0 < x \le 2). \tag{1.2.17}$$

$$\ln(1+x) = x - \frac{x^2}{2} + \frac{x^3}{3} - \frac{x^4}{4} + \cdots \quad (-1 < x \le 1). \tag{1.2.18}$$

$$\ln(1-x) = -\left[x + \frac{x^2}{2} + \frac{x^3}{3} + \frac{x^4}{4} + \cdots\right] \quad (-1 \le x < 1). \tag{1.2.19}$$

(ii) **Algebraic functions** (below, α is any positive real number)

$$\frac{1}{1-x} = 1 + x + x^2 + x^3 + \cdots + x^n + \cdots . \tag{1.2.20}$$

$$\frac{1}{1+x} = 1 - x + x^2 - x^3 + \cdots + (-1)^n x^n + \cdots . \tag{1.2.21}$$

$$\frac{1}{\sqrt{1 \pm x}} = 1 \mp \frac{1}{2} x + \frac{3}{2 \cdot 4} x^2 \mp \frac{3 \cdot 5}{2 \cdot 4 \cdot 6} x^3 + \frac{3 \cdot 5 \cdot 7}{2 \cdot 4 \cdot 6 \cdot 8} x^4 \mp \cdots . \tag{1.2.22}$$

$$\sqrt{1 \pm x} = 1 \pm \frac{1}{2} x - \frac{1}{2 \cdot 4} x^2 \pm \frac{3}{2 \cdot 4 \cdot 6} x^3 - \frac{3 \cdot 5}{2 \cdot 4 \cdot 6 \cdot 8} x^4 \pm \cdots . \tag{1.2.23}$$

$$(1 \pm x)^\alpha = 1 \pm \alpha x + \frac{\alpha(\alpha-1)}{2!} x^2 \pm \frac{\alpha(\alpha-1)(\alpha-2)}{3!} x^3 + \cdots . \tag{1.2.24}$$

$$(1 \pm x)^{-\alpha} = 1 \mp \alpha x + \frac{\alpha(\alpha+1)}{2!} x^2 \mp \frac{\alpha(\alpha+1)(\alpha+2)}{3!} x^3 + \cdots . \tag{1.2.25}$$

(iii) **Trigonometric and hyperbolic functions**

$$\sin x = x - \frac{x^3}{3!} + \frac{x^5}{5!} - \frac{x^7}{7!} + \cdots + (-1)^n \frac{x^{2n+1}}{(2n+1)!} + \cdots . \tag{1.2.26}$$

$$\cos x = 1 - \frac{x^2}{2!} + \frac{x^4}{4!} - \frac{x^6}{6!} + \cdots + (-1)^n \frac{x^{2n}}{(2n)!} + \cdots . \tag{1.2.27}$$

$$\sinh x = x + \frac{x^3}{3!} + \frac{x^5}{5!} + \frac{x^7}{7!} + \cdots + \frac{x^{2n+1}}{(2n+1)!} + \cdots . \tag{1.2.28}$$

$$\cosh x = 1 + \frac{x^2}{2!} + \frac{x^4}{4!} + \frac{x^6}{6!} + \cdots + \frac{x^{2n}}{(2n)!} + \cdots . \tag{1.2.29}$$

(iv) **Some non-elementary functions** (see Example 1.1.6)

$$\text{Si}(x) = x - \frac{x^3}{3 \cdot 3!} + \frac{x^5}{5 \cdot 5!} - \frac{x^7}{7 \cdot 7!} + \cdots . \tag{1.2.30}$$

$$\text{Ci}(x) = \gamma + \ln x - \frac{x^2}{2 \cdot 2!} + \frac{x^4}{4 \cdot 4!} - \frac{x^6}{6 \cdot 6!} + \cdots . \tag{1.2.31}$$

$$\text{Ei}(x) = \gamma + \ln |x| + x + \frac{x^2}{2 \cdot 2!} + \frac{x^3}{3 \cdot 3!} + \frac{x^4}{4 \cdot 4!} + \cdots . \tag{1.2.32}$$

$$\text{li}(x) = \ln |\ln x| + \ln x + \frac{(\ln x)^2}{2 \cdot 2!} + \frac{(\ln x)^3}{3 \cdot 3!} + \cdots . \tag{1.2.33}$$

$$\text{erf}(x) = \frac{2}{\sqrt{\pi}} \left(x - \frac{x^3}{3} + \frac{x^5}{5 \cdot 2!} - \frac{x^7}{7 \cdot 3!} + \cdots \right) . \tag{1.2.34}$$

where γ is Euler's constant defined in Example 1.1.1.

1.2.6   Complex variables

A complex number has the form

$$z = x + iy, \quad i = \sqrt{-1},$$

where x and y are real numbers termed the *real* and *imaginary* part of z, respectively. The *complex conjugate* to z is the complex number $\bar{z} = x - iy$.

Complex numbers are written also in the trigonometric representation

$$z = r(\cos\theta + i\sin\theta) \tag{1.2.35}$$

as well as in the equivalent polar representation (see further Euler's formula (1.2.41)):

$$z = re^{i\theta}, \tag{1.2.36}$$

where r and θ are real numbers termed the modulus and argument of z, respectively. It follows from (1.2.35) that the angle θ is determined only to within an arbitrary integer multiple of 2π. Taking this non-uniqueness into account, z is written, e.g., in the polar representation (1.2.36) in the form

$$z = re^{i(\theta+2\pi k)}, \quad k = 0, \pm 1, \pm 2, \ldots . \tag{1.2.37}$$

The value θ in the interval $-\pi < \theta \leq \pi$ is termed the *principal argument*.

The following functions of the complex variable are defined by extending the Maclaurin series (1.2.16), (1.2.26) and (1.2.27) to the complex domain:

$$e^z = 1 + z + \frac{z^2}{2!} + \cdots + \frac{z^n}{n!} + \cdots . \tag{1.2.38}$$

$$\sin z = z - \frac{z^3}{3!} + \frac{z^5}{5!} + \cdots + (-1)^n \frac{z^{2n+1}}{(2n+1)!} + \cdots . \tag{1.2.39}$$

$$\cos z = 1 - \frac{z^2}{2!} + \frac{z^4}{4!} + \cdots + (-1)^n \frac{z^{2n}}{(2n)!} + \cdots . \tag{1.2.40}$$

It follows that $e^{iz} = \cos z + i\sin z$, and hence *Euler's formula*:

$$e^{x+iy} = e^x (\cos y + i\sin y). \tag{1.2.41}$$

Using Euler's formula, one can readily show that

$$\cos z = \frac{e^{iz} + e^{-iz}}{2}, \quad \sin z = \frac{e^{iz} - e^{-iz}}{2i}. \tag{1.2.42}$$

Furthermore, Euler's formula (1.2.41) and the polar representation (1.2.36) of the complex number $z = r(\cos\theta + i\sin\theta)$ lead to de Moivre's formula

$$z^n = r^n[\cos(n\theta) + i\sin(n\theta)]. \tag{1.2.43}$$

Replacing in Eq. (1.2.43) n by $1/n$ and using representation (1.2.37) of z, one obtains n values of $\sqrt[n]{z}$:

$$\sqrt[n]{z} = \sqrt[n]{r}\left[\cos\left(\frac{\theta}{n} + \frac{2\pi}{n}k\right) + i\sin\left(\frac{\theta}{n} + \frac{2\pi}{n}k\right)\right], \quad k = 0, 1, \ldots, n-1. \quad (1.2.44)$$

Example 1.2.1. Let us obtain the cube roots of unity, i.e., solve the equation $w^3 = 1$. Writing the real number 1 in the trigonometric form:

$$1 = \cos(2\pi k) + i\sin(2\pi k),$$

and using Eq. (1.2.44) with $k = 0, 1$, and 2, we obtain the following three cube roots of unity:

$$w_1 = 1, \quad w_2 = \frac{-1 + i\sqrt{3}}{2}, \quad w_3 = \frac{-1 - i\sqrt{3}}{2}.$$

The hyperbolic functions (1.1.8) are also extended to the complex domain in a usual way:

$$\sinh z = \frac{e^z - e^{-z}}{2}, \quad \cosh z = \frac{e^z + e^{-z}}{2}.$$

The trigonometric and hyperbolic functions are connected by the following equations:

$$\sin z = -i\sinh(iz), \quad \cos z = \cosh(iz), \quad \tan z = -i\tanh(iz),$$
$$\sinh z = -i\sin(iz), \quad \cosh z = \cos(iz), \quad \tanh z = -i\tan(iz). \quad (1.2.45)$$

The derivatives of the complex exponential, trigonometric and hyperbolic functions are as follows:

$$(e^z)' = e^z, \quad (\sin z)' = \cos z, \quad (\cos z)' = -\sin z,$$
$$(\sinh z)' = \cosh z, \quad (\cosh z)' = \sinh z.$$

Using the polar representation (1.2.37) of z, one defines the (multivalued) *logarithmic function* of the complex variable as follows:

$$\ln z = \ln r + i(\theta + 2\pi k), \quad k = 0, \pm 1, \pm 2, \ldots \quad (1.2.46)$$

Finally, the exponential function a^z, where a is any complex number, is defined as follows:

$$a^z = e^{z\ln a}. \quad (1.2.47)$$

1.2.7   Approximate representation of functions

Definition 1.2.1. We say that a function $\alpha(x, \varepsilon)$ is of order less than ε^p (with an integer $p \geq 1$) and write

$$\alpha(x, \varepsilon) = o(\varepsilon^p), \quad \varepsilon \to 0, \quad (1.2.48)$$

if

$$\lim_{\varepsilon \to 0} \frac{\alpha(x, \varepsilon)}{\varepsilon^p} = 0. \tag{1.2 49}$$

Equation (1.2.49) is satisfied if e.g., any one of the following two conditions holds:

$$\alpha(x, \varepsilon) = \varepsilon^{p+1} \phi(x, \varepsilon), \quad \phi(x, \varepsilon) \neq \infty \text{ as } \varepsilon \to 0$$

and

$$|\alpha(x, \varepsilon)| \leq C \, |\varepsilon|^{p+1}, \quad C = \text{const.}$$

Definition 1.2.2. Functions $f(x, \varepsilon)$ and $g(x, \varepsilon)$ are said to be *approximately equal* with an error $o(\varepsilon^p)$ as $\varepsilon \to 0$ if

$$f(x, \varepsilon) - g(x, \varepsilon) = o(\varepsilon^p).$$

To designate the approximate equality, we use either the notation $g \approx f$ or more specifically

$$g(x, \varepsilon) = f(x, \varepsilon) + o(\varepsilon^p).$$

According to the Taylor expansion (1.2.14), an approximate representation of $f(x)$ with an error $o((x - a)^n)$ as $x \to a$ is as follows:

$$f(x) \approx f(a) + f'(a)\,(x - a) + \cdots + \frac{f^{(n)}(a)}{n!}\,(x - a)^n. \tag{1.2.50}$$

For example, expansion (1.2.34) hints, e.g., the following approximation to the error function:

$$\mathrm{erf}(x) = \frac{2}{\sqrt{\pi}} \left(x - \frac{x^3}{3} \right) + o(x^4).$$

Remark 1.2.1. Approximately equal functions f and g are often called *equivalent* functions and are denoted $f \sim g$.

1.2.8 Jacobian. Functional independence. Change of variables in multiple integrals

Definition 1.2.3. Let $x = (x^1, \ldots, x^n)$. Functions

$$u^\alpha = u^\alpha(x), \quad \alpha = 1, \ldots, m, \ (m \leq n) \tag{1.2.51}$$

are said to be functionally dependent if there exists at least one relation $\Phi(u^1(x), \ldots, u^m(x)) = 0$ and *functionally independent* otherwise.

A convenient test for the independence of functions (1.2.51) is formulated in terms of their *Jacobian matrix*

$$\left(\frac{\partial u^\alpha}{\partial x^i} \right)$$

here the indices α and i denote rows and columns, respectively. Namely, functions (1.2.51) are functionally independent if and only if the rank of the Jacobian matrix is equal to m. If $n = m$, the test for the functional independence is formulated in terms of the determinant

$$J = \det\left(\frac{\partial u^\alpha}{\partial x^i}\right)$$

known as the *Jacobian*.

Theorem 1.2.4. Functions $u^1(x), \ldots, u^n(x)$ of n variables $x = (x^1, \ldots, x^n)$ are functionally independent if and only if their Jacobian does not vanish.

The Jacobian is also useful in the generalization of the rule for the change of variables (1.2.11) to multiple integrals. Namely, let functions (1.2.51) be functionally independent and let $m = n$. Then (1.2.51) provides a change of variables $y = u(x)$ with the Jacobian $J \neq 0$. This change of variables leads to the following *rule for change of variables in multiple integrals:*

$$\int f(y^1, y^2, \ldots, y^n)\, dy^1 dy^2 \cdots dy^n \tag{1.2.52}$$
$$= \int f(u^1(x), u^2(x), \ldots, u^n(x))|J|\, dx^1 dx^2 \cdots dx^n.$$

1.2.9   Linear independence of functions. Wronskian

Definition 1.2.4. Consider m functions of one variable x :

$$y_1 = y_1(x), \; y_2 = y_2(x), \; \ldots, \; y_m = y_m(x). \tag{1.2.53}$$

Functions (1.2.53) are said to be linearly dependent if there exist constants $c_1, \ldots, c_m$, not all zero, such that

$$c_1 y_1(x) + c_2 y_2(x) + \cdots + c_m y_m(x) = 0, \tag{1.2.54}$$

and *linearly independent* if there is no relation of form (1.2.54).

A convenient test for the linear independence is formulated in terms of the $m \times m$ determinant

$$W[y_1, y_2, \ldots, y_m] = \begin{vmatrix} y_1 & y_2 & \cdots & y_m \\ y_1' & y_2' & \cdots & y_m' \\ \vdots & \vdots & & \vdots \\ y_1^{(m-1)} & y_2^{(m-1)} & \cdots & y_m^{(m-1)} \end{vmatrix} \tag{1.2.55}$$

called the *Wronskian* of functions (1.2.53).

Theorem 1.2.5. Functions (1.2.53) are linearly independent if and only if their Wronskian does not vanish,

$$W[y_1, y_2, \ldots, y_m] \neq 0. \tag{1.2.56}$$

1.2.10 Integration by quadrature

The practical integration of differential equations presumes knowledge of the general solution to the simplest differential equation

$$\frac{dy}{dx} = 0. \tag{1.2.57}$$

According to Theorem 1.2.2, the general solution of Eq. (1.2.57) is

$$y = C.$$

Let us dwell on the fundamental differential equation of the integral calculus solved in the pioneering works of Newton and Leibnitz. Suppose that we are given a continuous function $f(x)$ and are asked to find a function $y = F(x)$ whose derivative is equal to $f(x)$:

$$\frac{dF(x)}{dx} = f(x).$$

The function $F(x)$ is called an *integral of $f(x)$ with respect to x*. Thus, the problem is to solve the first-order *ordinary differential equation:*

$$y' = f(x). \tag{1.2.58}$$

Its *general solution* is written

$$y = \int f(x)dx + C, \tag{1.2.59}$$

where $\int f(x)dx = F(x)$ is any integral of $f(x)$ with respect to x, and C is an arbitrary constant known as the *constant of integration*. By specifying the constant of integration, one obtains a *particular solution*. Hence, Eq. (1.2.58) has an infinite number of solutions given by the one-parameter family of integrals (1.2.59).

Notation. In calculus, the symbol $\int f(x)dx$ of the *indefinite integral* is a standard notation for all solutions to Eq. (1.2.58), i.e., $\int f(x)dx = F(x) + C$ with $F(x)$ denoting a particular integral of $f(x)$. In the theory of differential equations, however, a different interpretation is commonly used. Namely, $\int f(x)dx$ is identified with any particular integral $F(x)$ of the function $f(x)$. This is the interpretation which is adopted in the present book.

In notation (1.2.59), the general solution, e.g., to a second-order equation,

$$y'' = f(x) \tag{1.2.60}$$

involves two arbitrary constants C_1 and C_2 and is written

$$y = \int dx \int f(x)dx + C_1 x + C_2, \tag{1.2.61}$$

where $\int dx \int f(x)dx = \int \left(\int f(x)dx \right) dx$.

Remark 1.2.2. In the classical literature, the integral formula (1.2.59) is termed the *quadrature*. Consequently, the differential equation (1.2.58) is said to be *integrable by quadrature*. The same terminology applies to the equation

$$y' = h(y),$$

since its solution is given by the integral formula similar to (1.2.59),

$$x = \int \frac{dy}{h(y)} + C \equiv H(y) + C,$$

whence $y = H^{-1}(x - C)$, where H^{-1} denotes the inverse to H.

1.2.11   Differential equations for families of curves

Consider, in the (x, y) plane, a family of curves,

$$y = f(x, C_1, \ldots, C_n) \tag{1.2.62}$$

given, in general, implicitly by

$$\Phi(x, y, C_1, \ldots, C_n) = 0. \tag{1.2.63}$$

By the definition of implicit functions, Eq. (1.2.63) with y replaced by the function f from (1.2.62) is satisfied identically in x (from some interval) for all admissible values of the parameters C_k, $k = 1, \ldots, n$. Consequently, one can differentiate this identity with respect to x. Iterating the procedure n times yields:

$$\frac{\partial \Phi}{\partial x} + \frac{\partial \Phi}{\partial y} y' = 0,$$

$$\frac{\partial^2 \Phi}{\partial x^2} + 2 \frac{\partial^2 \Phi}{\partial x \partial y} y' + \frac{\partial^2 \Phi}{\partial y^2} y'^2 + \frac{\partial \Phi}{\partial y} y'' = 0,$$

$$\cdots \cdots$$

$$\frac{\partial^n \Phi}{\partial x^n} + \cdots \cdots \cdots + \frac{\partial \Phi}{\partial y} y^{(n)} = 0. \tag{1.2.64}$$

Elimination of the parameters C_k from Eqs. (1.2.63) and (1.2.64) yields an nth-order ordinary differential equation

$$F(x, y, y', \ldots, y^{(n)}) = 0. \tag{1.2.65}$$

Function (1.2.62) provides a solution, depending on n arbitrary constants C_k, to the differential equation (1.2.65). Accordingly, (1.2.65) is termed the differential equation of the family of curves (1.2.62) or (1.2.63).

The above procedure can be simplified by using the following *differential algebraic* notation (see Section 1.4).

Notation. The *total differentiation* D_x of functions depending on a finite number of variables $x, y, y', y'', \ldots$ is defined by

$$D_x = \frac{\partial}{\partial x} + y'\frac{\partial}{\partial y} + y''\frac{\partial}{\partial y'} + \cdots + y^{(s+1)}\frac{\partial}{\partial y^{(s)}} + \cdots . \qquad (1.2.66)$$

In this notation, Eqs. (1.2.64) are written in a compact form:

$$D_x\Phi = 0, \quad D_x^2\Phi = 0, \quad \ldots, \quad D_x^n\Phi = 0. \qquad (1.2.67)$$

Moreover, in this way one eludes the necessity of invoking equation (1.2.63) n times while deriving equations (1.2.64).

Exercise 1.2.1. Find the differential equation of the family of straight lines, $y = ax + b$ containing two parameters, a and b.

Solution. By setting $\Phi = y - ax - b$, Eqs. (1.2.67) are written:

$$D_x\Phi \equiv y' - a = 0, \quad D_x^2\Phi \equiv y'' = 0.$$

Here the last equation does not contain the parameters. Hence, the differential equation (1.2.65) of straight lines is the simplest *linear* equation of the second order:

$$y'' = 0. \qquad (1.2.68)$$

Exercise 1.2.2. Find the differential equation of the family of parabolas given in the form $\Phi \equiv y - ax^2 - bx - c = 0$ and depending on three parameters. $a, b,$ and c.

Solution. Equations (1.2.67) are written:

$$D_x\Phi \equiv y' - 2ax - b = 0, \quad D_x^2\Phi \equiv y'' - 2a = 0, \quad D_x^3\Phi \equiv y''' = 0.$$

Thus the differential equation of the parabolas is the simplest linear equation of the third order:

$$y''' = 0. \qquad (1.2.69)$$

Exercise 1.2.3. Find the differential equation of the family of circles given in the form $\Phi \equiv (x - a)^2 + (y - b)^2 - c^2 = 0$.

Solution. Equations (1.2.67) yield:

$$x - a + (y - b)y' = 0, \quad 1 + y'^2 + (y - b)y'' = 0, \quad 3y'y'' + (y - b)y''' = 0.$$

Here the third equation does not contain the parameters a and c, and it suffices to substitute there the expression $y - b = -(1 + y'^2)/y''$ found from the second equation. Hence, the family of circles is described by the *nonlinear* equation:

$$y''' - 3\frac{y'y''^2}{1 + y'^2} = 0. \qquad (1.2.70)$$

Exercise 1.2.4. Find the differential equation of the family of hyperbolas given in the form $\Phi \equiv (y - a)(b - cx) - 1 = 0$.

Solution. Equations (1.2.67) yield:

$$(b - cx)y' - c(y - a) = 0, \quad (b - cx)y'' - 2cy' = 0, \quad (b - cx)y''' - 3cy'' = 0.$$

We find from the second equation that $b - cx = 2cy'/y''$ and substitute it into the third one to obtain the following differential equation of the hyperbolas:

$$y''' - \frac{3}{2}\frac{y''^2}{y'} = 0. \tag{1.2.71}$$

1.3   Vector analysis

Vector analysis provides a natural and concise notation for formulating geometrical and physical problems. Consequently, it has become a mainstay of undergraduate curricula in the sciences and engineering. It is widely used in Newtonian, continuum and relativistic mechanics, electrodynamics, etc. This section contains the basic notions and formulae from vector analysis.

1.3.1   Vector algebra

The *scalar product* $a \cdot b$ of vectors a and b is a scalar quantity (i.e., a real number) defined by

$$a \cdot b = |a||b| \cos \theta, \tag{1.3.1}$$

where θ ($0 \leq \theta \leq \pi$) is the angle between the vectors a and b. The scalar product is also written in the literature (a, b) or simply ab.

Let the vectors a and b be written in the Cartesian coordinates:

$$a = a^1 i + a^2 j + a^3 k = (a^1, a^2, a^3),$$
$$b = b^1 i + b^2 j + b^3 k = (b^1, b^2, b^3), \tag{1.3.2}$$

where i, j and k are unit vectors along the first, second and third coordinate axes, respectively. Then their scalar product is given by

$$a \cdot b = \sum_{i=1}^{3} a^i b^i. \tag{1.3.3}$$

An example of a scalar product is provided by mechanics: a force F having its point of application moved through a displacement x does the *work*

$$W = F \cdot x. \tag{1.3.4}$$

It follows from Eq. (1.3.4) that in the particular case when the displacement x is perpendicular to the force F then *no work is done*. For example, when

These chaps do no work!

He does work

one carries a load in the earth's gravitation field and moves along a horizontal plane, he does no work. Refute the opinion of the puppy.

The *vector product* $a \times b$ of vectors a and b is the vector defined as follows:

1. the magnitude of $a \times b$ is equal to $|a \times b| = |a||b| \sin \theta$, where θ $(0 \leq \theta \leq \pi)$ is the angle between the vectors a and b;

2. the vector $a \times b$ is perpendicular to the plane spanned by a and b and is such that the triplet a, b and $a \times b$ forms a right-handed system.

The vector product is also designated $[a, b]$ or $[a\,b]$. The magnitude $|a \times b|$ of the vector product is equal to the area of the parallelogram with the sides a and b. Hence, the vector product represents a directed area and therefore it is termed in the classical literature a *vectorial area*.

The *vector product* of two vectors (1.3.2) in the Cartesian coordinates is written

$$a \times b = \begin{vmatrix} i & j & k \\ a^1 & a^2 & a^3 \\ b^1 & b^2 & b^3 \end{vmatrix} \tag{1.3.5}$$

where

$$\begin{vmatrix} i & j & k \\ a^1 & a^2 & a^3 \\ b^1 & b^2 & b^3 \end{vmatrix} = (a^2 b^3 - a^3 b^2)i + (a^3 b^1 - a^1 b^3)j + (a^1 b^2 - a^2 b^1)k.$$

Eq. (1.3.5) shows that the vector product is anticommutative:

$$a \times b = -b \times a.$$

Given three vectors, a, b and c, one can produce the *triple products*

$$(a \cdot b)c, \quad a \times (b \times c), \quad a \cdot (b \times c).$$

The *vector triple product* $a \times (b \times c)$ is conveniently computed by

$$a \times (b \times c) = (a \cdot c)b - (a \cdot b)c. \qquad (1.3.6)$$

The *mixed product* $a \cdot (b \times c)$ is the *vector-scalar product* sometimes called the *scalar triple product*. It is scalar and is equal to the volume of the parallelepiped having a, b, c as its edges, provided that the triplet a, b, c forms a right-handed system. The mixed product satisfies the equations

$$a \cdot (b \times c) = b \cdot (c \times a) = c \cdot (a \times b). \qquad (1.3.7)$$

The mixed product is written in the Cartesian coordinates as follows. Let

$$a = (a^1, a^2, a^3), \quad b = (b^1, b^2, b^3), \quad c = (c^1, c^2, c^3).$$

Then

$$a \cdot (b \times c) = \begin{vmatrix} a^1 & a^2 & a^3 \\ b^1 & b^2 & b^3 \\ c^1 & c^2 & c^3 \end{vmatrix}. \qquad (1.3.8)$$

1.3.2 Vector functions

A *vector function* $a = f(t)$ of a *scalar variable* t is a variable vector depending on t. In coordinates, it can be identified with a triplet of scalar functions

$$a^1 = f_1(t), \quad a^2 = f_2(t), \quad a^3 = f_3(t)$$

and written as follows:

$$a = f_1(t)i + f_2(t)j + f_3(t)k.$$

The derivative of a vector function is defined by

$$a' \equiv \frac{d f(t)}{dt} = \lim_{\Delta t \to 0} \frac{f(t + \Delta t) - f(t)}{\Delta t}.$$

In coordinates:

$$a' = f_1'(t)i + f_2'(t)j + f_3'(t)k.$$

Likewise, the second-order derivative is given by

$$a'' = f_1''(t)i + f_2''(t)j + f_3''(t)k.$$

Vector functions obey the usual rules of differentiation, e.g.

$$(a + b)' = a' + b', \quad (\varphi a)' = \varphi' a + \varphi a',$$

$$(a \cdot b)' = a' \cdot b + a \cdot b', \quad (a \times b)' = a' \times b + a \times b'.$$

Let a vector a be a function of a scalar variable φ and let $\varphi = \varphi(t)$:

$$a = f(\varphi(t)).$$

Then the chain rule for vector functions is written in the usual form:

$$\frac{da}{dt} = \frac{df}{d\varphi} \frac{d\varphi}{dt}.$$

1.3.3 Vector fields

A rapid treatment of differential calculus of vector fields is given here. All calculations are given in a rectangular Cartesian reference frame. Consequently the independent variables are the coordinates x, y, z of the position vector $\boldsymbol{x} = (x, y, z)$. All functions under consideration are assumed to be continuously differentiable.

A *scalar field* ϕ is a function of the position vector $\boldsymbol{x} = (x, y, z)$:

$$\phi = \phi(x, y, z).$$

A *vector field* $\boldsymbol{a} = (a^1, a^2, a^3)$ is a vector function

$$a = a(x, y, z)$$

depending upon the position vector $\boldsymbol{x} = (x, y, z)$.

Partial derivatives of scalar and vector fields are defined in a usual way, as it was done above for vector functions of one variable.

Hamilton's operator or the *symbolic differential operator* ∇ is a vector given in the rectangular Cartesian coordinates (x, y, z) by

$$\nabla = i\frac{\partial}{\partial x} + j\frac{\partial}{\partial y} + k\frac{\partial}{\partial z}, \tag{1.3.9}$$

where i, j and k are unit vectors along the x, y and z axes, respectively. In other words, ∇ is the vector operator with the components

$$\nabla_x = \frac{\partial}{\partial x}, \quad \nabla_y = \frac{\partial}{\partial y}, \quad \nabla_z = \frac{\partial}{\partial z} \tag{1.3.10}$$

along the x, y, z axes. The familiar operations of gradient, divergence and curl are written via Hamilton's operator as follows.

The *gradient* of a scalar field $\phi = \phi(x, y, z)$ is the product of the vector ∇ by the scalar:

$$\operatorname{grad}\phi \overset{\text{def}}{=} \nabla\phi = \frac{\partial\phi}{\partial x}i + \frac{\partial\phi}{\partial y}j + \frac{\partial\phi}{\partial z}k. \tag{1.3.11}$$

The *divergence* of a vector field $\boldsymbol{a}$ is the scalar product of the vectors ∇ and $\boldsymbol{a} = (a^1, a^2, a^3)$:

$$\operatorname{div} \boldsymbol{a} \overset{\text{def}}{=} \nabla \cdot \boldsymbol{a} = \nabla_x a^1 + \nabla_y a^2 + \nabla_z a^3 \equiv \frac{\partial a^1}{\partial x} + \frac{\partial a^2}{\partial y} + \frac{\partial a^3}{\partial z}. \qquad (1.3.12)$$

The *curl* or *rotation* of a vector field $\boldsymbol{a} = \boldsymbol{a}(x, y, z)$ is the vector product of the vectors ∇ and $\boldsymbol{a}$:

$$\operatorname{curl} \boldsymbol{a} \equiv \operatorname{rot} \boldsymbol{a} \overset{\text{def}}{=} \nabla \times \boldsymbol{a} = \begin{vmatrix} \boldsymbol{i} & \boldsymbol{j} & \boldsymbol{k} \\ \dfrac{\partial}{\partial x} & \dfrac{\partial}{\partial y} & \dfrac{\partial}{\partial z} \\ a^1 & a^2 & a^3 \end{vmatrix} \qquad (1.3.13)$$

or

$$\operatorname{curl} \boldsymbol{a} = \left(\frac{\partial a^3}{\partial y} - \frac{\partial a^2}{\partial z} \right) \boldsymbol{i} + \left(\frac{\partial a^1}{\partial z} - \frac{\partial a^3}{\partial x} \right) \boldsymbol{j} + \left(\frac{\partial a^2}{\partial x} - \frac{\partial a^1}{\partial y} \right) \boldsymbol{k}. \qquad (1.3.14)$$

The operator ∇ together with formulae of *vector algebra* enables one to relate scalar and vector fields through differentiation. Let $\boldsymbol{a}$, $\boldsymbol{b}$ and ϕ, ψ be vector and scalar fields, respectively, and let α, β be arbitrary constants. The operator ∇ has the following properties:

1. $\nabla(\alpha\phi + \beta\psi) = \alpha\nabla\phi + \beta\nabla\psi$

2. $\nabla \cdot (\alpha\boldsymbol{a} + \beta\boldsymbol{b}) = \alpha\nabla \cdot \boldsymbol{a} + \beta\nabla \cdot \boldsymbol{b}$

3. $\nabla \times (\alpha\boldsymbol{a} + \beta\boldsymbol{b}) = \alpha\nabla \times \boldsymbol{a} + \beta\nabla \times \boldsymbol{b}$

4. $\nabla(\phi\psi) = \psi\nabla\phi + \phi\nabla\psi$ $\qquad\qquad\qquad\qquad\qquad (1.3.15)$

5. $\nabla \cdot (\phi\boldsymbol{a}) = (\nabla\phi) \cdot \boldsymbol{a} + \phi(\nabla \cdot \boldsymbol{a})$

6. $\nabla \times (\phi\boldsymbol{a}) = (\nabla\phi) \times \boldsymbol{a} + \phi\nabla \times \boldsymbol{a}$

7. $\nabla \cdot (\boldsymbol{a} \times \boldsymbol{b}) = \boldsymbol{b} \cdot (\nabla \times \boldsymbol{a}) - \boldsymbol{a} \cdot (\nabla \times \boldsymbol{b})$

8. $\nabla \cdot (\nabla\phi) \equiv \nabla^2\phi \equiv \Delta\phi = \dfrac{\partial^2\phi}{\partial x^2} + \dfrac{\partial^2\phi}{\partial y^2} + \dfrac{\partial^2\phi}{\partial z^2}$

9. $\nabla \cdot (\nabla \times \boldsymbol{a}) = 0, \quad$ 10. $\nabla \times (\nabla\phi) = 0.$

These properties are often written in notation (1.3.11)~(1.3.12). For example, properties 5, 9 and 10 can be written as:

$5'.$ $\operatorname{div}(\phi\boldsymbol{a}) = \phi\operatorname{div}\boldsymbol{a} + \boldsymbol{a} \cdot \operatorname{grad}\phi, \quad$ $9'.$ $\operatorname{div}\operatorname{rot}\boldsymbol{a} = 0, \quad$ $10'.$ $\operatorname{rot}\operatorname{grad}\phi = 0.$

1.3.4 Three classical integral theorems

Green's theorem: Let V be an arbitrary region in the (x, y) plane with the boundary ∂V. Then

$$\int_{\partial V} P dx + Q dy = \int_V \left(\frac{\partial Q}{\partial x} - \frac{\partial P}{\partial y} \right) dx dy \qquad (1.3.16)$$

for any (differentiable) functions $P(x, y)$ and $Q(x, y)$.

Stokes' theorem: Let V be an (orientable) surface in the space (x, y, z) with the boundary ∂V, and let $P(x, y, z), Q(x, y, z)$ and $R(x, y, z)$ be any (differentiable) functions. Then

$$\int_{\partial V} P dx + Q dy + R dz$$

$$= \int_V \left(\frac{\partial Q}{\partial x} - \frac{\partial P}{\partial y} \right) dx \, dy + \left(\frac{\partial R}{\partial y} - \frac{\partial Q}{\partial z} \right) dy \, dz + \left(\frac{\partial P}{\partial z} - \frac{\partial R}{\partial x} \right) dz \, dx. \quad (1.3.17)$$

Eq. (1.3.17) is written in the vector notation as follows:

$$\int_{\partial V} \boldsymbol{A} \cdot d\boldsymbol{x} = \int_V \operatorname{curl} \boldsymbol{A} \cdot d\boldsymbol{S}.$$

Here $\boldsymbol{A} = (P, Q, R)$, $d\boldsymbol{x} = (dx, dy, dz)$ and $d\boldsymbol{S} = \boldsymbol{\nu} dS$, where $\boldsymbol{\nu}$ is the unit outward normal to the surface V.

The divergence theorem (the Gauss-Ostrogradsky theorem): Let V be a volume in the space (x, y, z) with the closed boundary ∂V and $\boldsymbol{A}$ be any vector field. Then

$$\int_{\partial V} (\boldsymbol{A} \cdot \boldsymbol{\nu}) \, dS = \int_V (\nabla \cdot \boldsymbol{A}) \, dx \, dy \, dz \equiv \int_V \operatorname{div} \boldsymbol{A} \, dx \, dy \, dz, \quad (1.3.18)$$

where $\boldsymbol{\nu}$ is the unit outward normal to the boundary ∂V of V.

1.3.5 The Laplace equation

The differential operator of the second order

$$\Delta \equiv \nabla^2 = \frac{\partial^2}{\partial x^2} + \frac{\partial^2}{\partial y^2} + \frac{\partial^2}{\partial z^2} \quad (1.3.19)$$

is called the *Laplacian*. The *Laplace equation* in three variables,

$$\Delta \phi \equiv \frac{\partial^2 \phi}{\partial x^2} + \frac{\partial^2 \phi}{\partial y^2} + \frac{\partial^2 \phi}{\partial z^2} = 0, \quad (1.3.20)$$

is one of the most important equations of mathematical physics. It is a linear partial differential equation of the second order.

Example 1.3.1. The spherically symmetric solution, $\phi = \phi(r)$, to the Laplace equation (1.3.20) has the form (see Problem 1.21)

$$\phi(r) = \frac{C_1}{r} + C_2,$$

where $r = \sqrt{x^2 + y^2 + z^2}$ and $C_1, C_2 = \text{const}$.

1.3.6 Differentiation of determinants

The derivative of determinants is given by the following formula:

$$\frac{\mathrm{d}}{\mathrm{d}x}\begin{vmatrix} a_{11}(x) & a_{12}(x) & \cdots & a_{1n}(x) \\ \vdots & \vdots & & \vdots \\ a_{i1}(x) & a_{i2}(x) & \cdots & a_{in}(x) \\ \vdots & \vdots & & \vdots \\ a_{n1}(x) & a_{n2}(x) & \cdots & a_{nn}(x) \end{vmatrix} = \sum_{i=1}^{n} \begin{vmatrix} a_{11}(x) & a_{12}(x) & \cdots & a_{1n}(x) \\ \vdots & \vdots & & \vdots \\ a'_{i1}(x) & a'_{i2}(x) & \cdots & a'_{in}(x) \\ \vdots & \vdots & & \vdots \\ a_{n1}(x) & a_{n2}(x) & \cdots & a_{nn}(x) \end{vmatrix},$$

where $a'_{ij}(x) = \mathrm{d}a_{ij}(x)/\mathrm{d}x$.

1.4 Notation of differential algebra

The calculus of differential algebra furnishes us with a *convenient language and effective devices* for tackling differential equations. In the classical mathematical analysis, it is customary to deal with *functions* $u^\alpha(x)$, where $\alpha = 1, \ldots, m$, and $x = (x^1, \ldots, x^n)$. The derivatives

$$u_i^\alpha(x) = \frac{\partial u^\alpha(x)}{\partial x^i}, \quad u_{ij}^\alpha(x) = \frac{\partial^2 u^\alpha(x)}{\partial x^i \partial x^j}, \ldots$$

are also regarded as functions of x.

Differential algebra suggests to treat the quantities u^α, u_i^α, $u_{ij}^\alpha, \ldots$ as variables and to deal with composite functions $f(x, u(x), \partial u(x)/\partial x, \ldots)$ as with functions $f(x, u, u_{(1)}, \ldots)$ of the independent variables $x, u, u_{(1)}, \ldots$.

1.4.1 Differential variables. Total differentiation

Let us start with the one-dimensional case. Namely, let us consider one independent variable x and one dependent variable y with the successive derivatives $y', y'', \ldots, y^{(s)}, \ldots$. The *total derivative* (see (1.2.66))

$$D_x = \frac{\partial}{\partial x} + y'\frac{\partial}{\partial y} + y''\frac{\partial}{\partial y'} + \cdots + y^{(s+1)}\frac{\partial}{\partial y^{(s)}} + \cdots \qquad (1.4.1)$$

acts on *differential functions,* i.e., functions

$$f(x, y, y_{(1)}, \ldots) \qquad (1.4.2)$$

of any finite number of the *independent variables*

$$x, \ y, \ y_{(1)} = y', \ y_{(2)} = y'', \ldots. \qquad (1.4.3)$$

The set of all differential functions will be denoted by $\mathcal{A}$.

Let us illustrate the difference between D_x and the partial differentiation $\partial/\partial x$ with respect to x by considering the action of both operations to the following functions from $\mathcal{A}$:

$$f = x, \quad f = y, \quad f = xy'.$$

The total differentiation yields:

$$D_x(x) = 1, \quad D_x(y) = y', \quad D_x(xy') = y' + xy''$$

whereas the partial derivatives are:

$$\frac{\partial x}{\partial x} = 1, \quad \frac{\partial y}{\partial x} = 0, \quad \frac{\partial(xy')}{\partial x} = y'.$$

1.4.2 Higher derivatives of the product and of composite functions

The formula for higher derivatives of the product of functions applies to arbitrary differential functions as well. Namely, if $f, g \in \mathcal{A}$, then

$$D_x^k(fg) = D_x^k(f)g + \sum_{s=1}^{k-1} \frac{k!}{(k-s)!s!} D_x^{k-s}(f)D_x^s(g) + fD_x^k(g). \qquad (1.4.4)$$

Furthermore, there is a formula due to Faà de Bruno (1857) for computing higher derivatives of composite functions. It extends the chain rule (1.2.5). Namely, consider a differential function of the form $f(y)$. Then its kth-order derivative is given, in terms of $y', \ldots, y^{(k)}$ and

$$f' = \frac{df}{dy}, \quad f'' = \frac{d^2 f}{dy^2}, \ldots, \quad f^{(k)} = \frac{d^k f}{dy^k}$$

by the following formula:

$$D_x^k(f) = \sum \frac{k!}{l_1! l_2! \cdots l_k!} f^{(p)} \left(\frac{y'}{1!}\right)^{l_1} \left(\frac{y''}{2!}\right)^{l_2} \cdots \left(\frac{y^{(s)}}{s!}\right)^{l_s} \cdots \left(\frac{y^{(k)}}{k!}\right)^{l_k},$$
$$(1.4.5)$$

where the sum runs through all non-negative integers $l_1, \ldots, l_k$ such that

$$l_1 + 2l_2 + \cdots + kl_k = k, \qquad (1.4.6)$$

and p is the positive integer defined, for every solution set $l_1, \ldots, l_k$ of (1.4.6), by

$$p = l_1 + l_2 + \cdots + l_k. \qquad (1.4.7)$$

For example, if one needs the third derivative, one solves Eq. (1.4.6)—(1.4.7) with $k = 3$, i.e.,

$$l_1 + 2l_2 + 3l_3 = 3,$$

$$p = l_1 + l_2 + l_3.$$

They have three sets of solutions, given by the following values of l_1, l_2, l_3 (different from zero) and p :

$$1) \ l_1 = 3, \ p = 3; \quad 2) \ l_1 = 1, \ l_2 = 1, \ p = 2; \quad 3) \ l_3 = 1, \ p = 1.$$

Hence, we have

$$D_x^3(f) = f''' \, y'^3 + 3f'' \, y' \, y'' + f' \, y'''.$$

1.4.3 Differential functions with several variables

Now we deal with the following algebraically independent variables:

$$x = \{x^i\}, \ u = \{u^\alpha\}, \quad u_{(1)} = \{u_i^\alpha\}, \quad u_{(2)} = \{u_{ij}^\alpha\}, \ldots, \qquad (1.4.8)$$

where the index i runs over the values from 1 to n, and α from 1 to m. The variables u_{ij}^α, etc. are assumed to be symmetric in subscripts, i.e., $u_{ij}^\alpha = u_{ji}^\alpha$.

For every $i = 1, \ldots, n$, we introduce the operator

$$D_i = \frac{\partial}{\partial x^i} + u_i^\alpha \frac{\partial}{\partial u^\alpha} + u_{ij}^\alpha \frac{\partial}{\partial u_j^\alpha} + \cdots \qquad (1.4.9)$$

and call it the *total differentiation* with respect to x^i. The operator D_i is a formal sum of an infinite number of terms. However, it truncates when acting on any function of a finite number of the variables $x, u, u_{(1)}, \ldots$. In consequence, the total differentiations D_i are well defined on the set of all functions depending on a finite number of $x, u, u_{(1)}, \ldots$. For example, one can readily verify that

$$D_i(x^k) = \delta_i^k, \quad D_i(u^\beta) = u_i^\beta, \quad D_i(u_k^\beta) = u_{ik}^\beta, \quad D_i(f(u^1)) = \frac{df}{du^1} u_i^1,$$

where δ_i^k are the Kronecker symbols defined by

$$\delta_i^k = 1 \quad \text{if} \quad i = k; \qquad \delta_i^k = 0 \quad \text{if} \quad i \neq k.$$

Thus, though variables (1.4.8) are *algebraically independent*, they are connected by the following *differential relations*:

$$u_i^\alpha = D_i(u^\alpha), \quad u_{ij}^\alpha = D_j(u_i^\alpha) = D_j D_i(u^\alpha), \ldots. \qquad (1.4.10)$$

The quantities x^i are called *independent variables*, whereas u^α are termed *differential variables* with the successive *derivatives* $u_i^\alpha, u_{ij}^\alpha, \ldots$ of the first, second, etc. orders, due to relations (1.4.10). We call an analytic function of a finite set of the variables $x, u, u_{(1)}, \ldots$ a *differential function*. The maximal order p of a derivative involved in the differential function $f = f(x, u, u_{(1)}, \ldots, u_{(p)})$ is termed the *order* of this function and is denoted by $\mathrm{ord} f$. If f is a differential function of the pth order, then its total derivatives D_i are differential functions of order $p + 1$. The set of all differential functions of finite order endowed with the differentiations D_i (1.4.9) is called the *space of differential functions* and is denoted by $\mathcal{A}$.

1.4.4 The frame of differential equations

Let $F \in \mathcal{A}$ be a differential function of the order p. The equation

$$F(x, u, u_{(1)}, \ldots, u_{(p)}) = 0 \qquad (1.4.11)$$

defines a surface in the space of the variables $x, u, \ldots, u_{(p)}$.

The concept of a differential equation comprises two distinctly different ingredients, namely:

(1) a *surface* of form (1.4.11) called the *frame of a differential equation* (see Fig.1.1);

(2) a *class of solutions* suggested by mathematical or physical content of a differential equation.

In the classical literature, the solutions to differential equations were identified solely with sufficiently differentiable functions. Problems of modern mathematics and physics require that the concept of solutions was broadened by considering *generalized solutions* (distributions) instead of the classical solutions.

In integrating ordinary differential equations, a decisive step is that of simplifying the frame by a change of variables. Lie group analysis furnishes a method for determining a suitable change of variables. Provided that an infinitesimal symmetry is known, we merely introduce so-called *canonical variables*. This simplifies the equation by converting its frame into a cylinder, i.e., the explicit dependence of one of the variables x or y has been eliminated.

As an example, consider the Riccati equation from Fig. 1.1. We will straightened out the curved surface given in Fig. 1.1 into a cylinder given in Fig. 1.2.

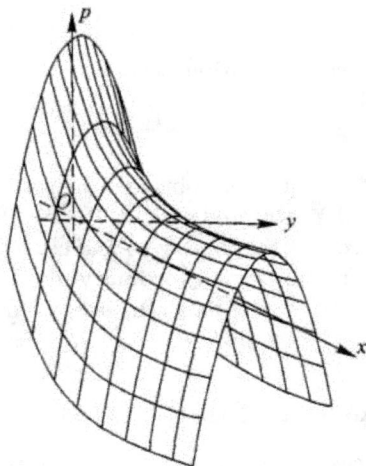

Figure 1.1: The frame of the Riccati equation $y' + y^2 - 2/x^2 = 0$, $p = y'$.

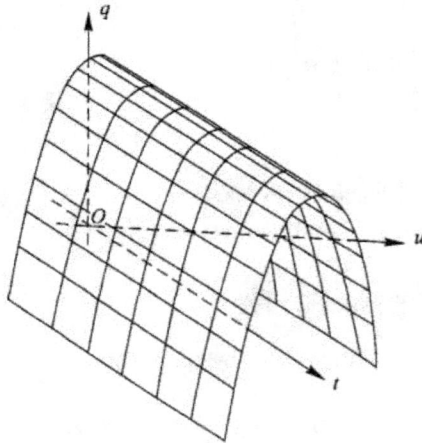

Figure 1.2: The frame of the equation $u' + u^2 - u - 2 = 0$, $q = u'$.

The Riccati equation

$$y' + y^2 - \frac{2}{x^2} = 0$$

is invariant under the dilation

$$\bar{x} = a\,x, \quad \bar{y} = \frac{y}{a}\,.$$

Lie group analysis readily gives the canonical variables t and u defined by

$$t = \ln x, \quad u = xy.$$

In these variables the Riccati equation becomes

$$u' + u^2 - u - 2 = 0.$$

The frame of the latter equation, obtained by setting $u' = q$, is a parabolic cylinder, $q + u^2 - u - 2 = 0$, protracted along t. Hence, the hyperbolic paraboloid of Fig. 1.1 is straightened out by passing to the canonical variables.

Similar approach is fruitful for all (ordinary and partial) differential equations with known symmetries. For this purpose, we use methods of Lie group analysis in this book.

1.4.5   Transformation of derivatives

Let us begin with the case of one independent variable x and one dependent variable y. Consider a change of variables

$$\bar{x} = \varphi(x, y), \quad \bar{y} = \psi(x, y). \tag{1.4.12}$$

Theorem 1.4.1. The change of variables (1.4.12) implies the change of the total differentiation:

$$D_x = D_x(\varphi)D_{\overline{x}}, \qquad (1.4.13)$$

where D_x and $D_{\overline{x}}$ are the total differentiations with respect to x and $\overline{x}$, respectively, and the following transformations of the successive derivatives:

$$\overline{y}' = \frac{D_x(\psi)}{D_x(\varphi)}, \quad \overline{y}'' = \frac{D_x(\varphi)D_x^2(\psi) - D_x(\psi)D_x^2(\varphi)}{[D_x(\varphi)]^3}, \dots \qquad (1.4.14)$$

where $\overline{y}' = d\overline{y}/d\overline{x}$, $\overline{y}'' = d\overline{y}'/d\overline{x}$.

Proof. The change of the total differentiation (1.4.13) is obtained from the first equation of (1.4.12) by invoking the chain rule. To prove the first equation of (1.4.14), it suffices to act by (1.4.13) to the second equation of (1.4.12) and obtain $D_x(\varphi)D_{\overline{x}}(\overline{y}) = D_x(\psi)$. Alternatively, one can use the following calculations:

$$\frac{d\overline{y}}{d\overline{x}} = \frac{\psi_x dx + \psi_y dy}{\varphi_x dx + \varphi_y dy} = \frac{(\psi_x + y'\psi_y)dx}{(\varphi_x + y'\varphi_y)dx} = \frac{\psi_x + y'\psi_y}{\varphi_x + y'\varphi_y} = \frac{D_x(\psi)}{D_x(\varphi)}.$$

The second equation of (1.4.14) and the higher derivatives are obtained by iterating the procedure:

$$\frac{d\overline{y}'}{d\overline{x}} = \frac{D_x(\overline{y}')}{D_x(\varphi)} = \frac{1}{D_x(\varphi)} D_x\left(\frac{D_x(\psi)}{D_x(\varphi)}\right).$$

Example 1.4.1. The rule for differentiation of inverse functions (1.2.4) follows simply from formula (1.4.14) applied to the change of variables

$$\overline{x} = y, \quad \overline{y} = x.$$

Indeed, since $\varphi = y$, $\psi = x$, Eqs. (1.4.13) and (1.4.14) are written

$$D_x = y' D_{\overline{x}}$$

and

$$\overline{y}' = \frac{1}{y'}, \quad \overline{y}'' = -\frac{y''}{y'^3},$$

respectively.

Let us turn now to the general case of several variables and consider a change of variables

$$\overline{x}^i = \varphi^i(x, u), \quad i = 1, \dots, n, \qquad (1.4.15)$$

$$\overline{u}^\alpha = \psi^\alpha(x, u), \quad \alpha = 1, \dots, m. \qquad (1.4.16)$$

In differential algebra, the corresponding change of derivatives are easily obtained as follows. Let us note that the change of independent variables (1.4.15) implies the following relation between total differentiations D_i and $\overline{D}_i$ with respect to the old and new variables, respectively (see (1.4.13)):

$$D_i = \sum_{j=1}^{n} D_i(\varphi^j)\overline{D}_j, \quad i = 1, \ldots, n. \tag{1.4.17}$$

Then we differentiate both sides of Eq. (1.4.16) by using the equation $\overline{u}_i^\alpha = \overline{D}_i(\overline{u}^\alpha)$ and, omitting the sign of summation, write the result in the form $D_i(\psi^\alpha) = D_i(\varphi^j)\overline{D}_j(\overline{u}^\alpha) = \overline{u}_j^\alpha D_i(\varphi^j)$. Thus, the change of the derivatives of the first order is determined by

$$\overline{u}_j^\alpha D_i(\varphi^j) = D_i(\psi^\alpha), \tag{1.4.18}$$

or

$$\left(\frac{\partial\varphi^j}{\partial x^i} + u_i^\beta\frac{\partial\varphi^j}{\partial u^\beta}\right)\overline{u}_j^\alpha = \frac{\partial\psi^\alpha}{\partial x^i} + u_i^\beta\frac{\partial\psi^\alpha}{\partial u^\beta}.$$

It remains to solve the latter equation with respect to $\overline{u}_j^\alpha$. The second differentiation of Eq. (1.4.18) yields transformations of derivatives of the second order, etc.

Example 1.4.2. Let t, x be two independent variables and u a dependent variable. Let us introduce the new independent variables τ, ξ and the dependent variable v defined by

$$\tau = t, \quad \xi = u, \quad v = x.$$

Eqs. (1.4.17) and (1.4.18) yield:

$$D_t = D_\tau + u_t D_\xi, \quad D_x = u_x D_\xi$$

and

$$0 = v_\tau + u_t v_\xi, \quad 1 = u_x v_\xi,$$

respectively. Hence, the change of the first-order derivatives has the form:

$$u_t = -\frac{v_\tau}{v_\xi}, \quad u_x = \frac{1}{v_\xi}.$$

1.5 Variational calculus

1.5.1 Principle of least action

Hamilton's variational principle or the *principle of least action* states: The motion of a mechanical system with a *kinetic energy* $T(t, q, v)$ and a *potential energy* $U(t, q)$ is determined by the requirement that the trajectories of the particles of the system provide an extremum for the *action integral*

$$S = \int_{t_1}^{t_2} L(t, q, v)\mathrm{d}t, \tag{1.5.1}$$

where $L(t, q, v) = T - U$ is called the *Lagrangian* of the system. Here t is time, $q = (q^1, \ldots, q^s)$ denote the coordinates of the particles of the system, and $v = \dot{q} \equiv dq/dt$ are their velocities. The action is defined on the set of functions $q^\alpha = q^\alpha(t)$ such that the integral exists in an interval $t_1 \leq t \leq t_2$.

Consider a variation of q when it is replaced by $q + \delta q$. It is assumed that the increment is a function $\delta q = \delta q(t)$ such that it is small everywhere in the interval $t_1 \leq t \leq t_2$ and vanishes at the boundary, $\delta q(t_1) = \delta q(t_2) = 0$. Differentiation yields $\delta v = d[\delta q(t)]/dt$. This causes the following variation of the action integral (1.5.1): $\int_{t_1}^{t_2} [L(t, q + \delta q, v + \delta v) - L(t, q, v)]dt$. Expansion of the integrand in powers of the increments δq and δv yields the linear principal part of δS (summation in $\alpha = 1, \ldots, s$):

$$\delta S = \int_{t_1}^{t_2} \left(\frac{\partial L}{\partial q^\alpha} \delta q^\alpha + \frac{\partial L}{\partial v^\alpha} \delta v^\alpha \right) dt.$$

Upon integrating the second term by parts, it is written:

$$\delta S = \int_{t_1}^{t_2} \left[\frac{\partial L}{\partial q^\alpha} - D_t \left(\frac{\partial L}{\partial v^\alpha} \right) \right] \delta q^\alpha dt + \left[\frac{\partial L}{\partial v^\alpha} \delta q^\alpha \right]_{t_1}^{t_2}$$

or, invoking the boundary conditions $\delta q(t_1) = \delta q(t_2) = 0$:

$$\delta S = \int_{t_1}^{t_2} \left[\frac{\partial L}{\partial q^\alpha} - D_t \left(\frac{\partial L}{\partial v^\alpha} \right) \right] \delta q^\alpha dt, \quad \text{where} \quad D_t = \frac{\partial}{\partial t} + v^\alpha \frac{\partial}{\partial q^\alpha} + \dot{v}^\alpha \frac{\partial}{\partial v^\alpha}.$$

The necessary condition for integral (1.5.1) to have an extremum is that $\delta S = 0$. Since the time interval $t_1 \leq t \leq t_2$ and the increments δq^α are arbitrary, it follows:

$$\frac{\partial L}{\partial q^\alpha} - D_t \left(\frac{\partial L}{\partial v^\alpha} \right) = 0, \quad \alpha = 1, \ldots, s. \tag{1.5.2}$$

Differential equations (1.5.2) are known as the *Euler-Lagrange equations*. Thus, the trajectory $q = q(t)$ of a mechanical system with the Lagrangian $L(t, q, v)$ solves the Euler-Lagrange equations (1.5.2).

1.5.2 Euler-Lagrange equations with several variables

The case several independent variables $x = (x^1, \ldots, x^n)$ and dependent (differential) variables $u = (u^1, \ldots, u^m)$ is treated similarly. Let us use the differential algebraic notation and terminology.

Let $L \in \mathcal{A}$ be a differential function of the first order. Let $V \subset \mathbb{R}^n$ be an arbitrary n-dimensional volume in the space of the independent variables x with the boundary ∂V. An *action* is the integral

$$l[u] = \int_V L(x, u, u_{(1)}) dx. \tag{1.5.3}$$

It is also termed a *variational integral*. The variation $\delta l[u]$ of the integral (1.5.3), caused by the variation $u + h(x)$ of u, is defined as the principal linear part (in h) of the integral $\int_V [L(x, u + h, u_{(1)} + h_{(1)}) - L(x, u, u_{(1)})]dx$ and has the form:

$$\delta l[u] = \int_V \left[\frac{\partial L}{\partial u^\alpha} h^\alpha + \frac{\partial L}{\partial u_i^\alpha} h_i^\alpha \right] dx.$$

On integrating the second term by parts, it is written:

$$\delta l[u] = \int_V \left[\frac{\partial L}{\partial u^\alpha} - D_i\left(\frac{\partial L}{\partial u_i^\alpha} \right) \right] h^\alpha dx + \int_V D_i\left(\frac{\partial L}{\partial u_i^\alpha} h^\alpha \right) dx.$$

Using the divergence theorem (1.3.18), we obtain

$$\delta l[u] = \int_V \left[\frac{\partial L}{\partial u^\alpha} - D_i\left(\frac{\partial L}{\partial u_i^\alpha} \right) \right] h^\alpha dx + \int_{\partial V} \frac{\partial L}{\partial u_i^\alpha} h^\alpha \nu^i dx,$$

where $\nu = (\nu^1, \ldots, \nu^n)$ is the unit outer normal to ∂V. Provided that the functions $h^\alpha(x)$ vanish on the boundary ∂V, we arrive at the following:

$$\delta l[u] = \int_V \left[\frac{\partial L}{\partial u^\alpha} - D_i\left(\frac{\partial L}{\partial u_i^\alpha} \right) \right] h^\alpha dx.$$

A function $u = u(x)$ is called an *extremum* of the variational integral (1.5.3) if $\delta l[u(x)] = 0$ for any volume V and any increment $h = h(x)$ vanishing on ∂V. It follows from the above expression for $\delta l[u]$ that a necessary condition for u to be an extremum is given by the *Euler-Lagrange equations:*

$$\frac{\delta L}{\delta u^\alpha} \equiv \frac{\partial L}{\partial u^\alpha} - D_i\left(\frac{\partial L}{\partial u_i^\alpha} \right) = 0, \quad \alpha = 1, \ldots, m. \tag{1.5.4}$$

Equations (1.5.4) provide, in general, a system of m partial differential equations of the second order. $\delta L/\delta u^\alpha$ is called the *variational derivative*.

Problems to Chapter 1

1.1. How much bigger is the surface area of the northern hemisphere of the Earth than the area of the equatorial section of the Earth?

1.2. Find the inverse hyperbolic functions

$$\text{(i)} \ t = \text{arcsinh}\, x, \quad \text{(ii)} \ t = \text{arctanh}\, x, \quad \text{(iii)} \ t = \text{arccosh}\, x$$

by solving the following equations with respect to t :

$$x = \sinh t = \frac{e^t - e^{-t}}{2}, \quad x = \tanh t = \frac{e^t - e^{-t}}{e^t + e^{-t}}, \quad x = \cosh t = \frac{e^t + e^{-t}}{2}.$$

1.3. Prove formula (1.1.7):

$$\arcsin x = \arctan \frac{x}{\sqrt{1 - x^2}} .$$

Find the relation, similar to (1.1.7), between the inverse hyperbolic functions $\operatorname{arcsinh} x$ and $\operatorname{arctanh} x$.

1.4. Solve the cubic equation $x^3 - 3x^2 + x + 5 = 0$.

1.5. Work out each of the following integrals ($k, m \geq 0$ are any integers):

$$\int_{-\pi}^{\pi} \sin(kx) \sin(mx)\, dx, \quad m \neq k,$$

$$\int_{-\pi}^{\pi} \cos(kx) \cos(mx)\, dx, \quad m \neq k,$$

$$\int_{-\pi}^{\pi} \cos(kx) \sin(mx)\, dx, \quad m, k = 0, 1, 2, \ldots ,$$

$$\int_{-\pi}^{\pi} \sin^2(kx)\, dx, \quad \int_{-\pi}^{\pi} \cos^2(kx)\, dx, \quad k = 1, 2, \ldots .$$

1.6. Obtain the Maclaurin expansion (1.2.20) of the function $1/(1 - x)$.

1.7. Obtain the Maclaurin expansion (1.2.34) of the error function $\operatorname{erf}(x)$ using its definition (1.1.12).

1.8. Check for the functional independence the following functions f, g and h of three variables x, y, z :

$$\text{(i)} \quad f = \sqrt{x^2 - y^2}, \quad g = \sqrt{y^2 - z^2}, \quad h = x^2 - z^2;$$

$$\text{(ii)} \quad f = \sqrt{x^2 + y^2}, \quad g = \sqrt{y^2 + z^2}, \quad h = x^2 + z^2;$$

$$\text{(iii)} \quad f = \sqrt{x^2 + y^2}, \quad g = \sqrt{y^2 + z^2}, \quad h = x^2 - z^2.$$

1.9. Differentiate the hyperbolic functions $\sinh x, \cosh x$ and $\tanh x$.

1.10. Check for the linear independence of the following functions:

$$\text{(i)} \quad y_1(x) = e^x, \quad y_2(x) = e^{-x};$$

$$\text{(ii)} \quad y_1(x) = e^x, \quad y_2(x) = \cosh x, \quad y_3(x) = \sinh x,$$

$$\text{(iii)} \quad y_1(x) = e^x, \quad y_2(x) = e^{-x}, \quad y_3(x) = \sinh x,$$

$$\text{(iv)} \quad y_1(x) = e^x, \quad y_2(x) = \cosh x, \quad y_3(x) = \tanh x,$$

$$\text{(v)} \quad y_1(x) = \sinh x, \quad y_2(x) = \cosh x, \quad y_3(x) = \tanh x,$$

1.11. Evaluate: (i) $e^{i\pi}$,   (ii) $e^{i(\pi/2)}$,   (iii) i^i.

1.12. Find the divergence $\nabla \cdot \boldsymbol{x}$ and the curl $(\nabla \times \boldsymbol{x})$ of the position vector $\boldsymbol{x} = (x, y, z)$.

1.13. Let ϕ, ψ and $\boldsymbol{a}$ be scalar and vector fields, respectively. Evaluate:

$$\begin{aligned}
&\text{(i)} \quad \nabla \times (\nabla \phi) \equiv \mathrm{curl}(\mathrm{grad}\phi), \\
&\text{(ii)} \quad \nabla \cdot (\nabla \times \boldsymbol{a}) \equiv \mathrm{div}(\mathrm{curl}\boldsymbol{a}), \\
&\text{(iii)} \quad \nabla \cdot (\boldsymbol{a} \times \boldsymbol{x}) \equiv \mathrm{div}(\boldsymbol{a} \times \boldsymbol{x}), \\
&\text{(iv)} \quad \nabla \times (\nabla \times \boldsymbol{a}) \equiv \mathrm{curl}(\mathrm{curl}\boldsymbol{a}), \\
&\text{(v)} \quad \nabla \cdot (\phi \nabla \psi - \psi \nabla \phi).
\end{aligned}$$

1.14. Convert the following volume integral into a surface integral:

$$\int_V (\phi \Delta \psi - \psi \Delta \phi) \, dx \, dy \, dz.$$

1.15. Work out expressions (1.4.14) for the change of the first and second derivatives.

1.16. Obtain the transformation of the first and second derivatives under the change of variables $\bar{x} = e^x$, $\bar{y} = 1/y$.

1.17. Obtain the transformation of the third derivative y''' under the change of variables $\bar{x} = y$, $\bar{y} = x$ (see Example (1.4.1)).

1.18. Solve the equation $w^3 + 1 = 0$.

1.19. Prove the mean value theorem for definite integrals:

$$\int_a^b f(x)dx = f(\xi)\,(b - a), \quad a \le \xi \le b.$$

1.20. Prove that if $g(x)$ does not change the sign when $a \le x \le b$, then

$$\int_a^b f(x)g(x)dx = f(\xi) \int_a^b g(x)dx, \quad a \le \xi \le b.$$

1.21. Obtain the spherically invariant solution to the Laplace equation given in Example 1.3.1 by looking for functions $\phi = \phi(r)$ satisfying the Laplace equation (1.3.20), $\Delta\phi = 0$.

1.22. Derive Eqs. (1.1.45).

1.23. Prove Lemma 1.1.1.

1.24. Show that transformation (1.1.54) maps Eq. (1.1.39) to an equation of form (1.1.55).

1.25. According to Remark 1.1.7, the elliptic and hyperbolic curves are connected by complex linear transformations. Hence, Eq. (1.2.70) for circles and Eq. (1.2.71) for hyperbolas should be connected by a complex transformation. Find this complex transformation.

1.26. Prove the rule for differentiating determinants given in Section 1.3.6.

1.27. According to Eqs. (1.1.16), the surface area ω_n of the unit sphere in n dimensions is written in terms of the Gamma function as follows:

$$\omega_n = \frac{2\sqrt{\pi^n}}{\Gamma\left(\frac{n}{2}\right)}.$$

Let here $n = 2$ and $n = 3$ and obtain the commonly known expressions for the circumference of the unit circle in the plane and the surface area of the unit sphere in the three-dimensional space, respectively.

1.28. Prove the first equation (1.1.15), $\Gamma(x + 1) = x\Gamma(x)$.

1.29. Evaluate $\Gamma(-1/2)$.

1.30. Evaluate the integral $\int_0^\infty e^{-s^2}ds$ by setting $t = s^2$ in definition (1.1.14) of $\Gamma(x)$ and letting $x = 1/2$.

1.31. Explain why formula (1.2.52) for the change of variables in multiple integrals contains the absolute value of the Jacobian J whereas the similar formula (1.2.11) in the one-dimensional case does not?

Chapter 2

Mathematical models

Differential equations, in a proper sense, have appeared in mathematics in the 1680s in the works of the creators of the differential and integral calculus. The term *differential equation* was mentioned by G.W. Leibnitz for the first time in his letter to I. Newton (1676) and then used in his publications after 1684. Newton's *Principles* [29] contains numerous differential equations formulated and integrated in the framework of elementary geometry. Since then, the formulation of fundamental natural laws and of technological problems in the form of rigorous mathematical models is given frequently, even prevalently, in terms of differential equations.

Additional reading: R. Courant and D. Hilbert [5]~[4], M.D. Greenberg [11], N.H. Ibragimov [21], J.D. Murray [27]~[28], G.F. Simmons [35].

2.1 Introduction

Differential equations involve independent variables and dependent variables together with their derivatives. A differential equation is said to be of the nth order if it involves derivatives of this order but not higher.

If the dependent variables are functions of a single independent variable, the equations are termed *ordinary differential equations* (ODE). The renowned Newton's second law for a particle in an external force field F,

$$\frac{\mathrm{d}(m\boldsymbol{v})}{\mathrm{d}t} = \boldsymbol{F}, \tag{2.1.1}$$

falls precisely into this category. Here time t is the independent variable, m and $\boldsymbol{v} = (v^1, v^2, v^3)$ denote the particle mass (in general, m is not constant) and its vector velocity, respectively. Newton's equation (2.1.1) is a system of three equations of the first order with respect to the velocity components v^i regarded as the dependent variables, provided that the force $\boldsymbol{F}$ depends upon t and $\boldsymbol{v}$ alone. However, if $\boldsymbol{F} = \boldsymbol{F}(t, \boldsymbol{x}, \boldsymbol{v})$, we have a second-order equation

(2.1.1) for the position vector $\boldsymbol{x} = (x^1, x^2, x^3)$ of the particle. Indeed consider, for simplicity sake, the case of constant mass m and substitute $\boldsymbol{v} = \boldsymbol{x}'$ where the prime denotes differentiation with respect to t. Then equation (2.1.1) is rewritten as the following system of three second-order equations:

$$m\frac{\mathrm{d}^2\boldsymbol{x}}{\mathrm{d}t^2} = \boldsymbol{F}(t, \boldsymbol{x}, \boldsymbol{x}'). \tag{2.1.2}$$

If, on the other hand, unknown functions depend on several independent variables, so that the equations in question relate the independent variables, the dependent variables and their partial derivatives, then one deals with *partial differential equations* (PDE). A celebrated representative of this category is d'Alembert's equation for small transversal vibrations of strings:

$$\frac{\partial^2 u}{\partial t^2} - k^2\frac{\partial^2 u}{\partial x^2} = 0, \tag{2.1.3}$$

where k^2 is a positive constant. This is a second-order partial differential equation with two independent variables, time t and the coordinate x along the string. It is also known as the one-dimensional wave equation.

A model of small transversal vibrations of uniform slender rods provides a partial differential equation of the fourth order:

$$\frac{\partial^2 u}{\partial t^2} + \mu\frac{\partial^4 u}{\partial x^4} = f, \tag{2.1.4}$$

where f is a total force acting on the rod and μ is a positive constant.

The mathematical model of thermal diffusion, due to J.B.J. Fourier (1811), provides a partial differential equation known as the *heat conduction equation*. The one-dimensional heat equation has the form

$$\frac{\partial u}{\partial t} - k^2\frac{\partial^2 u}{\partial x^2} = 0. \tag{2.1.5}$$

2.2   Natural phenomena

2.2.1   Population models

Thomas Robert Malthus, the pioneer in the mathematical treatment of demographic problems, suggested in *An essay on the principle of population as it affects the future improvement of society* (1798) his celebrated principle of population. His model is mathematically rather simple and is based on the natural assumption that the rate of population growth is proportional to the population P considered. Accordingly, it is formulated by the differential equation

$$\frac{\mathrm{d}P}{\mathrm{d}t} = \alpha P, \quad \alpha = \text{const.} > 0. \tag{2.2.1}$$

Hence the Malthusian principle provides *unlimited growth* of a population according to the exponential law:

$$P(t) = P_0 e^{\alpha(t-t_0)}, \tag{2.2.2}$$

where P_0 and $P(t)$ denote the population in millions at the initial time $t = t_0$ and at an arbitrary time t, respectively. The main consequence of the *Essay* was that the realization of a happy society will always be hindered by the universal tendency of the population to outrun the means of subsistence.

It was soon seen, however, that Malthus' model is unrealistic and requires a modification. Subsequently, several modified population models have been considered in an attempt to find more realistic laws of population. One of them, known as the *logistic law*, is described by the following nonlinear equation:

$$\frac{dP}{dt} = \alpha P - \beta P^2, \quad \alpha, \beta = \text{const.} \neq 0, \tag{2.2.3}$$

where the nonlinear term βP^2 can be interpreted as a kind of *social friction*. An analysis of this law shows that it is adequate to describe certain insect populations. However, its value as a general "law" of population growth is extremely limited as far as human population is concerned.

A model of predator and prey suggested by A.J. Lotka (1925) and V. Volterra (1926) is formulated by a system of nonlinear ordinary differential equations of the first order,

$$\frac{dx}{dt} = (a - by)x, \quad \frac{dy}{dt} = (kx - l)y, \tag{2.2.4}$$

where $a, b, k,$ and l are positive constants. Here y denotes a predator species and x its prey. It is assumed that the prey population provides the total food supply for the predators. A qualitative analysis of solutions of the above system shows, e.g., that any biological system described by the Lotka-Volterra equations (2.2.4) ultimately approaches either a constant or periodic population.

2.2.2 Ecology: Radioactive waste products

Radioactivity is a consequence of the breaking up of elements with high atomic weights such as uranium minerals. The discovery of radioactivity provided new means, e.g. of determining geological time, etc. Artificial radioactivity is widely used in practical affairs – chemistry, medicine, nuclear energetics, etc. However, an industrial use of nuclear energy requires an inexorable vigilance by the population because of the danger of pollution by radioactive waste products.

A mathematical description of radioactive decay assumes that the rate of decay is proportional to the amount of a radioactive substance. Therefore, the mathematical model has form (2.2.1):

$$\frac{dU}{dt} = -kU. \tag{2.2.5}$$

Here U is the amount of a radioactive substance present at time t and k is a positive constant. The solution of Eq. (2.2.5) has the form

$$U(t) = U_0 e^{-k(t-t_0)}, \tag{2.2.6}$$

where U_0 is an initial $(t = t_0)$ amount of the substance. The empirical constant k depends on the radioactive matter in question. Usually, it is determined in terms of the so-called *half-life* defined as the interval of time $\Delta t = t - t_0$ after which the substance will have diminished to half of its original amount.

Example 2.2.1. It is known that the half-life of *radium* is $\Delta t = 1600$ years. Therefore, according to formula (2.2.6), $U_0/2 = U_0 e^{-1600k}$, whence $k = (\ln 2)/1600 \approx 0.00043$. Thus, the radioactive disintegration of radium in t years is given by

$$U(t) = U_0 e^{-\frac{\ln 2}{1600} t}.$$

2.2.3   Kepler's laws. Newton's gravitation law

The apparent motions of the planets appear to be irregular and complicated. However, it was obvious in the remote past that the heavens ought to exemplify mathematical beauty. This would only be the case if the planets moved in circles. Indeed, in Greek science one can find a hypothesis that all the planets, including the earth, go round the sun in circles. J. Kepler discovered, however, that planets move in ellipses, not in circles, with the sun at a focus, not at the center. He formulated in 1609 two of the cardinal principles of modern astronomy: *Kepler's first law* (Fig. 2.1) and *Kepler's second law* (Fig. 2.2). Kepler's third law published in 1619 asserts that the ratio T^2/R^3 of the square of the period T and the cube of the mean distance R from the sun is the same for all planets.

Kepler's laws reduce the motion of planets to geometry and reveal, at a new level, a mathematical harmony in nature. From a practical point of view, it was important that Kepler gave an answer, based on empirical astronomy, to the question of *how* the planets move. The geometry of the heavens provided by Kepler's laws challenged scientists to answer the question of *why* the planets obey these laws. The question required an investigation of the dynamics of the Solar system. The necessary dynamics had been initiated by Galileo Galilei and developed into modern rational mechanics by Newton in his *Principles*.

According to Newton's gravitation law, the force of attraction between the sun and a planet has the form

$$\boldsymbol{F} = \frac{\alpha}{r^3} \boldsymbol{x}, \quad \alpha = -GmM, \tag{2.2.7}$$

where G is the universal constant of gravitation, m and M are the masses of a planet and the sun, respectively, $\boldsymbol{x} = (x^1, x^2, x^3)$ is the position vector of the planet considered as a particle, and $r = |\boldsymbol{x}|$ is the distance of the planet from

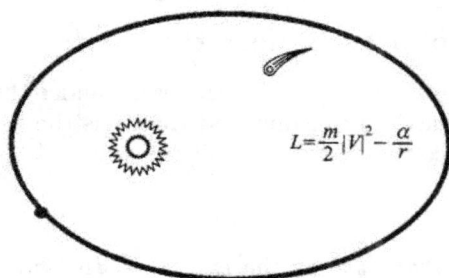

Figure 2.1: Kepler's first law: The orbit of a planet is an ellipse with the sun at one focus.

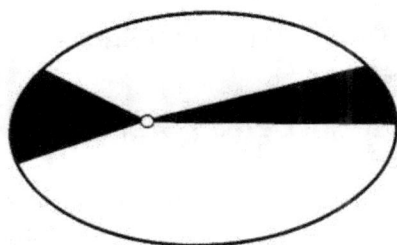

Figure 2.2: Kepler's second law: The areas swept out in equal times by the line joining the sun to a planet are equal.

the sun. Hence, ignoring the motion of the sun under a planet's attraction, Newton's second law (2.1.2) yields

$$m\frac{d^2\boldsymbol{x}}{dt^2} = \frac{\alpha}{r^3}\boldsymbol{x}, \quad \alpha = \text{const.} \tag{2.2.8}$$

The problem on integration of Eqs. (2.2.8) is referred to as *Kepler's problem*. Newton derived Kepler's laws by solving the differential equations (2.2.8). It can be shown however that the Kepler's laws are direct consequences of specific symmetries of Newton's gravitation force. Specifically, the first and second Kepler's laws can be derived, without integrating the nonlinear equations (2.2.8), from conservation of two vector fields, namely, conservation of the angular momentum

$$\boldsymbol{M} = m(\boldsymbol{x} \times \boldsymbol{v}), \tag{2.2.9}$$

where the vector $\boldsymbol{v} = \dot{\boldsymbol{x}} \equiv d\boldsymbol{x}/dt$ is the velocity, and conservation of what is known as the Laplace vector

$$\boldsymbol{A} = [\boldsymbol{v} \times \boldsymbol{M}] + \frac{\alpha}{r}\boldsymbol{x}. \tag{2.2.10}$$

2.2.4   Free fall of a body near the earth

Consider the free fall of a body toward the earth under the assumption that the gravity near the earth is constant and that it is the only force acting on the object. Let $m =$ const. be the mass of the object, h its height above the ground, t time, and

$$g \approx 9.81 \text{ m/s}^2$$

is the *acceleration of gravity* near the earth. In this notation, the force of gravity is $F = -mg$ and Newton's equation (2.1.2) is written

$$\frac{d^2 h}{dt^2} = -g.$$

This is an equation of form (1.2.60) with constant $f = -g$. Consequently, integration (1.2.61) yields

$$h = -\frac{g}{2} t^2 + C_1 t + C_2.$$

By letting $t = 0$ in this solution and in the velocity $v \equiv h' = -gt + C_1$, one obtains the physical meaning of the integration constants, namely $C_2 = h_0$ is the initial position of the body and $C_1 = v_0$ is its initial velocity. Thus, the trajectory of a falling body is given by

$$h = -\frac{g}{2} t^2 + v_0 t + h_0. \tag{2.2.11}$$

Exercise 2.2.1. A body at rest ($v_0 = 0$) falls from the height h_0. Find its terminal velocity v_*, i.e., the velocity when the body reaches the ground.

Solution. By formula (2.2.11), $h = h_0 - gt^2/2$, $v = -gt$. Denoting t_* the instant when the body reaches the ground ($h = 0$) and v_* its velocity at that instant, one obtains $v_* = -gt_*$, $h_0 = gt_*^2/2$, whence eliminating t_*:

$$v_* = -\sqrt{2gh_0}. \tag{2.2.12}$$

The minus sign appears owing to the fact that the h axis is directed upwards from the surface of the earth whereas the body falls toward the earth.

2.2.5   Meteoroid

The fall of a distant body (meteoroid) before entering the earth's atmosphere is defined by Newton's second law of dynamics (2.1.1) together with his *law of inverse squares* according to which a meteoroid and the earth attract each other by the force $F = GmM/r^2$, where

$$G = 6.67 \times 10^{-8} \text{cm}^3/(\text{g} \cdot \text{s}^2)$$

is the universal gravitational constant, m and M denote the masses of a meteoroid and the earth, and r the distance between their centers. Let R be the radius of the earth. Then the value of the force of attraction on the surface of the earth is $F = GmM/R^2$. On the other hand, the gravitation force near the earth (i.e., the *weight* of a body with mass m) is written mg. Whence the equation $mg = GmM/R^2$, or $GmM = mgR^2$. Hence, the object is attracted to the earth by the force $F = mgR^2/r^2$.

The mass m of a meteoroid is constant before entering the earth's atmosphere. Let us ignore the air resistance and assume that the mass does not change along the whole trajectory of a falling meteoroid. Then Eq. (2.1.2) is written:

$$\frac{d^2r}{dt^2} = -\frac{gR^2}{r^2}. \tag{2.2.13}$$

The minus sign appears since r is directed from the earth to the meteoroid and hence it is opposite to the direction of the force of the gravitational attraction.

Exercise 2.2.2. Reduce the order of Eq. (2.2.13).

Solution. By letting $dr/dt = v(r)$ and noting that

$$\frac{d^2r}{dt^2} = \frac{dv}{dr}\frac{dr}{dt} = v\frac{dv}{dr} \equiv \frac{1}{2}\frac{d(v^2)}{dr},$$

we rewrite Eq. (2.2.13) in the form

$$\frac{d(v^2)}{dr} = -\frac{2gR^2}{r^2}.$$

We integrate it, take into account that, in our notation, the velocity is negative and obtain:

$$v = -\sqrt{\frac{2gR^2}{r} + C}, \quad C = \text{const.} \tag{2.2.14}$$

Exercise 2.2.3. Find the terminal velocity v_* (i.e., the velocity on the surface of the earth) of a meteoroid falling from a point at infinity where it was in rest.

Solution. Let us first specify the constant of integration in (2.2.14) by assuming that initially ($t = 0$) the meteoroid rested ($v_0 = 0$) at a distance r_0 from the center of the earth. Letting $t = 0$ and hence $v = 0$ in (2.2.14) yields that $C = -2gR^2/r_0$, and formula (2.2.14) becomes

$$v = -R\sqrt{2g}\sqrt{\frac{1}{r} - \frac{1}{r_0}}.$$

Letting $r_0 = \infty$ and $r = R$, one obtains the terminal velocity:

$$v_* = -\sqrt{2gR}. \tag{2.2.15}$$

Hence the meteoroid reaches the ground with the same velocity as a body falling from the height h_0 equal to the radius R of the earth (compare (2.2.15) with (2.2.12)).

2.2.6   A model of rainfall

The idea to suggest a simple model of this natural phenomenon came to my mind while flying in a small airplane amongst whimsically shaped thick clouds over Africa.

To start with, let me give a piece of information about clouds relevant to the first stage of the suggested model. The typical thickness of clouds producing precipitation is from 100 m to 4 km, but very thick clouds (*cumulonimbus*) may reach 20 km. As an approximation to a mathematical model of rainfall, let us simulate two successive stages of the phenomenon, the first stage being the *development of raindrops in clouds* and the second one being the *fall of raindrops through the air*.

1. *Developing drops*:

The onset of raindrops in clouds is imitated here by the free fall toward the earth of a spherical mass of water in saturated atmosphere under the force of gravity.

The mass m of a drop increases owing to condensation, the increment being proportional to time and to the surface area of the drop, i.e., $dm = 4\pi k r^2 dt$, where r is the radius of the drop and k is an empirical constant. On the other hand, the mass of a spherical drop of water (with density $\rho = 1$) is $m = 4\pi r^3/3$, whence $dm = 4\pi r^2 dr$. Hence, $dr = kdt$, and Newton's second law (2.1.1) with $F = -mg$ is written:

$$k\frac{d(r^3 v)}{dr} = -gr^3. \tag{2.2.16}$$

The solution of this differential equation satisfying the initial condition, $v = v_0$ when $r = r_0$, has the form:

$$v = -\frac{gr}{4k}\left(1 - \frac{r_0^4}{r^4}\right) + \frac{r_0^3}{r^3}v_0. \tag{2.2.17}$$

Typical cloud droplets have the radius $r_0 \approx 10~\mu$m, while raindrops reach the earth with radii about 1 mm. Let us assume, in our simplified model, that the initial radius r_0 of a drop is infinitely small. Then we let $r_0 = 0$ in solution (2.2.17) to obtain $v = -gr/(4k)$. Invoking the equation $r = kt$, we can write

$$v = -\frac{1}{4}gt. \tag{2.2.18}$$

Hence the magnitude $|v|$ of the velocity of raindrops, at the stage of their developing in clouds, increases as a linear function of time.

2. *Falling rain*:

This stage is imitated by the fall of raindrops through the air toward the earth. It is assumed that gravity and air resistance are the only forces acting on the object, e.g. the evaporation of falling drops is ignored.

Let air resistance be a function, $f(v)$, of the velocity v of drops only. Let us denote by m the mass of a raindrop at the instant when the drop leaves the

clouds and assume that it remains unaltered during the fall, $m = $ const. Then
the velocity of the raindrop is determined, according to Newton's second law,
by a differential equation of the first order:

$$m\frac{dv}{dt} = -mg + f(v), \qquad (2.2.19)$$

together with the initial condition

$$v\big|_{t=t_*} = v_*, \qquad (2.2.20)$$

where the notation $|_{t=t_*}$ means evaluated at $t = t_*$. Here t_* is the instant when
the raindrop leaves the clouds and v_* is its terminal velocity in that instant.
Provided that t_* and v_* are found from the first stage, one obtains the velocity
of raindrops by solving the *initial value problem* (2.2.19)~(2.2.20).

Commonly, it is assumed that air resistance is proportional to the square of
the velocity of a falling object provided that the object is not "very small" and
that its velocity is less than that of sound but not infinitely small. However,
under certain conditions air resistance can be approximated by a linear function
of velocity as well. Thus, one can consider, as a reasonable model of rainfall,
the following simple form of Eq. (2.2.19):

$$m\frac{dv}{dt} = -mg - \alpha v + \beta v^2, \qquad (2.2.21)$$

where $\alpha \geq 0$ and $\beta \geq 0$ are empirical constants. The choice of the signs is in
accordance with the fact that the air resistance opposes the force of gravity
and that v is negative in our coordinate system which is directed upwards.

2.3 Physics and engineering sciences

2.3.1 Newton's model of cooling

The phenomenon of cooling (heating) by a surrounding medium is commonly
used in everyday life. One immerses a body, for cooling (heating) it, in a
medium of lower (higher) temperature than that of the body. The medium
may be the surrounding air, a large cold bath, a preheated oven, etc., while
the body in question may be a thermometer, a hot metal plate to be cooled,
blood plasma stored at low temperature to be warmed before using, milk and
other liquids. It is assumed that the temperature T of the surrounding bath is
unaffected by the immersed body, i.e., $T = $ const. or, in general, T is a given
function $T(t)$.

It is assumed further that the temperature τ of the immersed body is the
same in all its parts at each instant so that $\tau = \tau(t)$. Then what is known as
Newton's law of cooling states simply that the rate of change of τ is proportional

to the temperature difference $T - \tau$. Newton's law of cooling is written as the following ordinary differential equation of the first order:

$$\frac{d\tau}{dt} = k(T - \tau), \qquad (2.3.1)$$

where k is a positive constant depending on the substance of the immersed body and that of the surrounding medium.

Example 2.3.1. Pasteurization provides a good example. Recall that pasteurization is the partial sterilization of milk without boiling it and is based on Louis Pasteur's discovery that germs in milk temporarily stop functioning if every particle of the milk is heated to $64°\,C$ and then the milk is quickly cooled.

Let us imagine an educated farmer who decided to pasteurize milk for the first time but, unfortunately, found that his thermometer was broken. Since our farmer is an educated one, we can fancy that he would solve the problem of warming the milk precisely to $64°$ having at his disposal only an oven and his watch, by using Eq. (2.3.1) instead of the broken thermometer as follows.

The farmer has firstly to determine the coefficient k. To that end, he places a cup of milk stored at room temperature $\tau_0 = 25°\,C$ in the oven set, e.g. at $T = 250°\,C$ and waits until the milk boils. Suppose it took 15 min for the milk to boil. Now, letting the boiling temperature of milk be $90°\,C$, the farmer uses the solution to Eq. (2.3.1):

$$\tau = T - Be^{-kt}. \qquad (2.3.2)$$

At the initial moment ($t = 0$) this equation is written $25° = 250° - B$ and specifies the constant of integration, $B = 225°$. Then the solution at $t = 15$ min yields:

$$90 = 250 - 225e^{-15k},$$

where we consider only the numerical values. Hence, $15k = -\ln(160/225)$, or $k \approx 34/1500$. Thus, Eq. (2.3.1) yields the following formula for the temperature of the milk placed in the oven at $250°$:

$$\tau = 250 - 225e^{-34t/1500}.$$

For $\tau = 64$ it follows $-34t/1500 = \ln(186/225) \approx -0.19$, whence $t \approx 8.4$ min. Thus, the farmer should warm the milk for 8 min 24 s in the oven set at $250°\,C$.

Newton's cooling law, appropriately adapted to real situations, provides a good approximation to modelling, e.g. the temperature dynamics inside a building. Indeed, let the inside temperature τ be an unknown function of time t. Let $T = T(t)$ be the outside temperature considered as a given function. We firstly note that Newton's law (2.3.1), where k is a *positive* constant, is in agreement with the natural expectation that the inside temperature τ increases ($d\tau/dt > 0$) when $T > \tau$, and decreases ($d\tau/dt < 0$) when $T < \tau$. The constant

k has the dimension of t^{-1} and depends on the quality of the building, in particular on its thermal insulation. In common situations, $0 < k < 1$, and it is infinitely small in ideally insulated buildings.

Suppose that the building is supplied by a heater and by an air conditioner. Denote by $H(t)$ the rate of increase in temperature inside the building caused by the heater, and by $A(t)$ the rate of change (increase or decrease) in temperature caused by the air conditioner. Assuming that these are the only factors affecting the temperature in the building, we have the following modification of Newton's cooling law (2.3.1):

$$\frac{d\tau}{dt} = k[T(t) - \tau] + H(t) + A(t). \tag{2.3.3}$$

As an example of Eq. (2.3.3) with $A(t) \neq 0$, let us assume that a furnace supplies heating at a given rate $H(t) \geq 0$ and that the building is provided with a thermostat to keep the inside temperature around a desired (*critical*) temperature τ_c. If the actual temperature $\tau(t)$ is above τ_c, the air conditioner supplies cooling, otherwise it is off, then $A(t) = l(\tau_c - \tau)$, where l is a positive empirical parameter, and Eq. (2.3.3) is written as

$$\frac{d\tau}{dt} = k[T(t) - \tau] + H(t) + l(\tau_c - \tau), \tag{2.3.4}$$

with given functions $T(t)$ and $H(t)$ and the constants k, τ_c, l.

Imagine the following situation. In a still cold winter evening when the outside temperature stayed constant at $T_0 = -10°\,\text{C}$, the electricity was shut-down in your house at $t_0 = 6\,\text{pm}$. This caused cessation of operation of your heater and air conditioner for the whole night. Suppose that the inside temperature at $t_0 = 6\,\text{pm}$ was $\tau_0 = 25°\,\text{C}$. Unfortunately, however, the door and windows in your building were not well insulated. Therefore, it took only an hour for the inside temperature to drop to $\tau = 19.5°\,\text{C}$.

Exercise 2.3.1. What temperature do you expect in your bedroom at 6 am, provided that the outside temperature stays at $T_0 = -10°\,\text{C}$ the whole night?

Solution. According to the given conditions, we use Newton's cooling law (2.3.1). Its solution is given by (2.3.2),

$$\tau(t) = T_0 - Be^{-kt}.$$

Letting $t = t_0$, we have $25 = -10 - Be^{-kt_0}$ or $B = -35e^{kt_0}$. Hence (in ° C):

$$T(t) = -10 + 35e^{-k(t-t_0)}.$$

The condition $\tau|_{t=t_0+1} = 19.5$ yields that $19.5 = -10 + 35e^{-k}$, whence $k = \ln(1.18) \approx 1/6$. Thus, the inside temperature is given by

$$T(t) = -10 + 35e^{-(t-t_0)/6}, \quad t_0 \le t \le t_1.$$

In particular, the temperature at $t = t_1 = 6$ am is $-5.25°$ C. See Fig. 2.3.

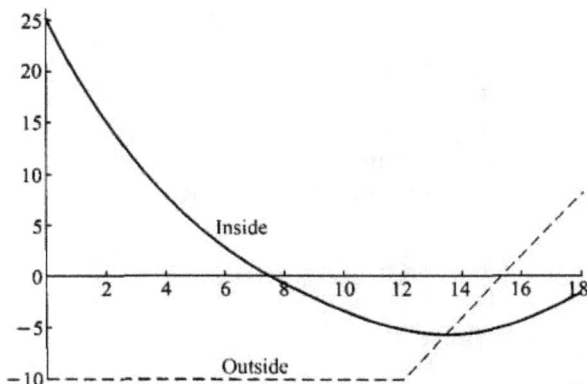

Figure 2.3: On Exercises 2.3.1 and 2.3.2.

Exercise 2.3.2. Let, in the conditions of Exercise 2.3.1, the outside temperature increase uniformly from $-10°$ C at 6 am to $+8°$ C at noon. Find the variation of the inside temperature during this time.

Solution. Invoking that the coefficient k of the building has the value $k = 1/6$ and that the variation of the outside temperature is given by

$$T(t) = -10 + 3(t - t_1), \quad t_1 \le t \le t_2,$$

where $t_1 = 6$, $t_2 = 12$, and solving the corresponding equation (2.3.1),

$$\frac{d\tau}{dt} = \frac{1}{6}[-10 + 3(t - t_1) - \tau],$$

we obtain the following variation of the inside temperature (see Fig. 2.3):

$$\tau = -28 + 3(t - t_1) + 22.75e^{-(t-t_1)/6}, \quad t_1 \le t \le t_2.$$

Exercise 2.3.3. Suppose now that the previous accident happened when the outside temperature was $+8°$ C at $t_0 = 6$ pm decreasing to $-10°$ C at $t_1 = 6$ am. Find the temperature variation in your house during the night.

Solution. Here, unlike the previous case, the outside temperature is unsteady and is given by $T(t) = 8 - \frac{3}{2}(t - t_0)$, where $t_0 \le t \le t_1$. Eq. (2.3.1):

$$\frac{d\tau}{dt} = k[8 - \frac{3}{2}(t - t_0) - \tau]$$

has the general solution

$$\tau(t) = 8 + \frac{3}{2k} - \frac{3}{2}(t - t_0) - Be^{-kt},$$

where $B = $ const. The initial condition $\tau\big|_{t=t_0} = 25$ yields $B = (\frac{3}{2k} - 17)e^{kt_0}$. We know from Exercise 2.3.3 that, for the building in question, $k = 1/6$. Hence, $B = -8e^{t_0/6}$. Thus, the variation of the inside temperature during the night is given by (in °C)

$$\tau(t) = 17 - \frac{3}{2}(t - t_0) + 8e^{-(t-t_0)/6}, \quad t_0 \le t \le t_1.$$

In particular, the temperature at $t_1 = 6\,\text{am}$ is around $0°$ C. See Fig. 2.4.

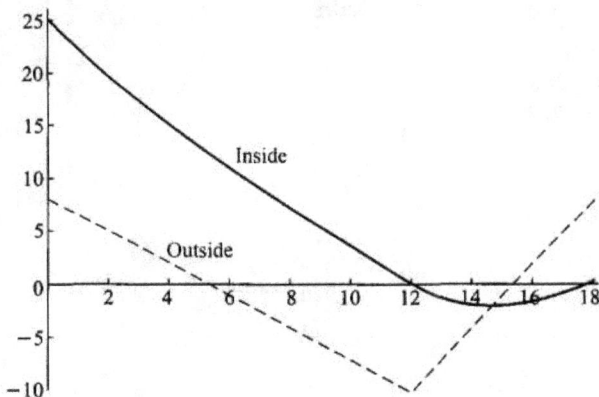

Figure 2.4: On Exercises 2.3.3 and 2.3.4.

Exercise 2.3.4. Solve Exercise 2.3.2 in the conditions of Exercise 2.3.3.

Solution. The solution has the form (see Fig. 2.4)

$$\tau = -28 + 3(t - t_1) + 28e^{-(t-t_1)/6}, \quad t_1 \le t \le t_2.$$

Exercise 2.3.5. In the 1970s, I attended a conference on differential equations organized by Sergey Sobolev at lake Baikal. We enjoyed, along with talks of brilliant Siberian mathematicians, surprises of the unstable Baikal weather. Though it was mid-summer, the temperature frequently varied during the day as drastically as from $+25°$ C to $+5°$ C, and vice versa. In this exercise, I invite you to imagine that you are near Baikal and answer the following question. Suppose that you are in a summer house with a bad insolation supplied by no heater. Let us assume that at a certain time t_0 (denote it by $t_0 = 0$), when the temperature indoors was $\tau_0 = +16°$ C, the whether suddenly changed and the outside temperature oscillated between $+26°$ C and $+6°$ C as follows (in $°$ C):

$$T(t) = 16 + A \sin(\pi t), \quad A = 10. \tag{2.3.5}$$

Will you prefer to stay at home or you believe that the temperature inside the house will be the same as outside because of the poor insulation?

Solution. The differential equation (2.3.1), with $T(t)$ given by (2.3.5), is written

$$\frac{d\tau}{dt} = k[16 + A \sin(\pi t) - \tau]. \tag{2.3.6}$$

Using the method of variation of the parameter in the solution $\tau = Ce^{-kt}$ of the homogeneous equation $\tau' = -k\tau$, we write

$$\tau = C(t)e^{-kt}$$

and substitute in Eq. (2.3.6) to obtain

$$C'(t) = k[16 + A \sin(\pi t)] e^{kt}.$$

The standard integral

$$\int e^{kx} \sin(lx) dx = \frac{1}{k^2 + l^2}[k \sin(lx) - l \cos(lx)]e^{kx} + B$$

yields

$$C(t) = 16e^{kt} + B + \frac{Ak}{\pi^2 + k^2}[k \sin(\pi t) - \pi \cos(\pi t)]e^{kt}.$$

Thus, the general solution of Eq. (2.3.6) has the form

$$\tau(t) = 16 + Be^{-kt} + \frac{Ak}{\pi^2 + k^2}[k \sin(\pi t) - \pi \cos(\pi t)].$$

The initial condition $\tau|_{t=0} = 16$ yields $B = Ak\pi/(\pi^2 + k^2)$, and ultimately we arrive at the following solution:

$$\tau(t) = 16 + \frac{Ak}{\pi^2 + k^2}[\pi e^{-kt} + k \sin(\pi t) - \pi \cos(\pi t)]. \tag{2.3.7}$$

Solution (2.3.7) manifests that it is better to stay inside the house than outside whatever the insulation is. See Figs. 2.5 and 2.6, Where the outside temperature (2.3.5) is given by the dotted line.

Likewise, one can solve a more general problem when the outside temperature $T(t)$ and the initial temperature $\tau|_{t=0}$ are

$$T(t) = T_0 + A \sin(\omega t) \tag{2.3.8}$$

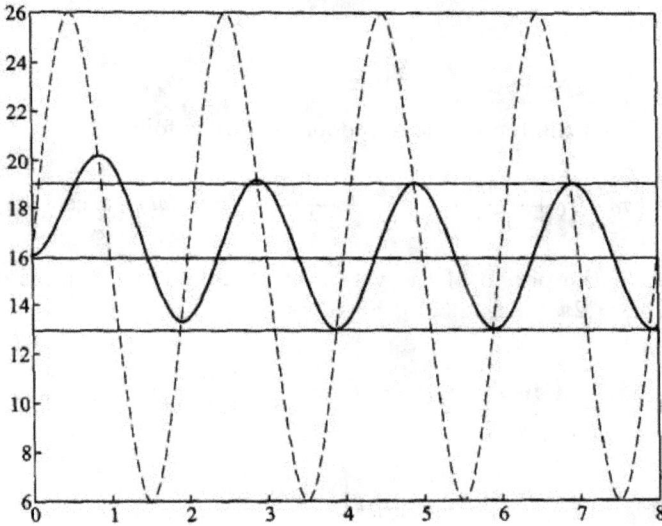

Figure 2.5: Surprising Baikal weather: $k = 1$, $A = 10$.

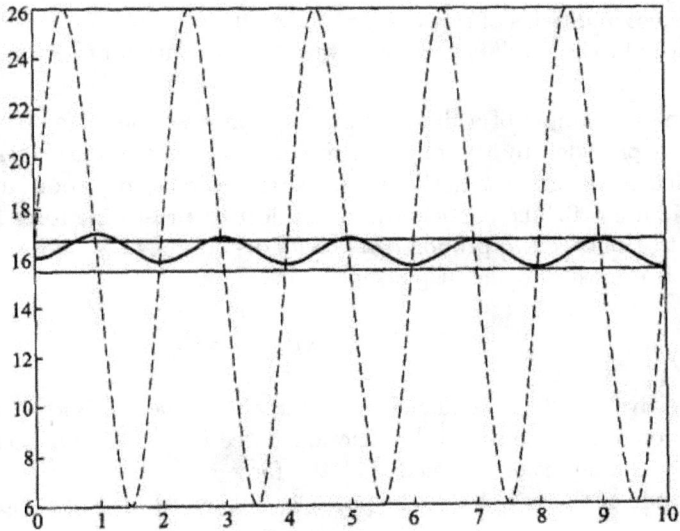

Figure 2.6: Surprising Baikal weather: $k = 1/6$, $A = 10$.

and

$$\tau\big|_{t=0} = \tau_0, \tag{2.3.9}$$

respectively. Then the temperature indoors varies as follows:

$$\tau(t) = T_0 + \left(\tau_0 - T_0 + \frac{Ak\omega}{\omega^2 + k^2}\right)e^{-kt} + \frac{Ak}{\omega^2 + k^2}[k\sin(\omega t) - \omega\cos(\omega t)]. \tag{2.3.10}$$

In particular, if the period of the variation of the outside temperature is 24 hours, i.e., $24\omega = 2\pi$, then (2.3.10) becomes

$$\tau(t) = T_0 + \left(\tau_0 - T_0 + \frac{12Ak\pi}{\pi^2 + 12^2 k^2}\right)e^{-kt}$$

$$+ \frac{12Ak}{\pi^2 + 12^2 k^2}\left[12k\sin\frac{\pi t}{12} - \pi\cos\frac{\pi t}{12}\right]. \tag{2.3.11}$$

2.3.2   Mechanical vibrations. Pendulum

In everyday life, one encounters many types of vibrations. These are, e.g., the rustle of leaves and twigs of trees caused by wind, the bouncing motion of a car due to cracks in the road, water waves and the oscillation of a ship on waves, etc.

A common example of a differential equation describing small mechanical vibrations is provided by the physical problem of a heavy particle suspended from a coiled spring and oscillating in the vertical direction y about its position of equilibrium $y = 0$. The particle will be subject to a restoring force F_1 that is, according to Hooke's law, proportional to its displacement y from the position $y = 0$ and is opposite to the displacement, i.e.,

$$F_1 = -ky,$$

with a positive constant coefficient k. In reality, when a body moves in a medium, there is also a damping (or friction) force F_2 which tends to retard the motion. Commonly, it is assumed that the force of friction is proportional to the magnitude of the velocity of the particle, but opposite in direction. Hence,

$$F_2 = -l\frac{dy}{dt},$$

where the independent variable is time t, and the coefficient of proportionality l is a positive parameter called the *damping constant*. We denote by $f(t)$ the total external force (due to wind, cracks in the road, etc.) regarded as a given function of time.

Thus, applying Newton's second law (2.1.2) with

$$F = F_1 + F_2 + f(t),$$

the equation for small oscillations of a particle with mass m is written

$$m\frac{d^2y}{dt^2} + l\frac{dy}{dt} + ky = f(t), \tag{2.3.12}$$

or

$$Ly = f(t), \quad \text{where } L = m\frac{d^2}{dt^2} + l\frac{d}{dt} + k.$$

The mechanical vibrations are said to be *damped* if $l \neq 0$, and *undamped* otherwise. The motion is said to be *free* if $f(t) \equiv 0$, and *forced* otherwise.

Damped oscillations of a mechanical system with n degrees of freedom are described by a system of ordinary differential equations:

$$\sum_{j=1}^{n}\left(m_{ij}\frac{d^2y^j}{dt^2} + l_{ij}\frac{dy^j}{dt} + k_{ij}y^j\right) = f^i(t), \quad i = 1, \ldots, n,$$

with constant coefficients m_{ij}, l_{ij}, and k_{ij}. It is written in form (2.3.12) after introducing the vector notation $y = (y^1, \ldots, y^n), f = (f^1, \ldots, f^n)$ and the matrix differential operator L as follows:

$$Ly = f(t), \qquad L = M\frac{d^2}{dt^2} + A\frac{d}{dt} + B,$$

where M, A, B are matrices with the entries m_{ij}, l_{ij}, and k_{ij}, respectively.

Remark 2.3.1. Throughout the book, vectors y etc. are regarded as *column vectors* though they are written as rows. Accordingly, in the matrix $M = (m_{ij})$, the indices i and j denote rows and columns, respectively.

The above linear models are legitimate for infinitesimal vibrations (called *harmonic oscillations*) when the amplitude of the vibrations is sufficiently small. Finite vibrations (or their higher-order approximations) are usually described by nonlinear differential equations and are known as *anharmonic oscillations*.

Example (Pendulum). A simple pendulum consists of a weight (a bob) mounted on the lower end of a vertical rod. The upper end of the rod is suspended from a flexible low-friction support. If one pulls the bob sidewise and releases it, the pendulum swings to and fro in a vertical plane under the influence of gravity. The use of a pendulum as a time measurer is based on the remarkable observation due to Galileo[1] that, within certain limits, its period of swing is constant provided that the length of the pendulum rod remains unaltered.

---

[1]Galileo discovered the principle of a pendulum in 1581. This principle showed that one can use an oscillating mass to keep time. The introduction of the pendulum was an important development in construction of mechanical clocks. Galileo's discovery can be compared with the invention of a wheel for its ingenuity.

The above property of a pendulum can be deduced from an appropriate mathematical model. Let m be the mass of the bob of a pendulum and let l be the length of the pendulum rod. The rod is a relatively light string so that its mass is negligible compare to m. Let us denote by y the angular displacement of the pendulum from its equilibrium position $y = 0$. The mathematical model of the pendulum is obtained from Newton's second law (2.1.1) by letting v be the linear velocity $v = l\,dy/dt$ and F be the restoring component of the gravitational force, $F = -mg\sin y$. Hence, the governing equation of the pendulum is written:

$$\frac{d^2y}{dt^2} + \omega^2 \sin y = 0, \qquad \text{where} \quad \omega^2 = \frac{g}{l}. \tag{2.3.13}$$

The equation of small oscillations is obtained from the *nonlinear* equation (2.3.13) by assuming $y \to 0$ and replacing $\sin y$ by its first approximation, $\sin y \approx y$. Then one arrives at the *linear* equation for free harmonic oscillations:

$$\frac{d^2y}{dt^2} + \omega^2 y = 0. \tag{2.3.14}$$

Its general solution has the form (see Section 3.3.3, Example 3.3.2)

$$y = C_1 \cos(\omega t) + C_2 \sin(\omega t), \quad C_1, C_2 = \text{const}.$$

The time taken for one complete cycle of swing of the pendulum is called the *period* and is denoted by τ. It can be formally determined by the condition that the above solution remains unaltered when t is replaced by $t + \tau$, in other words, if $\cos[\omega(t + \tau)] = \cos(\omega t)$ and $\sin[\omega(t + \tau)] = \sin(\omega t)$. Hence, $\omega(t + \tau) = \omega t + 2\pi$, or $\omega\tau = 2\pi$. Thus, invoking the definition of ω, one obtains the following expression for the period of free harmonic oscillations that accords with Galileo's observation:

$$\tau = \frac{2\pi}{\omega} = 2\pi\sqrt{\frac{l}{g}}, \tag{2.3.15}$$

where g is the gravitational constant. The quantity $\omega = 2\pi/\tau$ gives the number of oscillations in 2π units of time, and consequently it is known as the *angular frequency*.

Let us conclude this example by considering a pendulum of a practical interest. Namely, let us find the length of a pendulum which makes one swing per second (i.e., $\tau/2 = 1$ s). Eq. (2.3.15) yields

$$l = \frac{g}{\pi^2}\left(\frac{\tau}{2}\right)^2.$$

Substituting $g \approx 9.81$ m/s^2 and $\tau/2 = 1$ s, we find the desired length:

$$l \approx 1 \text{ m}.$$

Remark 2.3.2. The period τ (2.3.15) of harmonic oscillations does not depend on the initial displacement of the pendulum. Despite the fact that this conclusion provides a perfect mathematical explanation of Galileo's observation, it is, however, in a discord with the theory of free fall also based on Newton's law. Indeed, the time $t_* = \sqrt{2h_0}$ of a free fall (see Exercise 2.2.1), unlike period (2.3.15) of a free swing, depends on its initial height h_0. It is obvious that the effect is caused by the linear approximation (2.3.14) of the exact nonlinear model (2.3.13). Therefore, let us discuss the period of anharmonic oscillations governed by (2.3.13).

Exercise 2.3.6. Integrate Eq. (2.3.13) imposing the conditions $y = 0$ when $t = 0$ and $y = \alpha$ when $t = T/4$. Here T is the period of anharmonic oscillations of the pendulum and α is its angular *amplitude*, i.e., the maximum angular displacement, $\alpha = y|_{max}$. Evaluate the period T and compare it with the period τ of harmonic oscillations.

Solution. After multiplying Eq. (2.3.13) by dy/dt, one can integrate it once to obtain

$$\left(\frac{dy}{dt}\right)^2 - 2\omega^2 \cos y = C.$$

Since α is the maximum displacement, $(dy/dt)|_{y=\alpha} = 0$. Hence, $C = -2\omega^2 \cos \alpha$. Using the identity

$$\cos y - \cos \alpha = 2[\sin^2(\alpha/2) - \sin^2(y/2)],$$

one obtains:

$$\frac{dy}{dt} = 2\omega\sqrt{\sin^2(\alpha/2) - \sin^2(y/2)}.$$

The integration yields:

$$\int_0^y \left[\sin^2(\alpha/2) - \sin^2(u/2)\right]^{-1/2} d(u/2) = \omega t. \tag{2.3.16}$$

The integral in (2.3.16), known as the *elliptic integral* of the first kind, cannot be expressed in terms of elementary functions. Defining the new variable v by

$$\sin(u/2) = \sin(\alpha/2)\sin v$$

and denoting $\kappa = \sin(\alpha/2)$, one can rewrite the above integral in the standard form, viz.

$$\int \left[\sin^2(\alpha/2) - \sin^2(u/2)\right]^{-1/2} d(u/2) = \int \frac{dv}{\sqrt{1 - \kappa^2 \sin^2 v}}.$$

According to our formula for the change of variables, $u = \alpha$ implies $\sin v = 1$, i.e., $v = \pi/2$. Invoking the condition that $y = \alpha$ when $t = T/4$, we conclude that $v = \pi/2$ when $t = T/4$. Hence, one can readily determine the period by substituting $t = T/4$ in (2.3.16) and rewriting the elliptic integral in the standard form. Thus,

$$T = 4\sqrt{\frac{l}{g}} \int_0^{\pi/2} \frac{dv}{\sqrt{1 - \kappa^2 \sin^2 v}}, \quad \text{where} \quad \kappa = \sin(\alpha/2).$$

If the angular amplitude α is infinitesimally small, $\alpha \to 0$, then $k \to 0$, and hence T coincides with τ given by (2.3.15). The maximum difference of T from (2.3.15) occurs when $k \to 1$, i.e., when $\alpha \to \pi$ (the vertical displacement). If $0 < \alpha < \pi$, then $|\kappa^2 \sin^2 v| < 1$, and hence,

$$(1 - \kappa^2 \sin^2 v)^{-1/2} = 1 + (\kappa^2/2) \sin^2 v + (3\kappa^4/8) \sin^4 v + \cdots .$$

Integrating this series termwise by using the well-known integrals

$$\int \sin^2 v \, dv = (v/2) - (1/4) \sin(2v),$$

$$\int \sin^4 v \, dv = (3v/8) - (1/4) \sin(2v) + (1/32) \sin(4v),$$

etc., and finally substituting $\kappa = \sin(\alpha/2)$, we obtain the following expression for the period that is convenient for its numerical evaluation and, unlike (2.3.15), theoretically satisfactory:

$$T = 2\pi \sqrt{\frac{l}{g}} \left(1 + \frac{1}{4} \sin^2 \frac{\alpha}{2} + \frac{9}{64} \sin^4 \frac{\alpha}{2} + \cdots \right). \tag{2.3.17}$$

The first term coincides with (2.3.15) and defines the principal part τ of the period T, while the other terms of (2.3.17) are small. For example, high-class clocks employed in astronomy for their astonishing accuracy, are furnished by a pendulum with the amplitude $\alpha \approx 1.5°$. For such clocks, the use of the first correction term of (2.3.17) gives

$$T \approx \tau + \frac{\tau}{20000}.$$

2.3.3 Collapse of driving shafts

At the beginning of the 20th century, constructors of motor ships came across the troublesome phenomenon of a seemingly accidental "beating" and the possible collapse of shafts in power transmission systems (see Fig. 2.8). The strange phenomenon was explained by means of differential equations.

According to the model of the vibrations of rods (2.1.4), the positions of equilibrium of a uniformly rotating cylindrical shaft are given by the time-independent $(\partial u/\partial t = 0)$ solutions to Eq. (2.1.4), i.e., they are determined by the fourth-order ordinary differential equation

$$\mu \frac{d^4 u}{dx^4} = f,$$

where u is the shaft's displacement from its equilibrium position $u = 0$, and f is the density of the centrifugal force acting on the shaft. To find f, consider a small element of the shaft dx and denote by p the weight of the shaft per unit length. Then the mass of the element dx is

$$dm = \frac{p}{g}\, dx,$$

where g is the acceleration of gravity. The centrifugal force df acting on dx due to the rotation by a constant angular velocity ω is

$$df = \omega^2 u\, dm = \frac{p\omega^2}{g}\, u\, dx.$$

Hence,

$$f = \frac{p\omega^2}{g}\, u,$$

and the differential equation in question is written

$$\mu \frac{d^4 u}{dx^4} = \frac{p\omega^2}{g} u, \tag{2.3.18}$$

where the positive constant μ depends on the material of the shaft.

Let the shaft revolve at two bearings located at $x = 0$ and $x = l$ (Fig. 2.7). Then $u|_{x=0} = u|_{x=l} = 0$. Furthermore, it can be shown that bearings are points of inflection of the function $u = u(x)$. Thus, one arrives at the problem of investigating the solutions of the differential equation (2.3.18) satisfying four boundary conditions:

$$u\big|_{x=0} = 0, \quad u\big|_{x=l} = 0, \quad \frac{d^2 u}{dx^2}\bigg|_{x=0} = 0, \quad \frac{d^2 u}{dx^2}\bigg|_{x=l} = 0. \tag{2.3.19}$$

The phenomenon of "beating" occurs when the boundary value problem defined by Eqs. (2.3.18)~(2.3.19) has a "nontrivial solution", i.e., a solution $u = u(x)$ that does not vanish identically in the interval $0 \le x \le l$.

Integration of Eq. (2.3.18) rewritten in the form

$$\frac{d^4 u}{dx^4} = \alpha^4 u, \quad \text{where } \alpha^4 = \frac{p\omega^2}{g\mu} = \text{const.}, \tag{2.3.20}$$

provides its general solution

$$u = C_1 e^{\alpha x} + C_2 e^{-\alpha x} + C_3 \cos(\alpha x) + C_4 \sin(\alpha x), \quad C_i = \text{const.} \tag{2.3.21}$$

The boundary conditions (2.3.19) yield

$$C_1 + C_2 + C_3 = 0, \quad C_1 e^{\alpha l} + C_2 e^{-\alpha l} + C_3 \cos(\alpha l) + C_4 \sin(\alpha l) = 0,$$

$$C_1 + C_2 - C_3 = 0, \quad C_1 e^{\alpha l} + C_2 e^{-\alpha l} - C_3 \cos(\alpha l) - C_4 \sin(\alpha l) = 0.$$

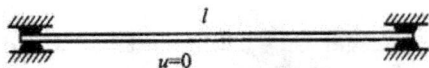

Figure 2.7: A stable shaft.

Figure 2.8: A beating shaft.

The reckoning shows that $C_1 = C_2 = C_3 = 0$ and $C_4 \sin(\alpha l) = 0$. If $C_4 = 0$, one arrives at the trivial solution, $u = 0$, so that the shaft is straight. On the other hand, by letting $\sin(\alpha l) = 0$, i.e., $\alpha = n\pi/l$, one obtains the cases when beating occurs. Then (2.3.21) yields

$$u = C_4 \sin(n\pi x/l).$$

According to the definition of α, the equation $\alpha = n\pi/l$ yields that a collapse of the shaft is possible whenever its angular velocity approaches any one of the following *critical values*:

$$\omega_n = \frac{n^2 \pi^2}{l^2} \sqrt{\frac{g\mu}{p}}, \quad n = 1, 2, \ldots. \tag{2.3.22}$$

2.3.4   The van der Pol equation

A common illustrative example is the discharge of an electrical condenser through an inductive coil of wire. According to elementary laws of electricity[2], the phenomenon is described by the equations

$$C\frac{dV}{dt} = -I, \quad V - L\frac{dI}{dt} = RI, \tag{2.3.23}$$

where I is the *current* of the discharge, V the *voltage* (the potential difference between the terminals of the condenser), R the *resistance*, C the condenser's *capacity*, and L the coil's *inductance*. Here I and V are functions of time t, whereas R, C, and L are regarded as given constants. Hence, (2.3.23) is a system of first-order ordinary differential equations with two dependent variables, I and V, considered as unknown functions of t.

Let us denote the dependent variable V by y and its first and second derivatives with respect to t by y' and y''. In this notation, the second equation of (2.3.23), upon substituting I from the first one, becomes a linear second-order ordinary differential equation:

$$ay'' + by' + cy = 0,$$

---

[2]Formulated by G.S. Ohm in 1827 and then generalized by G.R. Kirchhoff.

with constant coefficients $a = LC$, $b = RC$, and $c = 1$.

Replacing in the second equation of (2.3.23) the classical Ohm's law by the so-called *generalized Ohm's law*, one obtains a nonlinear system,

$$C\frac{dV}{dt} = -I, \quad L\frac{dI}{dt} = V - h(I),$$

or an equivalent single nonlinear equation of the second order:

$$ay'' + y = -f(y'), \quad a = \text{const}.$$

By letting $f(y') = \varepsilon(y'^3 - y')$, one arrives at the van der Pol equation used in the theory of triodes:

$$ay'' + y = \varepsilon(y' - y'^3), \quad \varepsilon = \text{const}. \tag{2.3.24}$$

In fact, van der Pol's equation was the first nonlinear differential equation of real physical significance having periodic solutions, the latter property being originally recognized by Balth van der Pol from his experiences in studying oscillations in electrical circuits and in connection with an electrical model of the beating of the heart[3].

2.3.5   Telegraph equation

It is common knowledge that in electrodynamics current streaming along a cable is well described by the following set of equations:

$$Cw_t + Gw + j_x = 0, \quad Lj_t + Rj + w_x = 0, \tag{2.3.25}$$

where the dependent variables are voltage w and current intensity j, considered as functions of time t and the coordinate x along the cable. Coefficients involved in the equations are constant and they characterize physical properties of the cable, namely C is capacity, L is self-induction, R is resistance and G is leakage, defined as loss of current divided by voltage. Excluding one of the dependent variables, w or j, from Eqs. (2.3.25) by means of differentiation and denoting the remaining variable by v, one obtains a linear second-order differential equation (for intensity or voltage) known as the *telegraph equation*

$$v_{tt} - c^2 v_{xx} + (a + b)v_t + abv = 0.$$

The constant factors involved here have the following physical meaning: c is light velocity, a and b are capacity and inductive dumping factors. They are connected with coefficients of the initial set of equations by the formulae

$$c^2 = \frac{1}{CL}, \quad a = \frac{G}{C}, \quad b = \frac{R}{L}.$$

---

[3]B. van der Pol, *Philosophical Magazine*, vol. 2, 1926, pp. 978-992; B. van der Pol and J. van der Mark, *Philosophical Magazine*, vol. 6, 1928, pp. 763-775. See also [11], Section 7.5.

The first derivative in the telegraph equation can be excluded by setting

$$u = e^{\frac{a+b}{2}t}v.$$

Then, upon designating the resulting positive constant $(a - b)^2/4$ by k^2, one obtains the following form of the telegraph equation:

$$u_{tt} - c^2 u_{xx} - k^2 u = 0, \quad c, k = \text{const.} \tag{2.3.26}$$

2.3.6 Electrodynamics

An electromagnetic field has two components, namely, the vector E of the electric field and the vector H of the magnetic field. The theory of electromagnetic waves, or *electrodynamics* is based on the Maxwell equations

$$\frac{\partial E}{\partial t} = c(\nabla \times H) - 4\pi j, \quad \nabla \cdot E = 4\pi \rho,$$

$$\frac{\partial H}{\partial t} = -c(\nabla \times E), \qquad \nabla \cdot H = 0. \tag{2.3.27}$$

Here j and ρ are the electric current density and the electric charge density, respectively, and $c \approx 3 \times 10^{10}$ cm/s is the velocity of light.   The Maxwell equations have four independent variables, namely, the time t and the position vector $x = (x, y, z)$. The dependent variables are the vectors E and H. The current j and the charge ρ are given functions, $j = j(t, x)$, $\rho = \rho(t, x)$. Thus, (2.3.27) is an *over-determined system* of first-order partial differential equations: it contains eight equations for six components of E and H.

The Maxwell equations (2.3.27) are often written in physics in the form

$$\frac{1}{c}\frac{\partial E}{\partial t} = \text{curl } H - \frac{4\pi}{c}j, \quad \text{div } E = 4\pi \rho,$$

$$\frac{1}{c}\frac{\partial H}{\partial t} = -\text{curl } E, \qquad \text{div } H = 0. \tag{2.3.28}$$

In the simplest case of propagation of electromagnetic waves in vacuum the Maxwell equations become

$$\frac{1}{c}\frac{\partial E}{\partial t} = \text{curl } H, \quad \text{div } E = 0,$$

$$\frac{1}{c}\frac{\partial H}{\partial t} = -\text{curl } E, \quad \text{div } H = 0. \tag{2.3.29}$$

In this case, one can consider the *determined system* of differential equations

$$\frac{1}{c}\frac{\partial E}{\partial t} = \text{curl } H,$$

$$\frac{1}{c}\frac{\partial H}{\partial t} = -\text{curl } E. \tag{2.3.30}$$

Indeed one can show (see Problem 2.7) that the relations

$$\text{div } \boldsymbol{E} = 0, \quad \text{div } \boldsymbol{H} = 0 \tag{2.3.31}$$

hold at any time if they are satisfied at an initial time $t = t_0$, and hence they are merely initial conditions. This statement applies to more general situation when Eqs. (2.3.31) are replaced by (see [21], Section 10.5)

$$\text{div } \boldsymbol{E} = f(\boldsymbol{x}), \quad \text{div } \boldsymbol{H} = g(\boldsymbol{x}),$$

in particular to Eqs. (2.3.28) when $\boldsymbol{j}$ and ρ do not depend on time.

2.3.7 The Dirac equation

One of the fundamental equations in quantum mechanics is the Dirac equation

$$\gamma^k \frac{\partial \psi}{\partial x^k} + m\psi = 0, \quad m = \text{const.} \tag{2.3.32}$$

Eq. (2.3.32) is used for the study of relativistic particles with a mass m and spin $1/2$, such as electron, neutron, proton and neutrino (when $m = 0$).

Here the dependent variable ψ is a 4-dimensional column vector with complex valued components $\psi^1, \psi^2, \psi^3, \psi^4$. In quantum mechanics, dependent variables are usually called *wave functions*. The wave function ψ satisfying the Dirac equation is called a *spinor* due to its specific transformation properties under the Lorentz group (see Section 7.3.8). The independent variable is the four-dimensional vector $x = (x^1, x^2, x^3, x^4)$, where x^1, x^2, x^3 are the real valued spatial variables and x^4 is the complex variable defined by $x^4 = ict$ with t being time and c the light velocity. Furthermore, γ^k are the following 4×4 complex matrices called the Dirac matrices:

$$\gamma^1 = \begin{pmatrix} 0 & 0 & 0 & -i \\ 0 & 0 & -i & 0 \\ 0 & i & 0 & 0 \\ i & 0 & 0 & 0 \end{pmatrix}, \quad \gamma^2 = \begin{pmatrix} 0 & 0 & 0 & -1 \\ 0 & 0 & 1 & 0 \\ 0 & 1 & 0 & 0 \\ -1 & 0 & 0 & 0 \end{pmatrix},$$

$$\gamma^3 = \begin{pmatrix} 0 & 0 & -i & 0 \\ 0 & 0 & 0 & i \\ i & 0 & 0 & 0 \\ 0 & -i & 0 & 0 \end{pmatrix}, \quad \gamma^4 = \begin{pmatrix} 1 & 0 & 0 & 0 \\ 0 & 1 & 0 & 0 \\ 0 & 0 & -1 & 0 \\ 0 & 0 & 0 & -1 \end{pmatrix}.$$

2.3.8 Fluid dynamics

The fundamental mathematical model in fluid dynamics is provided by the following system of nonlinear partial differential equations of the first order describing motions of a compressible fluid (gas):

$$\rho_t + \boldsymbol{v} \cdot \nabla\rho + \rho\,\mathrm{div}\,\boldsymbol{v} \quad = 0,$$
$$\rho\left[\boldsymbol{v}_t + (\boldsymbol{v} \cdot \nabla)\boldsymbol{v}\right] + \nabla p \quad = 0, \tag{2.3.33}$$
$$p_t + \boldsymbol{v} \cdot \nabla p + A(p,\rho)\,\mathrm{div}\,\boldsymbol{v} = 0,$$

where $A(p,\rho)$ is an arbitrary function connected with the entropy $S(p,\rho)$ by the equation

$$A = -\rho\,\frac{\partial S/\partial\rho}{\partial S/\partial p}. \tag{2.3.34}$$

The dependent variables are the velocity $\boldsymbol{v}$, the pressure p and the density ρ of the fluid. The independent variables are the time t and the position vector $\boldsymbol{x} = (x, y, z)$.

If the entropy in the liquid is constant, $S = \mathrm{const.}$, the flow is said to be *isentropic*.

In the case of so-called *polytropic flows*, function (2.3.34) has the form $A = \gamma p$, where γ is a constant known as an *adiabatic (polytropic) exponent*. The case $\gamma = 5/3$ corresponds to the flow of a monatomic gas. Since the solar neighborhood contains mainly monatomic gases, this case is important. Thus, monatomic gases are described by the equations

$$\rho_t + \boldsymbol{v} \cdot \nabla\rho + \rho\,\mathrm{div}\,\boldsymbol{v} = 0,$$
$$\rho\left[\boldsymbol{v}_t + (\boldsymbol{v} \cdot \nabla)\boldsymbol{v}\right] + \nabla p = 0, \tag{2.3.35}$$
$$p_t + \boldsymbol{v} \cdot \nabla p + \frac{5}{3}p\,\mathrm{div}\,\boldsymbol{v} = 0.$$

Another physically significant case corresponds to a planar isentropic flow (i.e., $S = \mathrm{const.}$) of a gas with the adiabatic exponent $\gamma = 2$. The condition that the flow is isentropic implies that one has to drop the last equation of the gasdynamic system (2.3.33). Then, setting

$$p = \frac{1}{2}\rho^2, \quad \rho = gh,$$

where g is the acceleration of gravity, one reduces the first two equations in (2.3.33) to the following system:

$$h_t + \boldsymbol{v} \cdot \nabla h + h\,\mathrm{div}\,\boldsymbol{v} = 0,$$
$$\boldsymbol{v}_t + (\boldsymbol{v} \cdot \nabla)\boldsymbol{v} + g\,\nabla h = 0. \tag{2.3.36}$$

Here $\boldsymbol{v}$ is a two-dimensional vector and ∇ is Hamilton's operator with two components, ∇_x and ∇_y (see (1.3.10)). Eqs. (2.3.36) describe the flow of a shallow water over a flat solid wall in the (x, y) plane, where h is the height of the water surface above the wall.

The planar non-steady-state potential gas flow with transonic speeds is described by the equation

$$2u_{tx} + u_x u_{xx} - u_{yy} = 0. \tag{2.3.37}$$

2.3.9 The Navier-Stokes equations

The incompressible flow of a viscous fluid is governed by the Navier-Stokes equations

$$v_t + (v \cdot \nabla)v + \frac{1}{\rho}\nabla p = \nu\Delta v, \quad \operatorname{div} v = 0. \qquad (2.3.38)$$

Here the dependent variables are the velocity $v = (v^1, v^2, v^3)$ and the pressure p, whereas the density ρ is assumed to be a given constant. The parameter ν is the viscosity of the fluid.

2.3.10 A model of an irrigation system

A mathematical model for investigating certain irrigation systems is given by the following nonlinear partial differential equation (see [19], Section 9.8 and the references therein):

$$C(\psi)\psi_t = [K(\psi)\psi_x]_x + [K(\psi)(\psi_z - 1)]_z - S(\psi). \qquad (2.3.39)$$

Here ψ is the soil moisture pressure head, $C(\psi)$ is the specific water capacity, $K(\psi)$ is the unsaturated hydraulic conductivity, $S(\psi)$ is a source term, t is the time, x is the horizontal axis and z is the vertical axis which is considered positive downward. This equation may be used for describing of soil infiltration, redistribution and extraction in a bedded non-deformable soil profile overlaying a shallow water table and irrigated by a line source drip irrigation system. Line source drip systems produce a continuous wetted band along the length of the lateral (the y-axis), and hence the phenomenon actually involves all three space coordinates, x, y, and z.

2.3.11 Magnetohydrodynamics

Magnetohydrodynamics deals with significant physical and engineering problems arising in investigating the motion of ionized fluids in the presence of electromagnetic forces. Let us consider the mathematical model describing the motion of a *perfectly conducting* fluid in a magnetic field. We assume that the magnetic permeability $\mu = 1$. Let H denote the magnetic vector field and v the flow velocity of the fluid. Taking into account the assumption of infinite electrical conductivity of the fluid, one has the expressions

$$j = \operatorname{curl} H$$

and

$$E = H \times v$$

for the vector j of the electric current density and the electric vector field E, respectively.

The equations of magnetohydrodynamics are obtained by combining the Maxwell equations (2.3.28) with equations (2.3.33) of hydrodynamics. Using the above expressions for j and E, one obtains the following equations (see, e.g. [4], Chapter VI, §3a.6):

$$\frac{1}{c}\frac{\partial H}{\partial t} + \operatorname{curl}(H \times v) = 0, \quad \operatorname{div} H = 0,$$

$$\rho_t + v \cdot \nabla\rho + \rho \operatorname{div} v = 0, \tag{2.3.40}$$

$$\rho\left[v_t + (v \cdot \nabla)v\right] + \nabla p - (\operatorname{curl} H) \times H = 0.$$

Here the term $(\operatorname{curl} H) \times H$ is due to the force $j \times H$ exerted by the magnetic field on a unit volume of the fluid. Since the equation $\operatorname{div} H = 0$ is merely an initial condition, Eqs. (2.3.40) provide a *sub-definite system:* they contain 7 equations for 8 unknown functions $H^1, H^2, H^3; v^1, v^2, v^3; \rho$ and p. An additional equation should be added in accordance with physical requirements in the problem.

2.4   Diffusion phenomena

2.4.1   Linear heat equation

The behavior of physical systems in diffusion processes is approximately described by neglecting the molecular character of the system. The elements of this idealized system are assumed to be unaffected by molecular fluctuations regardless of how small a volume is being considered.

Let us derive the differential equation governing a steady heat diffusion in a homogeneous material, where homogeneity means that the *mass density* ρ of the material, its *specific heat* c_* and *thermal conductivity* k are positive constants. We isolate in the material an arbitrary volume Ω and denote by $\partial\Omega$ its boundary. Let ν be the unit outward normal to the surface $\partial\Omega$. We denote by u the absolute temperature, so that $u = u(t, x)$ is the temperature field to be determined for any time t and $x \in \Omega$.

After J.B.J. Fourier's paper (1811) on the theory of heat conduction and his famous book *Théorie analitique de la chaleur* (1822), the mathematical model of thermal diffusion is usually based on the following physical principles of heat balance known as *Fourier's law of heat conduction.*

(i) The quantity of heat Q in Ω is proportional to the mass of Ω and to its temperature:

$$Q(t) = \int_\Omega \rho\, c_*\, u \, dx \, dy \, dz. \tag{2.4.1}$$

(ii) Heat diffuses from a higher to a lower temperature, and the heat flow is proportional to the gradient of temperature, i.e., the heat flux in the volume Ω through its surface $\partial\Omega$ is given by

$$\int_{\partial\Omega} (k \, \nabla u \cdot \nu) dS. \tag{2.4.2}$$

(iii) The rate of change of the heat content (2.4.1) within Ω, i.e., the quantity

$$\frac{dQ}{dt} = \int_\Omega \rho\, c_* \frac{\partial u}{\partial t}\, dx\, dy\, dz$$

is equal to the rate of heat entering through the surface $\partial\Omega$ in accordance with (2.4.2). Hence, we have the following balance equation:

$$\int_\Omega \rho\, c_* \frac{\partial u}{\partial t}\, dx\, dy\, dz = \int_{\partial\Omega} (k\,\nabla u \cdot \boldsymbol{\nu})dS. \tag{2.4.3}$$

Using the divergence theorem (1.3.18), one can convert the surface integral in the right-hand side of Eq. (2.4.3) to a volume integral:

$$\int_{\partial\Omega} (k\,\nabla u \cdot \boldsymbol{\nu})dS = \int_\Omega \nabla \cdot (k\,\nabla u)\, dx\, dy\, dz.$$

Consequently, the integral equation (2.4.3) becomes

$$\int_\Omega \rho\, c_* \frac{\partial u}{\partial t}\, dx\, dy\, dz = \int_\Omega \nabla \cdot (k\,\nabla u)\, dx\, dy\, dz. \tag{2.4.4}$$

Since Ω is arbitrary, the integral equation (2.4.4) is equivalent to the differential equation

$$\rho\, c_* \frac{\partial u}{\partial t} = \nabla \cdot (k\,\nabla u). \tag{2.4.5}$$

Since the thermal conductivity k is constant, we have

$$\nabla \cdot (k\,\nabla u) = k\,\nabla \cdot (\nabla u) = k\,\Delta u,$$

where Δ is Laplacian (1.3.19). Thus, we arrive at the linear *heat equation*

$$u_t = a^2 \Delta u, \tag{2.4.6}$$

where the positive constant $a^2 = k/(\rho\, c_*)$ is the *diffusivity* of the material. In the one-dimensional case, when the temperature depends on time t and one spatial variable x, the heat equation (2.4.6) has the form

$$\frac{\partial u}{\partial t} - a^2 \frac{\partial^2 u}{\partial x^2} = 0. \tag{2.4.7}$$

A physical realization of the one-dimensional heat diffusion is as follows. Consider an infinite uniform rectangular bar of a cross-sectional area S protracted along the x axis, with the sides perfectly insulated. Assume that the temperature T is uniform in any cross-section of the bar. That is, $T = T(t, x)$ is the temperature of the bar at time t in the section parallel to the (y, z) plane at a distance x from the origin O of the rectangular axes Ox, Oy, Oz. Let the

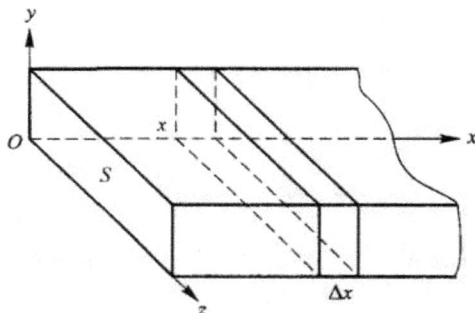

Figure 2.9: One-dimensional heat flow.

domain Ω be the slice of the bar of thickness Δx at a distance x from O (see Fig. 2.9). The balance equation (2.4.3) is written approximately as

$$\frac{\partial(S\Delta xT)}{\partial t} = S\left(\left.\frac{\partial T}{\partial x}\right|_{x+\Delta x} - \left.\frac{\partial T}{\partial x}\right|_x\right).$$

Dividing this equation by $S\Delta x$, passing to the limit $\Delta x \to 0$ and denoting the temperature T by u, we obtain the one-dimensional heat equation (2.4.7). It is linear and has constant coefficients, i.e., it is invariant under the t- and x-translations, because of our consideration of steady diffusion processes and the assumption on uniformity of the material of the bar.

2.4.2   Nonlinear heat equation

Our speculations in the previous section are based, in fact, on the assumption that a change of temperature does not affect the physical characteristics ρ, c_* and k of the material. This assumption is reasonable if the change of temperature is not high. Furthermore, it is also reasonable to assume that the density and the specific heat will keep their initial constant values even under the high temperature. The thermal conductivity will be affected, however, if the temperature varies considerably.

Therefore, let us consider the balance equation (2.4.3) under the assumption that ρ and c_* are positive constants as before, but k depends on the temperature, $k = k(u)$. Furthermore, one can let $\rho c_* = 1$ by using an appropriate scaling of time and write Eq. (2.4.5) in the following form:

$$\frac{\partial u}{\partial t} = \nabla \cdot [k(u)\nabla u]. \tag{2.4.8}$$

Eq. (2.4.8) is called a *nonlinear heat equation*. It is often written in the form

$$\frac{\partial u}{\partial t} = \text{div}\,[k(u)\,\text{grad}\,u]. \tag{2.4.9}$$

In the one-dimensional case, the nonlinear heat equation has the form

$$\frac{\partial u}{\partial t} = \frac{\partial}{\partial x}\left[k(u)\,\frac{\partial u}{\partial x}\right] \qquad (2.4.10)$$

or

$$u_t = [k(u)\,u_x]_x = k(u)\,u_{xx} + k'(u)\,(u_x)^2. \qquad (2.4.11)$$

2.4.3 The Burgers and Korteweg-de Vries equations

The Burgers equation

$$u_t = uu_x + \nu\,u_{xx} \qquad (2.4.12)$$

is widely used in fluid mechanics, nonlinear acoustics, etc. It is used, e.g. to model the formation and decay of non-plane shock waves where the variable x is a coordinate moving with the wave at the speed of the sound, and the dependent variable u represents the velocity fluctuations.

The coefficient ν in the Burgers equation (2.4.12) is usually considered as a constant. However, it is actually a function of the time, and hence there is merit in studying the generalized Burgers equation

$$u_t = uu_x + \nu(t)\,u_{xx}. \qquad (2.4.13)$$

The Korteweg-de Vries equation

$$u_t = uu_x + \mu\,u_{xxx}, \qquad \mu = \text{const.}, \qquad (2.4.14)$$

is used, e.g. in mathematical description of propagation of long water waves in channels.

The Burgers and Korteweg-de Vries equations are distinguished among nonlinear partial differential equation due to their remarkable mathematical properties.

2.4.4 Mathematical modelling in finance

The mathematics of finance is aimed at studying stock price fluctuation as a diffusion process in a random environment. Accordingly, time and uncertainty are central elements in modelling the financial behaviour of economic agents. Therefore, the basic mathematical models in finance are formulated in terms of stochastic processes thus leading to *stochastic differential equations*. However, under certain simplifying assumptions, the models often can be approximated by usual differential equations.

A well-known equation of this type is provided by the Black-Scholes model (1973) used in stock option pricing. The model is approximated by the following linear equation with variable coefficients:

$$u_t + \frac{1}{2}A^2 x^2 u_{xx} + Bx u_x - Cu = 0, \qquad (2.4.15)$$

where $A, B,$ and C are constant coefficients connected with characteristics of the model. Note that the Black-Scholes equation (2.4.15) can be transformed to the heat equation by a rather complicated change of variables.

2.5   Biomathematics

2.5.1   Smart mushrooms

It is natural to assume that growing mushrooms strive to minimize the waste of moisture. Consequently, they should grow so that their surface area is minimal thus reducing the amount of evaporation.

Starting from this assumption, let us find the optimal form of the mushrooms by solving the following simple mathematical problem. We will then compare the result with real mushrooms. Consider curves $y = y(x)$ in the (x, y) plane connecting two fixed points, $P_1 = (x_1, y_1)$ and $P_2 = (x_2, y_2)$. One revolves the curves about the y axis to obtain surfaces.

The problem is to find that curve for which the surface of revolution has a minimum area. Let us solve the problem. Consider a narrow strip of the surface obtained when the variable x is between the values x and $x + dx$. The area of the strip is

$$2\pi x ds = 2\pi x \sqrt{1 + y'^2} dx$$

since

$$(ds)^2 = (dx)^2 + (dy)^2$$

and hence

$$ds = \sqrt{1 + y'^2}\, dx.$$

Therefore, the total area of the surface of revolution is given by the integral

$$S = 2\pi \int_{x_1}^{x_2} x\sqrt{1 + y'^2} dx.$$

Hence, one arrives at the following variational formulation of the problem: find the curve for which the variational integral

$$\int L(x, y, y') dx$$

with the Lagrangian

$$L = x\sqrt{1 + y'^2}$$

has a stationary value. The condition for a stationary value is equivalent to the Euler-Lagrange equation (1.5.2):

$$\frac{\partial L}{\partial y} - D_x \left(\frac{\partial L}{\partial y'} \right) = 0. \tag{2.5.1}$$

Since in our example we have

$$\frac{\partial L}{\partial y} = 0, \quad \frac{\partial L}{\partial y'} = \frac{xy'}{\sqrt{1 + y'^2}},$$

Eq. (2.5.1) is written in the form of a *conservation law* (see Section 7.3):

$$D_x \left(\frac{xy'}{\sqrt{1 + y'^2}} \right) = 0. \tag{2.5.2}$$

Whence, upon differentiation, one obtains the following nonlinear differential equation of the second order:

$$y'' + \frac{1}{x}(y' + y'^3) = 0. \tag{2.5.3}$$

The conservation law (2.5.2) yields the following first integral for Eq. (2.5.3):

$$\frac{xy'}{\sqrt{1 + y'^2}} = A = \text{const.}$$

Now we solve the above equation with respect to y', integrate it and obtain the general solution

$$y = B + k \operatorname{arccosh} \frac{x}{k},$$

with two constants of integration, B and k. Evaluating the inverse to the hyperbolic cosine (see (1.1.8)), we can write the solution in the form

$$y = B + k \ln \left| \frac{x + \sqrt{x^2 - k^2}}{k} \right| = C + k \ln \left| x + \sqrt{x^2 - k^2} \right|$$

where $C = B - k \ln |k|$.

Thus, the desired curve is given by the solution

$$y = C + k \ln \left| x + \sqrt{x^2 - k^2} \right|$$

of (2.5.3) satisfying the boundary conditions $y(x_1) = y_1$, $y(x_2) = y_2$.

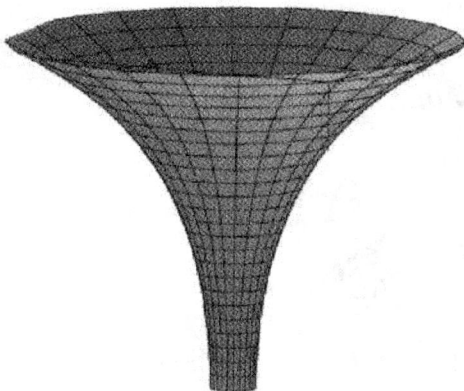

Figure 2.10: Growing by the law
$y = C + k \ln |x + \sqrt{x^2 - k^2}|$.

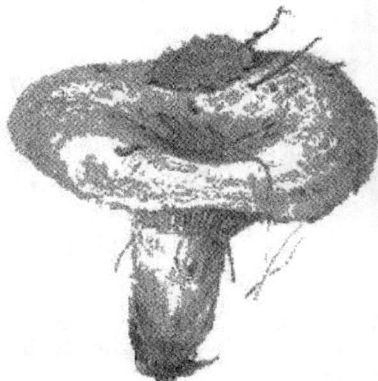

Figure 2.11: A real mushroom.

2.5.2 A tumour growth model

Recently, several mathematical models appeared in the literature for describing spread of malignant tumours. These models are formulated as systems of nonlinear partial differential equations. One of these models is formulated here.

In healthy tissue, balance is preserved between cellular reproduction and cell death. A change of DNA caused by genetic, chemical or other environmental reasons, can give rise to a malignant tumour cell which disrupts this balance and causes an uncontrolled reproduction of cells followed by infiltration into neighboring or remote tissues (metastasis).

Several authors[4] investigated the problem of invasion of malignant cells into surrounding tissue neglecting cellular diffusion. Motivated by several important observations in tumour biology, they suggested a mathematical model appropriate for studying the averaged one-dimensional spatial dynamics of malignant cells by ignoring variations in the plane perpendicular to the direction of invasion. The model is formulated in terms of nonlinear partial differential equations as the following system:

$$u_t = f(u) - (uc_x)_x ,$$
$$c_t = -g(c, p),$$
$$p_t = h(u, c) - Kp.$$

Here u, c and p, depend on time t and one space coordinate x and represent the concentrations of invasive cells, extracellular matrix (e.g. type IV collagen) and protease, respectively. To describe the dynamics of a specific biological

---

[4]A.J. Perumpanani, J.A. Sherrat, J. Norbury, and H.M. Byrne, *Physica D*, 126, 1999.

system, the authors of this model introduced *arbitrary elements* $f(u)$, $g(c,p)$ and $h(u,c)$ that are supposed to be increasing functions of the dependent variables u, c, p. For example, the function $h(u,c)$ in the last equation of the above system represents the dependence of the protease production on local concentrations of malignant cells and collagen, while the term $-Kp$ is based on the assumption that the protease decays linearly, where K is a positive constant to be determined experimentally via half-life.

By observing that the timescales associated with the protease production and decay are considerably shorter than for the invading cells, the above model can be reduced to the following system of two equations:

$$u_t = f(u) - (uc_x)_x,$$
$$c_t = -g(c,u), \tag{2.5.4}$$

where $f(u)$ and $g(c,u)$ are arbitrary functions satisfying the conditions

$$f(u) > 0, \quad g_c(c,u) > 0, \quad g_u(c,u) > 0.$$

2.6 Wave phenomena

Mathematical models of vibration phenomena are most simply derived by using *Hamilton's variational principle* or the *principle of least action* (see Section 1.5). In classical mechanics, any mechanical system is described by a finite number of variables (coordinates of the system) considered as unknown functions of time. Consequently, the motion of the system is governed by *ordinary* differential equations (1.5.2) discussed in Section 1.5.1.

In continuum mechanics, the position of a continuous system no longer can be characterized by a finite number of variables of time. In this case, the kinetic and potential energies are represented by integrals from functions of several variables. This leads to *partial* differential equations obtained via Hamilton's variational principle discussed in Section 1.5.2. Mathematical models have particularly simple (linear) form if the motions are confined to the vicinity of an equilibrium position of a continuous system.

2.6.1 Small vibrations of a string

A *string*, considered as a thin elastic thread spanned along the x-axis, provides a simple example of a one-dimensional continuous system.

Consider the following physical problem. Let us disturb a string from its equilibrium position and release. Evidently, the string will seek to revert to its original position due to the force caused by extension of the string when disturbed. However, when the equilibrium position is reached the string does not stop, inertia deviates it from the x-axis in the opposite direction, and the process repeats. Thus, assuming stretching is the only force acting upon the string, one has free vibrations of the string about its equilibrium position.

Let us obtain the differential equation for determining the perpendicular deviation $u = u(x, t)$ of a point x on the string from the equilibrium position at the time t. Thus we consider the *small vibrations* of the string by assuming that higher powers of the function $u(x, t)$ and of its derivatives can be neglected compared with lower powers.

We use lower indices for partial derivatives along with the standard notation. For example the derivatives of the function $u(x, t)$ with respect to the variables x and t are denoted by u_x and u_t, respectively.

Let $\rho(x)$ be a line density of the string. Then the mass of a small part of the string with the length dx in the interval $(x, x + dx)$ equals $\rho(x)dx$. Accordingly, the density of the kinetic energy T of the string at the point x at a time t is written in the form

$$T = \frac{1}{2} \rho(x) u_t^2. \tag{2.6.1}$$

The potential energy is proportional to the increase in length of the string compared with its length at rest. The factor of proportionality is a positive number $\mu > 0$ called the *tension*. Since the length of a string element dx upon deviation becomes

$$ds = \sqrt{(dx)^2 + (u_x dx)^2} = \sqrt{1 + u_x^2}\ dx \approx [1 + \frac{1}{2} u_x^2]\ dx,$$

the increase in length of the element of the string has the form

$$\sqrt{1 + u_x^2}\ dx - dx \approx \frac{1}{2} u_x^2\ dx.$$

Hence, the density of the potential energy is given by

$$U = \frac{1}{2} \mu u_x^2. \tag{2.6.2}$$

One can apply to the continuous system the concept of an *action* S introduced in Section 1.5.1. Then the action integral for the string is written

$$S = \int L dx dt,$$

where, according to Eqs. (2.6.1) and (2.6.2), the Lagrangian

$$L = T - U$$

has the form

$$L = \frac{1}{2} \left[\rho(x) u_t^2 - \mu u_x^2 \right]. \tag{2.6.3}$$

Thus, the Euler-Lagrange equation (1.5.4) has the form:

$$\frac{\delta L}{\delta u} \equiv \frac{\partial L}{\partial u} - D_t \left(\frac{\partial L}{\partial u_t} \right) - D_x \left(\frac{\partial L}{\partial u_x} \right) = 0. \tag{2.6.4}$$

Here D_t and D_x are the total differentiations (see (1.4.9))

$$D_t = \frac{\partial}{\partial t} + u_t \frac{\partial}{\partial u} + u_{tt} \frac{\partial}{\partial u_t} + u_{tx} \frac{\partial}{\partial u_x}$$

and

$$D_x = \frac{\partial}{\partial x} + u_x \frac{\partial}{\partial u} + u_{tx} \frac{\partial}{\partial u_t} + u_{xx} \frac{\partial}{\partial u_x}$$

with respect to t and x, respectively.

For Lagrangian (2.6.3), we have

$$\frac{\partial L}{\partial u} = 0, \quad \frac{\partial L}{\partial u_t} = \rho(x)u_t, \quad \frac{\partial L}{\partial u_x} = -\mu u_x,$$

and hence,

$$D_t \left(\frac{\partial L}{\partial u_t} \right) = \rho(x)u_{tt}, \quad D_x \left(\frac{\partial L}{\partial u_x} \right) = -\mu u_{xx}.$$

Substituting the above expressions in Eq. (2.6.4), we obtain the following linear partial differential equation of the second order:

$$-\rho(x)\, u_{tt} + \mu\, u_{xx} = 0.$$

Dividing by $-\rho(x)$, we obtain the *wave equation* for free small transverse vibrations of a string:

$$u_{tt} - k^2(x)u_{xx} = 0, \quad \text{where} \quad k^2(x) = \frac{\mu}{\rho(x)},$$

or

$$\frac{\partial^2 u}{\partial t^2} - k^2 \frac{\partial^2 u}{\partial x^2} = 0. \tag{2.6.5}$$

If the string is subjected to the action of an external force $f(x,t)$ perpendicular to the string, the kinetic energy (2.6.1) remains the same, but the potential energy (2.6.2) takes the form

$$U = \frac{1}{2}\, \mu\, u_x^2 - f(x,t)\, u. \tag{2.6.6}$$

Now we have

$$L = \frac{1}{2} \left[\rho(x)u_t^2 - \mu u_x^2 \right] + f(x,t)\, u \tag{2.6.7}$$

instead of (2.6.3) and therefore,

$$\frac{\partial L}{\partial u} = f(x,t), \quad \frac{\partial L}{\partial u_t} = \rho(x)u_t, \quad \frac{\partial L}{\partial u_x} = -\mu u_x.$$

Accordingly, Eq. (2.6.5) is replaced by the following equation for *forced vibrations* of the string:

$$\rho(x)\, u_{tt} - \mu\, u_{xx} = f(x,t). \tag{2.6.8}$$

Likewise, one can obtain the equation for *longitudinal* small vibrations of an elastic rod:

$$\rho(x)u_{tt} - [E(x)u_x]_x = f(x,t), \tag{2.6.9}$$

where $E(x)$ is Young's modulus, i.e., the modulus of elongation of the rod.

2.6.2   Vibrating membrane

A *membrane* is a portion of a two-dimensional surface made from an elastic material whose potential energy is proportional to change in the surface area. The positive constant $\mu > 0$ (factor of proportionality) is called the *tension*. The density of the membrane will be denoted by $\rho(x, y)$.

Let the membrane at rest occupy a region of the plane (x, y), and let $u(x, y, t)$ be the deformation of the membrane normal to its equilibrium position. We again suppose that the deformations are small, and hence higher powers of u, u_x, u_y are negligible compared with lower ones. Then the area of an element of the deformed membrane is written

$$\sqrt{1 + u_x^2 + u_y^2}\, dxdy \approx \left[1 + \frac{1}{2}(u_x^2 + u_y^2)\right] dxdy. \qquad (2.6.10)$$

Subtracting the area $dxdy$ of the element before deformation, one obtains the following change of the area:

$$\frac{1}{2}\left(u_x^2 + u_y^2\right) dxdy.$$

Hence, the density of the potential energy is

$$U = \frac{\mu}{2}\left(u_x^2 + u_y^2\right),$$

while the kinetic energy is similar to that of string (2.6.1), namely:

$$T = \frac{1}{2}\,\rho(x, y)\,u_t^2.$$

Thus, the Lagrangian for the membrane has the form

$$L = \frac{1}{2}\left[\rho(x, y)u_t^2 - \mu\left(u_x^2 + u_y^2\right)\right], \qquad (2.6.11)$$

and the corresponding Euler-Lagrange equation is written

$$\frac{\delta L}{\delta u} \equiv \frac{\partial L}{\partial u} - D_t\left(\frac{\partial L}{\partial u_t}\right) - D_x\left(\frac{\partial L}{\partial u_x}\right) - D_y\left(\frac{\partial L}{\partial u_y}\right) = 0, \qquad (2.6.12)$$

where D_t, D_x and D_y are the total differentiations with respect to t, x and y, respectively:

$$D_t = \frac{\partial}{\partial t} + u_t\frac{\partial}{\partial u} + u_{tt}\frac{\partial}{\partial u_t} + u_{tx}\frac{\partial}{\partial u_x} + u_{ty}\frac{\partial}{\partial u_y},$$

$$D_x = \frac{\partial}{\partial x} + u_x\frac{\partial}{\partial u} + u_{tx}\frac{\partial}{\partial u_t} + u_{xx}\frac{\partial}{\partial u_x} + u_{xy}\frac{\partial}{\partial u_y},$$

$$D_y = \frac{\partial}{\partial y} + u_y\frac{\partial}{\partial u} + u_{ty}\frac{\partial}{\partial u_t} + u_{xy}\frac{\partial}{\partial u_x} + u_{yy}\frac{\partial}{\partial u_y}.$$

Substituting Lagrangian (2.6.11) in (2.6.12), proceeding as in the case of the string and using again the notation

$$k^2(x, y) = \frac{\mu}{\rho(x, y)},$$

one obtains the two-dimensional *wave equation for a vibrating membrane:*

$$\frac{\partial^2 u}{\partial t^2} - k^2 \left(\frac{\partial^2 u}{\partial x^2} + \frac{\partial^2 u}{\partial y^2} \right) = 0. \tag{2.6.13}$$

The expression in the brackets is the two-dimensional version of Laplacian (1.3.19), and Eq. (2.6.13) is often written in the form

$$u_{tt} - k^2(x, y)\Delta u = 0. \tag{2.6.14}$$

In the presence of an external force $f(x, y, t)$ normal to the plane (x, y) the equation for forced vibrations of the membrane is written

$$u_{tt} - k^2(x, y)\Delta u = F(x, y, t), \tag{2.6.15}$$

where $F(x, y, t) = f(x, y, t)/\rho(x, y)$.

Most frequently, the wave equation is considered when the density ρ is constant. Then the coefficient k^2 is also constant. This assumption will also be used in the three-dimensional wave equation:

$$u_{tt} - k^2\Delta u = F(x, y, z, t), \quad k^2 = \text{const.}, \tag{2.6.16}$$

where Δ is the Laplace operator (1.3.19):

$$\Delta = \frac{\partial^2}{\partial x^2} + \frac{\partial^2}{\partial y^2} + \frac{\partial^2}{\partial z^2}.$$

The wave equation (2.6.16) is called *homogeneous* if $F = 0$, and *non-homogeneous* otherwise (see Section 5.1).

Along with the one-dimensional (2.6.5) and two-dimensional (2.6.13) wave equations, we also consider the three-dimensional linear wave equation

$$\frac{\partial^2 u}{\partial t^2} - k^2 \left(\frac{\partial^2 u}{\partial x^2} + \frac{\partial^2 u}{\partial y^2} + \frac{\partial^2 u}{\partial z^2} \right) = 0. \tag{2.6.17}$$

This is one of the basic equations of mathematical physics. For example, it describes propagation of light waves; then the coefficient k^2 in Eq. (2.6.17) is identical with c^2, where c is the velocity of light in vacuum.

Note that any wave equation with a constant coefficient k^2 can be reduced to an equation with $k^2 = 1$ by means of an appropriate dilation transformation. Therefore, we shall also use the following equivalent form, e.g. of the free wave equation:

$$u_{tt} - \Delta u = 0. \tag{2.6.18}$$

2.6.3   Minimal surfaces

We derived the wave equation for small vibrations of membranes using approximation (2.6.10) for the variation of the surface area of the membrane. The problem on minimal surfaces[5] requires determination of all possible configurations of the membrane when its surface area

$$\int_V \sqrt{1 + u_x^2 + u_y^2}\, dxdy$$

has a minimum value. Hence, the corresponding differential equation for minimal surfaces is the Euler-Lagrange equation with the Lagrangian

$$L = \sqrt{1 + u_x^2 + u_y^2}. \tag{2.6.19}$$

Since Lagrangian (2.6.19) does not involve u and u_t, and since

$$\frac{\partial L}{\partial u_x} = \frac{u_x}{\sqrt{1 + u_x^2 + u_y^2}}, \quad \frac{\partial L}{\partial u_y} = \frac{u_y}{\sqrt{1 + u_x^2 + u_y^2}},$$

the Euler-Lagrange equation (2.6.12) is written

$$D_x\left(\frac{u_x}{\sqrt{1 + u_x^2 + u_y^2}}\right) + D_y\left(\frac{u_y}{\sqrt{1 + u_x^2 + u_y^2}}\right) = 0 \tag{2.6.20}$$

and leads to the following nonlinear equation:

$$(1 + u_y^2)u_{xx} - 2u_x u_y u_{xy} + (1 + u_x^2)u_{yy} = 0. \tag{2.6.21}$$

The linearization of Eq. (2.6.21) gives the Laplace equation

$$\Delta u \equiv u_{xx} + u_{yy} = 0 \tag{2.6.22}$$

for the *equilibrium problem* for the membrane.

2.6.4   Vibrating slender rods and plates

A physical slender rod is a thin wire which resists bending unlike the string which resists elongation. A mathematical rod is a one-dimensional continuum, lying at the straight line when at rest, which being bent gains potential energy

---

[5]The problem was first formulated by L. Euler. Some hundred years later, a Belgian physicist J. Plateau suggested experiments for obtaining minimal surfaces and described them in 1873. Since then, the problem of minimal surfaces became known as *Plateau's problem*, see. [35], p.534. A profound mathematical investigation of the problem is due to R. Courant. I remember his lecture at the Russian-American conference held in 1963 at Novosibirsk University, where I was a student. His talk was enlivened by an illustration where he repeated Plateau's experiments and obtained minimal surfaces by dipping in a soap solution pieces of wire bent into closed curves of various forms.

with a density proportional to the square of the curvature. We denote by $u(x, t)$ the deviation of the rod from its position of equilibrium. We consider again small vibrations defined as in the case of strings. The deformed rod is a curve $u = u(x, t)$ in the x, u plane, where time t is regarded as a parameter.

Recall that the curvature of a curve is *the rate of change of its direction* (i.e., the tangent to the curve) and describes the flatness or sharpness of the curve. For a plane curve $u = u(x, t)$, the square of curvature K at the point x is given by

$$K^2 = \frac{u_{xx}^2}{(1 + u_x^2)^3}$$

and is approximated by

$$K^2 \approx u_{xx}^2.$$

Hence, the density of the potential energy is given by

$$U = \frac{\mu}{2} u_{xx}^2,$$

while the density of the kinetic energy has again form (2.6.1),

$$T = \frac{1}{2} \rho u_t^2,$$

where $\rho = \rho(x)$, $\mu = \text{const}$. Thus, the Lagrangian has the form

$$L = \frac{1}{2} \left(\rho u_t^2 - \mu u_{xx}^2 \right). \tag{2.6.23}$$

In the case of the Lagrangians involving the second-order derivatives:

$$L = L(t, x, u, u_x, u_t, u_{xx}, u_{xt}, u_{tt}),$$

Hamilton's variational principle provides the following Euler-Lagrange equation (cf. (1.5.4)):

$$\frac{\delta L}{\delta u} \equiv \frac{\partial L}{\partial u} - D_t \left(\frac{\partial L}{\partial u_t} \right) - D_x \left(\frac{\partial L}{\partial u_x} \right)$$

$$+ D_t^2 \left(\frac{\partial L}{\partial u_{tt}} \right) + D_t D_x \left(\frac{\partial L}{\partial u_{tx}} \right) + D_x^2 \left(\frac{\partial L}{\partial u_{xx}} \right) = 0. \tag{2.6.24}$$

For Lagrangian (2.6.23), Eq. (2.6.24) is written as follows:

$$-D_t(\rho u_t) - D_x^2(\mu u_{xx}) = -\rho u_{tt} - \mu u_{xxxx} = 0.$$

We can also take into account external forces as we did for strings. Thus, small transversal vibrations of slender rods are governed by the following partial differential equation of the fourth order:

$$\rho(x) \frac{\partial^2 u}{\partial t^2} + \mu \frac{\partial^4 u}{\partial x^4} = f, \tag{2.6.25}$$

where f is a total force acting on the rod and μ is a positive constant.

Derivation of the differential equation for vibrating plates is similar to that for rods. A plate is an elastic two-dimensional surface, plane when at rest, the density of whose potential energy U after deformation is proportional to a quadratic form in the *principal curvatures* K and H of the plate:

$$U = \alpha H^2 + \beta K, \quad \alpha, \beta = \text{const.}$$

The expressions for K and H are given in any textbook of differential geometry. In the case of two-dimensional surfaces given by the equation $u = u(x, y, t)$ with time t considered as a parameter, we have

$$K = \frac{u_{xx}u_{yy} - u_{xy}^2}{\left(1 + u_x^2 + u_y^2\right)^2}, \quad H = \text{div } \frac{\nabla u}{\sqrt{1 + u_x^2 + u_y^2}},$$

where $\nabla u = (u_x, u_y)$. In the approximation of small vibrations, we have

$$K \approx u_{xx}u_{yy} - u_{xy}^2, \quad H \approx \text{div } (\nabla u) = \Delta u \equiv u_{xx} + u_{yy}.$$

Hence, setting $\alpha = \mu/2$, we have

$$U \approx \frac{\mu}{2} (\Delta u)^2 + \beta(u_{xx}u_{yy} - u_{xy}^2).$$

Exercise 2.6.1. Prove that

$$\frac{\delta}{\delta u} (u_{xx}u_{yy} - u_{xy}^2) = 0. \tag{2.6.26}$$

In view of Eq. (2.6.26), we take the Lagrangian in the form

$$L = \frac{1}{2} [\rho u_t^2 - \mu(\Delta u)^2]. \tag{2.6.27}$$

With this Lagrangian, the two-dimensional version of the Euler-Lagrange equation (2.6.24) gives the following fourth-order partial differential equation for vibrating plates:

$$\rho u_{tt} + \mu(u_{xxxx} + 2u_{xxyy} + u_{yyyy}) = 0. \tag{2.6.28}$$

2.6.5   Nonlinear waves

Let us consider a uniform membrane whose tension varies during deformations, i.e., we assume that $\mu = \phi(u) > 0$ and $\rho = \text{const.}$ We set $\rho = 1$ for the simplicity sake. Then Lagrangian (2.6.11) is replaced by

$$L = \frac{1}{2} \left[u_t^2 - \phi(u)\left(u_x^2 + u_y^2\right) \right], \tag{2.6.29}$$

and the corresponding Euler-Lagrange equation

$$\frac{\partial L}{\partial u} - D_t\left(\frac{\partial L}{\partial u_t}\right) - D_x\left(\frac{\partial L}{\partial u_x}\right) - D_y\left(\frac{\partial L}{\partial u_y}\right) = 0$$

is written

$$-\frac{1}{2}\phi'(u)\,u_x^2 - \frac{1}{2}\phi'(u)\,u_y^2 - D_t[u_t] + D_x[\phi(u)u_x] + D_y[\phi(u)u_y] = 0$$

or

$$-\frac{1}{2}\phi'(u)\,u_x^2 - \frac{1}{2}\phi'(u)\,u_y^2 - u_{tt} + \phi(u)[u_{xx} + u_{yy}] + \phi'(u)[u_x^2 + u_y^2] = 0,$$

whence, upon collecting the like terms:

$$-u_{tt} + \phi(u)[u_{xx} + u_{yy}] + \frac{1}{2}\phi'(u)[u_x^2 + u_y^2] = 0.$$

Thus, we have the following *nonlinear wave equation:*

$$u_{tt} = \phi(u)\,\Delta u + \frac{1}{2}\phi'(u)|\nabla u|^2. \tag{2.6.30}$$

The nonlinear wave equation (2.6.30) can be written for any number of variables $x^1, \ldots, x^n$. For example, the one-dimensional case yields:

$$u_{tt} = \phi(u)\,u_{xx} + \frac{1}{2}\phi'(u)u_x^2. \tag{2.6.31}$$

The following nonlinear differential equations different from (2.6.31) are also used in studying nonlinear wave phenomena:

$$\begin{aligned}
u_{tt} &= [f(u)u_x]_x, \\
u_{tt} &= [f(x,u)u_x]_x, \\
u_{tt} &= [f(u)u_x + g(x,u)]_x.
\end{aligned} \tag{2.6.32}$$

Defining the potential v by the equation $u = v_x$, Eqs. (2.6.32) are written, respectively, as follows:

$$\begin{aligned}
v_{tt} &= f(v_x)v_{xx}, \\
v_{tt} &= f(x,v_x)v_{xx}, \\
v_{tt} &= f(v_x)v_{xx} + g(x,v_x).
\end{aligned} \tag{2.6.33}$$

The latter equations are encapsulated in the following reasonably general class of nonlinear one-dimensional wave equations:

$$v_{tt} = f(x,v_x)v_{xx} + g(x,v_x). \tag{2.6.34}$$

Another type of nonlinear wave phenomena of practical interest are known in gas dynamics as "short waves". They are described by the system

$$u_y - 2v_t - 2(v - x)v_x - 2kv = 0,$$

$$v_y + u_x = 0, \quad k = \text{const.}$$

This system of two first-order equations can be reduced by the substitution $u = w_y, v = -w_x$ to one second-order equation, namely:

$$2w_{tx} + 2(x + w_x)w_{xx} + w_{yy} + 2kw_x = 0. \tag{2.6.35}$$

2.6.6   The Chaplygin and Tricomi equations

The Chaplygin equation has the form

$$\varphi(x)u_{yy} + u_{xx} = 0. \tag{2.6.36}$$

It plays a significant role in problems of high velocity aerodynamics and was suggested by S.A. Chaplygin in 1902 in his dissertation "On gas jets". The Chaplygin equation is used for the study of the two-dimensional steady transonic flow and has practical applications, e.g. in aircraft engineering for modelling the flow of gas jets past a wing when the flight speed is close to the speed of sound.

A good approximation of the Chaplygin equation is the Tricomi equation

$$xu_{yy} + u_{xx} = 0. \tag{2.6.37}$$

Equations (2.6.36) and (2.6.37) provide examples of partial differential equations of so-called *mixed elliptic-hyperbolic type*, e.g. (2.6.37) is elliptic when $x > 0$ and hyperbolic when $x < 0$ (see Section 5.2.5). As a matter of fact, Eq. (2.6.37) was suggested by F.G. Tricomi in 1923 in his study of linear second-order partial differential equations of mixed type.

Problems to Chapter 2

2.1. Derive the Euler-Lagrange equations for the following Lagrangians:

(i) $L = \dfrac{1}{2}\left[u_x^2 + u_y^2 + u_z^2 - u_t^2\right] - f(t, x, y, z)u,$

(ii) $L = \dfrac{1}{2}u_y^2 - u_t u_x - \dfrac{1}{6}u_x^3,$

(iii) $L = \dfrac{1}{2}\left(-u_t^2 + \mu u_{xx}^2\right) - f(t, x)u, \quad \mu = \text{const.},$

(iv) $L = \dfrac{1}{2}\left[-u_t^2 + (u_{xx} + u_{yy})^2\right] - f(t, x, y)u.$

2.2. Give a detailed derivation of Eq. (2.6.21).

2.3. The gravitational force of the Sun is spherically symmetric. Therefore, one may naturally conclude that the motion of the planets should also be spherically symmetric. This would only be the case if the planets moved in circles on surfaces of spheres. J. Kepler discovered (1609), however, that planets move in ellipses not in circles, on fixed planes not spheres, with the Sun at a focus not at the centre. Explain what violates the symmetry in the motion of planets.

2.4. Let us generalize Kepler's problem and consider the motion of a particle with mass m in an arbitrary central potential field

$$U = U(r), \quad r = |\boldsymbol{x}| \equiv \sqrt{(x^1)^2 + (x^2)^2 + (x^3)^2}.$$

According to the principle of least action (see Section 1.5.1), the motion of the particle is determined by the Lagrangian

$$L = \frac{m}{2} \sum_{i=1}^{3} (v^i)^2 - U(r).$$

Find the corresponding Euler-Lagrange equations (1.5.2).

2.5. The Dirac equation (2.3.32) is a vector equation. Its components provide four equations. Write down these four equations explicitly.

2.6. Integrate Eq. (2.2.3): $dP/dt = \alpha P - \beta P^2$ $(\alpha, \beta = \text{const.} \neq 0)$.

2.7. Derive from Eqs. (2.3.30) that $D_t(\text{div } \boldsymbol{E}) = 0$, $D_t(\text{div } \boldsymbol{H}) = 0$, and hence Eq. (2.3.31), div $\boldsymbol{E} = 0$, div $\boldsymbol{H} = 0$, hold at any time if they are satisfied at an initial time $t = t_0$.

Chapter 3

Ordinary differential equations: Traditional approach

This chapter is designed as a short account of basic traditional methods invented mainly in the 17th and 18th centuries. These classical devices are simple and therefore commonly used in the practice of integration of special types of ordinary differential equations by means of *ad hoc* methods.

Additional reading: E. Goursat [9], G.F. Simmons [35].

3.1 Introduction and elementary methods

3.1.1 Differential equations. Initial value problem

An nth-order ordinary differential equation (ODE) is a relation

$$F(x, y, y', \ldots, y^{(n)}) = 0 \qquad (3.1.1)$$

connecting the single independent variable x, the dependent variable y and its derivatives $y', \ldots, y^{(n)}$.

The classical definition of solutions of differential equations is as follows.

Definition 3.1.1. A function $y = \phi(x)$, defined in a neighborhood of x_0 and continuously differentiable n times, is said to be a solution of a differential equation (3.1.1) if

$$F(x, \phi(x), \phi'(x), \ldots, \phi^{(n)}(x)) = 0$$

identically in x from a certain interval

$$(x_0 - \varepsilon, x_0 + \varepsilon), \quad \varepsilon > 0.$$

Since any function $y = y(x)$ represents a curve in the (x, y) plane, solutions of ordinary differential equations are also termed *integral curves*.

Existence theorems furnish the core of the general theory of differential equations, in particular, in Lie group analysis.

The first systematic investigations on the existence of solutions of differential equations are due to Cauchy (1845). Note that, e.g. in the case of Eq. (1.2.58),

$$\frac{dy}{dx} = f(x),$$

with continuous $f(x)$, one can readily obtain the solution that assumes a given value y_0 at $x = x_0$, by means of formula (1.2.59). The solution of this *initial value problem* is unique, is defined in a neighborhood of the point x_0 and is given by

$$y(x) = y_0 + \int_{x_0}^{x} f(t)dt.$$

Cauchy extended this result by proving the existence of solutions of the initial value problem for the general first-order equation:

$$\frac{dy}{dx} = f(x,y), \quad y\big|_{x=x_0} = y_0, \tag{3.1.2}$$

where $f(x,y)$ is a continuous function in a neighborhood of the point (x_0, y_0) in the (x,y) plane; the notation $\big|_{x=x_0}$ means evaluated at $x = x_0$. Consequently, initial value problems are often referred to as the *Cauchy problem*.

Thus, Cauchy's result states the existence of integral curves passing through any given point (x_0, y_0). However, the solution need not be unique if only the continuity of the right-hand side, $f(x,y)$, is required. For example, the initial value problem

$$\frac{dy}{dx} = 2\sqrt{|y|}, \quad y\big|_{x=x_0} = 0,$$

has two solutions, namely:

$$y = 0 \quad \text{and} \quad y = |x - x_0|(x - x_0).$$

Therefore, Cauchy's investigations were continued and led to the general theorems on existence and uniqueness of the solution of the Cauchy problem. For our purposes, it suffices to use the following simple version of the existence and uniqueness theorem.

Theorem 3.1.1. Let $f(x,y)$ be a continuously differentiable function in a neighborhood of the point (x_0, y_0). than the initial value problem (3.1.2) has one and only one solution $y = \phi(x)$ defined in a neighborhood of x_0.

Remark 3.1.1. A more general version (though not the most general one) of the theorem requires a weaker condition than the continuous differentiability, namely so-called *Lipschitz condition*.

The existence and uniqueness theorem for higher-order equations has the following form.

Theorem 3.1.2. Given an nth-order equation

$$y^{(n)} = f(x, y, y', \ldots, y^{(n-1)}),\tag{3.1.3}$$

the Cauchy problem is to find the solution to Eq. (3.1.3) satisfying the following initial conditions:

$$y\big|_{x=x_0} = y_0, \quad \frac{dy}{dx}\bigg|_{x=x_0} = y_0', \quad \ldots, \quad \frac{d^{n-1}y}{dx^{n-1}}\bigg|_{x=x_0} = y_0^{(n-1)}.\tag{3.1.4}$$

Let the function f in Eq. (3.1.3) be continuously differentiable in a neighborhood of $x_0, y_0, y_0', \ldots, y_0^{(n-1)}$. Then the Cauchy problem (3.1.3)~(3.1.4) has a unique solution defined in a neighborhood of x_0.

Remark 3.1.2. It follows that the general solution of nth-order differential equations (3.1.3) depends precisely on n arbitrary constants $C_1, \ldots, C_n$.

3.1.2 Integration of the equation $y^{(n)} = f(x)$

The solution of the equation

$$y^{(n)} = f(x)$$

is similar to solution (1.2.61) of Eq. (1.2.60). Namely, the consecutive integration yields:

$$y^{(n-1)} = \int f(x)dx + C_1, \quad y^{(n-2)} = \int dx \int f(x)dx + C_1 x + C_2, \quad \ldots,$$

whence finally the solution formula similar to (1.2.61):

$$y = \int dx \int dx \ldots \int f(x)dx + C_1 \frac{x^{n-1}}{(n-1)!} + C_2 \frac{x^{n-2}}{(n-2)!} + \cdots + C_{n-1}x + C_n,$$

where $C_1, \ldots, C_n$ are arbitrary constants of integration.

3.1.3 Homogeneous equations

Any homogeneous equation of order n can be integrated by quadrature if $n = 1$, and reduced to an equation of order $n-1$ if $n > 1$ (see Chapter 6). The general homogeneity of differential equations is defined as follows.

Definition 3.1.2. An ordinary differential equation of an arbitrary order

$$F(x, y, y', \ldots, y^{(n)}) = 0\tag{3.1.5}$$

is said to be *homogeneous* if it is invariant under a scaling transformation (dilation) of the independent and dependent variables (cf. (1.1.35)):

$$\bar{x} = a^k x, \quad \bar{y} = a^l y,\tag{3.1.6}$$

where $a > 0$ is a parameter not identical with 1, and k and l are any fixed real numbers. The invariance means that

$$F(\overline{x}, \overline{y}, \overline{y}', \ldots, \overline{y}^{(n)}) = 0, \tag{3.1.7}$$

where $\overline{y}' = d\overline{y}/d\overline{x}$, etc. In particular, in the case of a first-order equation

$$y' = f(x, y) \tag{3.1.8}$$

the homogeneity means that after dilation (3.1.6), Eq. (3.1.8) becomes

$$\frac{d\overline{y}}{d\overline{x}} = f(\overline{x}, \overline{y}). \tag{3.1.9}$$

Example 3.1.1. The first-order equation (cf. Example 3.2.2)

$$y' - \frac{2xy}{3x^2 - y^2} = 0$$

is homogeneous since it is invariant under the dilation $\overline{x} = a\,x$, $\overline{y} = a\,y$. Indeed,

$$\frac{d\overline{y}}{d\overline{x}} = \frac{ady}{adx} = \frac{dy}{dx} \equiv y', \qquad \frac{2\overline{x}\,\overline{y}}{3\overline{x}^2 - \overline{y}^2} = \frac{2a^2xy}{a^2(3x^2 - y^2)} = \frac{2xy}{3x^2 - y^2}$$

and hence condition (3.1.9) is satisfied:

$$\frac{d\overline{y}}{d\overline{x}} - \frac{2\overline{x}\,\overline{y}}{3\overline{x}^2 - \overline{y}^2} = y' - \frac{2xy}{3x^2 - y^2} = 0.$$

Example 3.1.2. The second-order equation

$$y'' - \frac{2xy}{3x^2 - y^2} = 0 \tag{3.1.10}$$

is not homogeneous. Indeed, consider the general dilation (3.1.6) replacing it, for the sake of convenience of calculations, by $\overline{x} = a\,x$, $\overline{y} = b\,y$ with positive parameters a and b. Then we have

$$\frac{d^2\overline{y}}{d\overline{x}^2} = \frac{b}{a^2}\frac{d^2y}{dx^2}, \qquad \frac{2\overline{x}\,\overline{y}}{3\overline{x}^2 - \overline{y}^2} = \frac{ab(2xy)}{3a^2x^2 - b^2y^2}.$$

Hence, the equation

$$\frac{d^2\overline{y}}{d\overline{x}^2} = \frac{2\overline{x}\,\overline{y}}{3\overline{x}^2 - \overline{y}^2}$$

is written

$$\frac{b}{a^2}\frac{d^2y}{dx^2} = \frac{ab(2xy)}{3a^2x^2 - b^2y^2}.$$

The invariance condition requires that

$$\text{(i) } 3a^2x^2 - b^2y^2 = c(3x^2 - y^2), \quad \text{(ii) } \frac{b}{a^2} = \frac{ab}{c}. \tag{3.1.11}$$

Since Eq. (3.1.11)(i) should hold identically in x and y, it follows that $a^2 = b^2 = c$. Now Eq. (3.1.11)(ii) is written

$$\frac{b}{a^2} = \frac{ab}{a^2}$$

and yields $a = 1$. Furthermore, since a and b are positive, it follows from $b^2 = a^2 = 1$ that $b = 1$. Thus, the dilation $\bar{x} = ax$, $\bar{y} = by$ reduces to the identity transformation $\bar{x} = x$, $\bar{y} = y$ and hence, by Definition 3.1.2, Eq. (3.1.10) is not homogeneous.

Example 3.1.3. The equation

$$y' + y^2 = \frac{C}{x^2}, \qquad C = \text{const.}, \tag{3.1.12}$$

is homogeneous since it is invariant under the dilation $\bar{x} = ax$, $\bar{y} = a^{-1}y$ (see also Problem 3.4 and Example 6.3.3).

Definition 3.1.3. One calls Eq. (3.1.5) *double homogeneous* if it is invariant with respect to the independent dilations of the independent and dependent variables, i.e., if it does not alter under the transformations

$$\bar{x} = ax, \quad \bar{y} = y \tag{3.1.13}$$

and

$$\bar{x} = x, \quad \bar{y} = by \tag{3.1.14}$$

with independent positive parameters a and b, respectively.

Example 3.1.4. The linear equations

$$xy' + Cy = 0, \quad C = \text{const.}, \tag{3.1.15}$$

and

$$x^2 y'' + C_1 xy' + C_2 y = 0, \quad C_1, C_2 = \text{const.}, \tag{3.1.16}$$

provide examples of double homogeneous equations of the first- and second-order, respectively. They are known as Euler's equations of the first and second order, respectively (see Section 3.4.4). As a matter of fact, Eq. (3.1.15) is the most general double homogeneous equation of the first order (see Problem 6.11). On the other hand, the most general double homogeneous equation of the second order has the form (see Problem 6.12)

$$y'' = \frac{y}{x^2} H\left(\frac{xy'}{y}\right), \tag{3.1.17}$$

where H is an arbitrary function. Eq. (3.1.16) is a particular case of (3.1.17) and is obtained by setting

$$H\left(\frac{xy'}{y}\right) = -C_1\left(\frac{xy'}{y}\right) + C_2.$$

Remark 3.1.3. The single homogeneity considered in Definition 3.1.2, unlike the double homogeneity, deals with an invariance under dilations depending on one parameter only. It can be obtained in practice by seeking a scaling transformation $\bar{x} = a\,x$, $\bar{y} = b\,y$ (see (1.1.35)) involving two parameters, a and b. In most of applications, the calculations will end up with either the identity transformation corresponding to $a = b = 1$, or certain one-parameter dilation (3.1.6). See, e.g. Example 3.1.2 and Problem 3.4.

3.1.4   Different types of homogeneity

The following two types of homogeneity corresponding to special types of dilations (3.1.6) are of particular interest and usually (even mostly) are considered in standard texts. An exception is provided by linear partial differential equations of the first order (see Section 4.1).

Type 1: Uniform homogeneity. The uniformly homogeneous equations are invariant under the *uniform scaling*

$$\bar{x} = a\,x, \quad \bar{y} = a\,y \tag{3.1.18}$$

obtained from (3.1.6) by letting $k = l = 1$. Since the uniform scaling (3.1.18) leaves unaltered the first derivative, $\bar{y}' = y'$, Eqs. (3.1.8) and (3.1.9) yield that $f(ax, ay) = f(x, y)$.

Example 3.1.5. The following first and second order equations with arbitrary constant coefficients A, B and C are uniformly homogeneous:

$$y' + \frac{y}{x} = C, \quad y'' + \frac{A}{x} y' + \frac{B}{x^2} y = \frac{C}{x},$$

$$y' + \frac{x}{y} = C, \quad y'' + \frac{A}{xy'} + \frac{B}{y} = \frac{C}{x}.$$

The standard form of the uniformly homogeneous equations of the first order is (see Problem 6.2 (i))

$$y' = \varphi\left(\frac{y}{x}\right). \tag{3.1.19}$$

Eq. (3.1.19) can be solved by considering the invariant y/x under dilation (3.1.18) as a new dependent variable, i.e., by setting

$$\frac{y}{x} = u, \quad \text{or} \quad y = xu(x).$$

Indeed, Eq. (3.1.19) takes the form

$$xu' + u = \varphi(u)$$

and can be solved by separation of variables:

$$\int \frac{du}{\varphi(u) - u} = \int \frac{dx}{x} \equiv \ln x + C.$$

Type 2: Homogeneity by function. This type of homogeneity designates the invariance with respect to transformation (3.1.6) with $k = 0$, $l = 1$, i.e., with respect to the dilation of y only:

$$\bar{x} = x, \quad \bar{y} = a\,y. \tag{3.1.20}$$

The homogeneity by function is commonly employed in the case of linear ordinary differential equations (Sections 3.2.6, 3.3, 3.4) as well as linear partial differential equations (see Sections 4.1 and 5.1).

The linear ordinary differential equation of the form

$$y' + P(x)y = 0$$

is homogeneous by function and, moreover, furnishes the general form of first-order equation that homogeneous by function (see Problem 6.2 (ii)). The higher-order equations of the form

$$y^{(n)} + a_1(x)y^{(n-1)} + \cdots + a_{n-1}(x)y' + a_n(x)y = 0, \quad n \geq 2,$$

are also homogeneous by function. However, in the case of higher-order equations, unlike the first-order equations, homogeneity by function does not imply the linearity. For example, the general form of second-order equations homogeneous by function is (see Problem 6.2 (ii))

$$y'' = yF\left(x, \frac{y'}{y}\right).$$

Consider an example of general homogeneity different from the above two types. Let us take dilation (3.1.6) with $k = \sqrt{2}$, $l = 1$:

$$\bar{x} = a^{\sqrt{2}}x, \quad \bar{y} = a\,y.$$

The corresponding general first-order homogeneous equations has the form

$$\frac{dy}{dx} = \frac{y}{x}F\left(\frac{y^{\sqrt{2}}}{x}\right). \tag{3.1.21}$$

Its integration is discussed further in Problem 6.10.

3.1.5 Reduction of order

Any second-order equation of the form

$$y'' = f(y, y') \tag{3.1.22}$$

can be reduced to a first-order equation by the substitution

$$y' = p(y). \tag{3.1.23}$$

Indeed, employing the chain rule we have from Eq. (3.1.23):

$$y'' = y'p'(y) \equiv pp'.$$

Now (3.1.22) becomes the first-order equation:

$$pp' = f(y,p) \tag{3.1.24}$$

for the new unknown function $p(y)$ with the independent variable y.

Provided that the general solution $p = \phi(y, C_1)$ to Eq. (3.1.24) is known, the solution of the original equation (3.1.22) is obtained via Eq. (3.1.23),

$$\frac{dy}{dx} = \phi(y, C_1),$$

by one quadrature:

$$\int \frac{dy}{\phi(y, C_1)} = x + C_2.$$

Likewise, substitution (3.1.23) reduces the order by one of any higher-order equation not explicitly involving the independent variable x, i.e., of equations of the form

$$y^{(n)} = f(y, y', \ldots, y^{(n-1)}).$$

In this case we have

$$y'' = y'p' = pp', \quad y''' = y'(pp')' = p(pp')' = p(p')^2 + p^2 p'', \ldots$$

and hence our equation becomes an equation of order $n - 1$ for $p(y)$:

$$p^{(n-1)} = F\left(y, p, p', \ldots, p^{(n-2)}\right).$$

3.1.6   Linearization through differentiation

Sometimes, nonlinear equations can be linearized through differentiation. The following example explains the idea.

Example 3.1.6. Consider the following nonlinear second-order equation:

$$2yy'' - y'^2 = 0. \tag{3.1.25}$$

Differentiation yields $2yy''' = 0$, whence $y = 0$ (a trivial solution to Eq. (3.1.25)) or $y''' = 0$. The equation $y''' = 0$ yields that $y = ax^2 + bx + c$ with arbitrary constants a, b, c. To determine these constants, we substitute the expression for y in Eq. (3.1.25) and obtain $4ac - b^2 = 0$. It follows that either $a \neq 0$ and then $c = b^2/(4a)$, or $a = b = 0$. Accordingly, the general solution to Eq. (3.1.25) has the form:

$$y = ax^2 + bx + \frac{b^2}{4a} = \frac{1}{4a}(2ax + b)^2 \quad (a \neq 0), \quad \text{and} \quad y = c.$$

3.2 First-order equations

3.2.1 Separable equations

The technique of *separation of variables* is applicable to first-order ordinary differential equations of the type

$$y' = p(x)q(y).$$ (3.2.1)

We rewrite Eq. (3.2.1) in the differential form

$$\frac{1}{q(y)}\frac{dy}{dx}\, dx = p(x)dx,$$

and integrate its both sides with respect to x :

$$\int \frac{1}{q(y)}\frac{dy}{dx}\, dx = \int p(x)dx + C.$$

Here we change the variable of integration in the left-hand side from x to y by using the rule for a change of variables in integrals (see Section 1.2.4) and the invariance of differential (1.2.7), and rewrite the above integral equation in the form

$$\int \frac{dy}{q(y)} = \int p(x)dx + C.$$ (3.2.2)

Evaluating the integrals in both sides and solving with respect to y, one obtains the general solution involving the constant of integration C.

3.2.2 Exact equations

Definition 3.2.1. A first-order differential equation of the form

$$M(x,y)dx + N(x,y)dy = 0$$ (3.2.3)

is said to be *exact* if its left-hand side is the differential, i.e., if

$$M\,dx + N\,dy = d\Phi \equiv \frac{\partial \Phi}{\partial x}dx + \frac{\partial \Phi}{\partial y}dy$$ (3.2.4)

with some function $\Phi(x,y)$.

For an exact equation (3.2.3), the function Φ is found from Eq. (3.2.4) rewritten as a system of differential equations for unknown Φ:

$$\frac{\partial \Phi}{\partial x} = M(x,y), \quad \frac{\partial \Phi}{\partial y} = N(x,y).$$ (3.2.5)

This *over-determined* system (namely, two equations for one unknown function Φ) is integrable (i.e., has a solution) if and only if the following holds:

$$\frac{\partial N}{\partial x} = \frac{\partial M}{\partial y}.$$ (3.2.6)

To solve Eqs. (3.2.5), let us integrate, e.g. the first equation of (3.2.5) with respect to x,

$$\Phi(x,y) = \int M(x,y)\mathrm{d}x + g(y), \tag{3.2.7}$$

and substitute into the second equation of (3.2.5):

$$\frac{\partial}{\partial y}\int M(x,y)\mathrm{d}x + g'(y) = N(x,y). \tag{3.2.8}$$

Solving (3.2.8) for $g'(y)$ and integrating, we find $g(y)$, substitute it into (3.2.7) and ultimately obtain $\Phi(x,y)$. The solution $y = f(x,C)$ of the exact equation (3.2.3) is given implicitly by

$$\Phi(x,y) = C, \tag{3.2.9}$$

where C is an arbitrary constant.

One can also begin with integrating the second equation of (3.2.5) with respect to y. Then Eqs. (3.2.7) and (3.2.8) are replaced by the equations

$$\Phi(x,y) = \int N(x,y)\mathrm{d}y + h(x) \tag{3.2.10}$$

and

$$\frac{\partial}{\partial x}\int N(x,y)\mathrm{d}y + h'(x) = M(x,y), \tag{3.2.11}$$

respectively.

Remark 3.2.1. See also a simple method given in Section 6.6.2.

Example 3.2.1. Consider the equation $(y\,\mathrm{e}^{xy} + \cos x)\mathrm{d}x + x\,\mathrm{e}^{xy}\mathrm{d}y = 0$. The functions $M = y\,\mathrm{e}^{xy} + \cos x$ and $N = x\,\mathrm{e}^{xy}$ obey condition (3.2.6):

$$\frac{\partial N}{\partial x} = \frac{\partial M}{\partial y} = (1 + xy)\,\mathrm{e}^{xy}.$$

Equation (3.2.10) yields $\Phi(x,y) = \int x\,\mathrm{e}^{xy}\mathrm{d}y + h(x) = \mathrm{e}^{xy} + h(x)$ and (3.2.11) is written $y\,\mathrm{e}^{xy} + h'(x) = y\,\mathrm{e}^{xy} + \cos x$, whence $h'(x) = \cos x$. Thus,

$$\Phi(x,y) = \mathrm{e}^{xy} + \sin x.$$

Equation (3.2.9), $\mathrm{e}^{xy} + \sin x = C$, yields the following general solution to the equation in question:

$$y = \frac{1}{x}\,\ln|C - \sin x|.$$

3.2.3 Integrating factor (A. Clairaut, 1739)

If Eq. (3.2.3) is not exact, it can be converted into an exact equation by multiplying by an appropriate function. Namely, it was first shown by Clairaut in 1739 that for any equation (3.2.3), there exists a function $\mu(x, y)$, called an *integrating factor*, such that the equivalent equation

$$\mu(M\,dx + N\,dy) = 0$$

is exact. By the definition of exact equations, the integrating factor satisfies the equation (see Eq. (3.2.6))

$$\frac{\partial(\mu N)}{\partial x} = \frac{\partial(\mu M)}{\partial y}. \tag{3.2.12}$$

The solution of this equation with respect to $\mu(x, y)$ is not usually simpler than the integration of the original equation (3.2.3). However, integrating factors may be guessed and used in particular cases[1]. For example, it is widely used in modern text for integration of non-homogeneous first-order linear equations instead of the simple, effective and more general method of *variation of parameters* (see Section 3.2.7). The following theorem is useful.

Theorem 3.2.1. If two linearly independent integrating factors, $\mu_1(x, y)$ and $\mu_2(x, y)$, are known for Eq. (3.2.3), then its general solution is obtained without integration by the equation

$$\frac{\mu_1(x, y)}{\mu_2(x, y)} = C. \tag{3.2.13}$$

Example 3.2.2. The equation

$$2xy\,dx + (y^2 - 3x^2)dy = 0 \tag{3.2.14}$$

is not exact since its coefficients $M = 2xy$ and $N = y^2 - 3x^2$ do not satisfy (3.2.6). Let us check that $\mu = 1/y^4$ is an integrating factor. Indeed,

$$\frac{\partial(\mu N)}{\partial x} = \frac{\partial}{\partial x}\left(\frac{1}{y^2} - \frac{3x^2}{y^4}\right) = -6\frac{x}{y^4}, \qquad \frac{\partial(\mu M)}{\partial y} = \frac{\partial}{\partial y}\left(\frac{2x}{y^3}\right) = -6\frac{x}{y^4}.$$

Let us integrate the corresponding exact equation

$$\frac{2x}{y^3}\,dx + \left(\frac{1}{y^2} - \frac{3x^2}{y^4}\right)dy = 0. \tag{3.2.15}$$

Equation (3.2.7) yields

$$\Phi(x, y) = \int \frac{2x}{y^3}\,dx + g(y) = \frac{x^2}{y^3} + g(y)$$

---

[1] Lie group analysis provides a general formula for an integrating factor for first-order equations with known infinitesimal symmetries (see Section 6.4.1).

and Eq. (3.2.8) is written

$$-\frac{3x^2}{y^4} + g'(y) = \frac{1}{y^2} - \frac{3x^2}{y^4},$$

whence $g'(y) = y^{-2}$. Thus, $g(y) = -1/y$ and we obtain finally (see also Example 6.6.3 in Section 6.6.2)

$$\Phi(x,y) = \frac{x^2}{y^3} - \frac{1}{y}.$$

The solution of our differential equation is given implicitly by

$$\frac{x^2}{y^3} - \frac{1}{y} = C, \quad \text{or} \quad x^2 - y^2 = Cy^3.$$

The solution can be obtained without integration by using, along with $\mu_1 = 1/y^4$, the second integrating factor $\mu_2 = 1/(y^3 - x^2 y)$ (it is obtained in Section 6.4.1, Example 6.4.1). Eq. (3.2.13) yields

$$\frac{\mu_1}{\mu_2} = \frac{y^3 - x^2 y}{y^4} = \frac{y^2 - x^2}{y^3} = C.$$

3.2.4   The Riccati equation

The general *Riccati equation* is a first-order equation with the quadratic non-linearity:

$$y' = P(x) + Q(x)y + R(x)y^2. \tag{3.2.16}$$

The remarkable property of the Riccati equation is that it admits a *nonlinear superposition*. Namely, the cross-ratio of any four solutions

$$y_1(x), \quad y_2(x), \quad y_3(x), \quad y_4(x)$$

of Eq. (3.2.16) does not depend on x, i.e.,

$$\frac{y_4(x) - y_2(x)}{y_4(x) - y_1(x)} : \frac{y_3(x) - y_2(x)}{y_3(x) - y_1(x)} = C, \quad C = \text{const.} \tag{3.2.17}$$

It follows that one can obtain the general solution to Eq. (3.2.16) provided that one knows three solutions. Indeed, let us fix in (3.2.17) any three distinct particular solutions $y_1(x), y_2(x), y_3(x)$ and vary y_4 to obtain general solution y of Eq. (3.2.16). Then (3.2.17) assumes the form

$$\frac{y - y_2(x)}{y - y_1(x)} : \frac{y_3(x) - y_2(x)}{y_3(x) - y_1(x)} = C,$$

with an arbitrary constant C. Solving the above equation with respect to y, one arrives at the following representation of the general solution to Eq. (3.2.16):

$$y = \frac{C\psi_1(x) + \psi_2(x)}{C\varphi_1(x) + \varphi_2(x)}. \tag{3.2.18}$$

Here,

$$\varphi_1(x) = y_2(x) - y_3(x), \quad \varphi_2(x) = y_3(x) - y_1(x),$$
$$\psi_1(x) = y_1(x)\,\varphi_1(x), \quad \psi_2(x) = y_2(x)\,\varphi_2(x).$$

Thus, the general solution of the Riccati equation is a linear-rational function (3.2.18) of an arbitrary constant C. Conversely, if the general solution of a first-order differential equation is a linear-rational function of an arbitrary constant, then the differential equation is a Riccati equation.

The arbitrary Riccati equation (3.2.16) can be reduced by a substitution $y \mapsto \alpha(x)y$ to the form

$$y' + y^2 = Q(x)y + P(x). \tag{3.2.19}$$

Indeed, let $\bar{y} = \alpha(x)y$. Then

$$y = \frac{1}{\alpha}\,\bar{y}, \quad y' = \frac{1}{\alpha}\,\bar{y}' - \frac{\alpha'}{\alpha^2}\,\bar{y}$$

and

$$y' - Ry^2 - Qy - P = \frac{1}{\alpha}\left[\bar{y}' - \frac{R}{\alpha}\,\bar{y}^2 - \left(Q + \frac{\alpha'}{\alpha}\right)\bar{y} - \alpha P\right].$$

Hence, taking $\alpha(x) = -R(x)$ we map Eq. (3.2.16) to

$$\bar{y}' + \bar{y}^2 = \overline{Q}(x)\,\bar{y} + \overline{P}(x),$$

where $\overline{Q} = Q + (\alpha'/\alpha)$, $\overline{P} = \alpha P$. Denoting $\bar{y}$, $\overline{Q}$ and $\overline{P}$ again by y, Q and P, respectively, we obtain (3.2.19).

Furthermore, Eq. (3.2.19) can be transformed by a substitution $y \mapsto y + \beta(x)$ to the form

$$y' + y^2 = P(x) \tag{3.2 20}$$

referred to as the *canonical form* of the Riccati equation. Indeed, we let $\bar{y} = y + \beta(x)$ and have

$$y = \bar{y} - \beta(x), \quad y' = \bar{y}' - \beta'(x),$$
$$y' + y^2 - Qy - P = \bar{y}' + \bar{y}^2 - (Q + 2\beta)\,\bar{y} - (P + \beta' - \beta^2 - Q\beta).$$

Hence, taking $\beta(x) = -\frac{1}{2}Q(x)$ we map Eq. (3.2.19) to

$$\bar{y}' + \bar{y}^2 = \overline{P}(x),$$

where $\overline{P}(x) = P(x) - \frac{1}{2}Q'(x) + \frac{1}{4}Q^2(x)$. Denoting $\bar{y}$ and $\overline{P}$ again by y and P, respectively, we obtain (3.2.20).

In general, a change of variables $(x, y) \mapsto (\bar{x}, \bar{y})$ is called an *equivalence transformation* of the Riccati equation if any equation of form (3.2.16) is transformed into an equation of the same type with possibly different coefficients. Equations related by an equivalence transformation are said to be *equivalent*.

The set of all equivalence transformations of the Riccati equation (3.2.16) comprise

(i) an arbitrary change of the independent variable:

$$\bar{x} = \phi(x), \qquad \phi'(x) \neq 0; \tag{3.2.21}$$

(ii) linear-rational transformations of the dependent variable:

$$\bar{y} = \frac{\alpha(x)y + \beta(x)}{\gamma(x)y + \delta(x)}, \qquad \alpha\delta - \beta\gamma \neq 0. \tag{3.2.22}$$

The Riccati equation is a *first-order nonlinear* equation. It can be rewritten as a *second-order linear* equation. Namely, one can first reduce Eq. (3.2.16) to form (3.2.19) and then set

$$y = \frac{u'}{u}$$

to obtain the linear second-order equation

$$u'' = Q(x)u' + P(x)u. \tag{3.2.23}$$

The linearization, even by raising the order, may be useful for integration.

Example 3.2.3. Consider the Riccati equation (3.2.19) with $P(x) = 0$:

$$y' + y^2 = Q(x)y.$$

The associated linear equation (3.2.19),

$$u'' = Q(x)u',$$

can be readily integrated by quadrature. Indeed, setting $u' = z$, one has the first-order equation $z' = Q(x)z$. Hence,

$$z = Ae^{\int Q(x)dx}.$$

Substituting $z = u'$ we have the equation

$$u' = Ae^{\int Q(x)dx}$$

whence upon integration:

$$u = A \int e^{\int Q(x)dx}dx + B.$$

Invoking that $y = u'/u$ we have

$$y = \frac{e^{\int Q(x)dx}}{C + \int e^{\int Q(x)dx}dx}, \qquad C = \text{const.}$$

Example 3.2.4. Riccati himself discovered and investigated in 1724 particular case of Eq. (3.2.16), namely the equation

$$y' = ay^2 + bx^\alpha, \quad a, b, \alpha = \text{const.} \tag{3.2.24}$$

known today as the *special Riccati equation*. The general Riccati equation (3.2.16) was introduced and studied for the first time by d'Alembert in 1763. Francesco Riccati and Daniel Bernoulli noted independently that Eq. (3.2.24) is integrable in finite form in terms of elementary functions if

$$\alpha = -\frac{4k}{2k \pm 1} \quad \text{with} \quad k = 0, \pm 1, \pm 2, \dots. \tag{3.2.25}$$

Josef Liouville showed in 1841 that the solution to the special Riccati equation (3.2.24) cannot be expressed via integration of elementary functions if α is different from (3.2.25).

The author found in 1989 all linearizable Riccati equations [15]. The following statement extracted from [15], Section 4.2 (see also [21], Section 11.2.5) provides a simple practical test for linearization.

Theorem 3.2.2. The Riccati equation (3.2.16),

$$y' = P(x) + Q(x)y + R(x)y^2,$$

is linearizable by a change of the dependent variable y if and only if it obeys any of the following two equivalent conditions (A) or (B):

(A) Equation (3.2.16) has either the form

$$y' = Q(x)y + R(x)y^2 \tag{3.2.26}$$

with two arbitrary functions $Q(x)$ and $R(x)$, or the form

$$y' = P(x) + Q(x)y + k[Q(x) - kP(x)]y^2 \tag{3.2.27}$$

with two arbitrary functions $P(x)$, $Q(x)$ and a constant (in general complex) coefficient k;

(B) Equation (3.2.16) has a constant (in general complex) solution.

Remark 3.2.2. Equation (3.2.26) has a constant solution $y = 0$. On the other hand, Eq. (3.2.27) has a constant solution $y = -1/k$. Therefore, the linear equation $y' = P(x) + Q(x)y$, which is a particular case of (3.2.27) for $k = 0$, may be regarded as a Riccati equation having $y = \infty$ as its particular constant solution.

3.2.5   The Bernoulli equation

The nonlinear equation

$$y' + P(x)y = Q(x)y^n, \quad n \neq 0 \text{ and } n \neq 1,$$

is known as the Bernoulli equation[2]. It can be reduced to a linear equation and solved by quadrature. Indeed, dividing both sides of the Bernoulli equation by y^n we have $y^{-n}y' + P(x)y^{1-n} = Q(x)$, or

$$\frac{1}{1-n} \frac{dy^{1-n}}{dx} + P(x)y^{1-n} = Q(x).$$

Hence, the substitution $z = y^{1-n}$ reduces the Bernoulli equation to the linear equation

$$z' + (1-n)P(x)z = (1-n)Q(x).$$

3.2.6   Homogeneous linear equations

The general linear equation of the first order has the form

$$y' + P(x)y = Q(x). \tag{3.2.28}$$

Equation (3.2.28) is homogeneous by function if and only if $Q(x) = 0$ (see Section 3.1.4). Consequently, the following nomenclature is commonly used in textbooks: Equation (3.2.28) is called *homogeneous* if $Q(x) = 0$, and *non-homogeneous* otherwise.

Remark 3.2.3. For the sake of brevity, the general homogeneity is identified in this terminology with the homogeneity by function. The convention has not lead to confusion in the past. But nowadays when there is a tendency to replace knowledge of mathematics by computer manipulations, more and more students and teachers understand homogeneity formally as a mere statement that the right-hand side of a differential equation is equal to zero, and apply it erroneously to nonlinear equations as well. From their point of view, Eq. (3.1.10)

$$y'' - \frac{2xy}{3x^2 - y^2} = 0$$

would be homogeneous while the equations from Example 3.1.5, e.g.

$$y'' + \frac{A}{x} y' + \frac{B}{x^2} y = \frac{C}{x}, \quad C \neq 0,$$

would be non-homogeneous.

---

[2]It was discovered by Jacques Bernoulli in 1695 and solved by Leibnitz in 1696.

Consider the homogeneous linear equation

$$y' + P(x)y = 0. \tag{3.2.29}$$

We separate the variables:

$$\frac{dy}{y} + P(x)dx = 0$$

and integrate to obtain

$$\ln y + \int P(x)dx = \text{const.}$$

Hence, the solution is

$$y = Ce^{-\int Pdx}, \quad C = \text{const.} \tag{3.2.30}$$

3.2.7 Non-homogeneous linear equations. Variation of the parameter

The simplest way for solving non-homogeneous linear equations (3.2.28) is provided by the method of *variation of parameters* suggested by Jean Bernoulli in 1697. Let us begin with an example.

Example 3.2.5. Let us solve the non-homogeneous equation

$$y' - y = x. \tag{3.2.31}$$

We first consider the homogeneous equation of Eq. (3.2.31):

$$y' - y = 0.$$

Its general solution is

$$y = Ce^x, \quad C = \text{const.}$$

Now we replace the constant of integration C by an unknown function $u(x)$ and look for the solution of the non-homogeneous equation in the form

$$y = u(x)e^x. \tag{3.2.32}$$

Substituting $y' = u'e^x + ue^x$ into Eq. (3.2.31), we obtain the following separable equation for $u(x)$:

$$u' = xe^{-x}.$$

Integrating it and denoting the constant of integration again by C, we have

$$u = \int xe^{-x}dx + C = -(x+1)e^{-x} + C, \quad C = \text{const.}$$

Finally, we substitute the expression for $u(x)$ into (3.2.32) and obtain the following general solution for Eq. (3.2.31):

$$y = Ce^x - x - 1. \tag{3.2.33}$$

In the case of the general non-homogeneous equation (3.2.28),

$$y' + P(x)y = Q(x),$$

we proceed likewise. Namely, we solve the homogeneous equation of Eq. (3.2.28),

$$y' + P(x)y = 0,$$

and replace in its general solution (3.2.30),

$$y = Ce^{-\int P\,dx}, \quad C = \text{const.},$$

the constant of integration C by an unknown function $u(x)$. In other words, we look for the solution of the non-homogeneous equation in the form

$$y = u(x)e^{-\int P\,dx}. \tag{3.2.34}$$

We have

$$y' = u'(x)e^{-\int P\,dx} - u(x)P(x)e^{-\int P\,dx}.$$

Substitution of this expression into Eq. (3.2.28) yields

$$u'(x)\,e^{-\int P\,dx} = Q(x), \quad \text{or} \quad u'(x) = Q(x)\,e^{\int P\,dx},$$

whence

$$u(x) = \int Q\,e^{\int P\,dx}\,dx + C, \quad C = \text{const.}$$

Inserting the expression for $u(x)$ into (3.2.34), we obtain the general solution to the non-homogeneous linear equation (3.2.28) given by two quadratures:

$$y = \left(C + \int Q\,e^{\int P\,dx}\,dx\right)e^{-\int P\,dx}. \tag{3.2.35}$$

3.3 Second-order linear equations

The general linear equation of the second order has the form

$$y'' + a(x)\,y' + b(x)\,y = f(x). \tag{3.3.1}$$

It is called *homogeneous* if $f(x) = 0$ and *non-homogeneous* otherwise (cf. Section 3.2.6).

Equation (3.3.1) is often written in the form

$$L_2[y] = f(x). \tag{3.3.2}$$

Here L_2 is the following linear differential operator of the second order:

$$L_2 = D^2 + a(x)\,D + b, \tag{3.3.3}$$

where

$$D = \frac{\mathrm{d}}{\mathrm{d}x}, \quad D^2 = \frac{\mathrm{d}^2}{\mathrm{d}x^2}.$$

Thus,

$$L_2[y] = D^2 y + a(x) D y + by \equiv y'' + a(x) y' + b(x) y. \tag{3.3.4}$$

The term *linear* refers to the following fundamental property of the operator L_2 :

$$L_2[C_1 y_1 + C_2 y_2] = C_1 L_2[y_1] + C_2 L_2[y_2], \quad C_1, C_2 = \text{const.} \tag{3.3.5}$$

3.3.1 Homogeneous equation: Superposition

The homogeneous linear equation

$$y'' + a(x) y' + b(x) y = 0 \tag{3.3.6}$$

or

$$L_2[y] = 0$$

possesses the remarkable property called the *superposition principle* or more specifically, the *linear superposition*. This principle follows from property (3.3.5) of the linear differential operator L_2 and states that if $y_1(x)$ and $y_2(x)$ are solutions of the homogeneous equation (3.3.6), then their linear combination with arbitrary constant coefficients,

$$y = C_1 y_1(x) + C_2 y_2(x),$$

is also a solution. Indeed, since $L_2[y_1(x)] = 0$, $L_2[y_2(x)] = 0$, Equation (3.3.5) yields:

$$L_2[y] = C_1 L_2[y_1(x)] + C_2 L_2[y_2(x)] = 0.$$

Since the general solution of any second-order ordinary differential equation involves two arbitrary constants, the superposition principle shows that the general solution to Eq. (3.3.6) is given by

$$y = C_1 y_1(x) + C_2 y_2(x), \tag{3.3.7}$$

where $y_1(x)$ and $y_2(x)$ are linearly independent solutions of Eq. (3.3.6). Therefore, in order to construct the general solution of the homogeneous equation, it suffices to find two independent solutions only. Consequently, we say that the pair of linearly independent solutions, $y_1(x)$ and $y_2(x)$, furnish a *fundamental system of solutions* for Eq. (3.3.6).

3.3.2   Homogeneous equation: Equivalence properties

Equivalence properties are useful in practical integration of differential equations. An *equivalence transformation* of the homogeneous linear equations is a change of variables preserving the linearity and homogeneity of Eqs. (3.3.6). The set of all equivalence transformations comprises an arbitrary change of the independent variable (cf. (3.2.21)):

$$\bar{x} = \phi(x), \qquad \phi'(x) \neq 0, \tag{3.3.8}$$

and the linear substitution of the dependent variable:

$$y = \sigma(x)\,\bar{y}, \quad \sigma \neq 0. \tag{3.3.9}$$

Definition 3.3.1. Two equations of form (3.3.6) are said to be *equivalent* if they are connected by a combination of transformations (3.3.8)—(3.3.9). Furthermore, two equations are termed *equivalent by function* if they can be mapped into each other by a linear substitution (3.3.9).

Theorem 3.3.1. Any homogeneous linear equation (3.3.6),

$$y'' + a(x)y' + b(x)y = 0,$$

is equivalent to the simplest linear equation

$$\bar{y}'' = 0, \tag{3.3.10}$$

where $\bar{y}'' = \mathrm{d}^2\bar{y}/\mathrm{d}\bar{x}^2$.

Proof. Equation (3.3.6) is reduced to form (3.3.10) by the transformation

$$\bar{x} = \int \frac{e^{-\int a(x)\mathrm{d}x}}{z^2(x)}\,\mathrm{d}x, \quad \bar{y} = \frac{y}{z(x)}, \tag{3.3.11}$$

where $z(x)$ is any solution of Eq. (3.3.6), i.e., $z'' + a(x)z' + b(x)z = 0$. For verification that transformation (3.3.11) maps the general linear equation into Eq. (3.3.10), see the solution to Problem 6.9.

According to Theorem 3.3.1, one can map into another any two equations of form (3.3.6) by using both transformations (3.3.8) and (3.3.9). In other words, all equations of (3.3.6) are equivalent. However, (3.3.11) shows the computation of an appropriate equivalence requires knowledge of particular solutions of the equations which one wants to transform into another.

Therefore, we will use here the equivalence by function which employs only the linear substitution (3.3.9) and provides a simple and constructive way of integration of a wide class of equations.

Lemma 3.3.1. The general homogeneous linear equation (3.3.6),

$$y'' + a(x)\,y' + b(x)\,y = 0,$$

is *equivalent by function* to the equation

$$y'' + \alpha(x)\,y = 0. \tag{3.3.12}$$

Namely, the linear substitution

$$y = \overline{y}\,e^{-\frac{1}{2}\int a(x)dx} \tag{3.3.13}$$

reduces it to the equation

$$\overline{y}'' + J(x)\,\overline{y} = 0, \tag{3.3.14}$$

where

$$J(x) = b(x) - \frac{1}{4}\,a^2(x) - \frac{1}{2}\,a'(x). \tag{3.3.15}$$

Proof. We take an arbitrary linear substitution (3.3.9) and have

$$y = \sigma(x)\,\overline{y}, \quad y' = \sigma(x)\,\overline{y}' + \sigma'(x)\,\overline{y},$$
$$y'' = \sigma(x)\,\overline{y}'' + 2\sigma'(x)\,\overline{y}' + \sigma''(x)\,\overline{y}.$$

Hence, Eq. (3.3.6) becomes

$$\sigma\,\overline{y}'' + [2\sigma' + a\,\sigma]\,\overline{y}' + [\sigma'' + a\,\sigma' + b\,\sigma]\,\overline{y} = 0. \tag{3.3.16}$$

We annul the term with $\overline{y}'$ by letting

$$2\sigma' + a\,\sigma = 0,$$

whence

$$\sigma = e^{-\frac{1}{2}\int a(x)dx}. \tag{3.3.17}$$

Now we substitute function (3.3.17) and its derivatives

$$\sigma' = -\frac{1}{2}\,a\,e^{-\frac{1}{2}\int a\,dx}, \quad \sigma'' = \left(\frac{1}{4}\,a^2 - \frac{1}{2}\,a'\right)e^{-\frac{1}{2}\int a\,dx}$$

into Eq. (3.3.16), multiply the result by $e^{\frac{1}{2}\int a(x)dx}$ and arrive at Eq. (3.3.14):

$$\overline{y}'' + \left(b - \frac{1}{4}\,a^2 - \frac{1}{2}\,a'\right)\overline{y} = 0. \tag{3.3.18}$$

It is manifest from the proof that the function $J(x)$ given by (3.3.15) remains unaltered under the transformations of Eq. (3.3.6) via any substitution (3.3.9). Therefore, J is called the *invariant* of Eq. (3.3.6). The invariance of $J(x)$ and Lemma 3.3.1 lead to the following result.

Theorem 3.3.2. Two homogeneous linear equations,

$$y'' + a(x)\, y' + b(x)\, y = 0 \tag{3.3.19}$$

and

$$\bar{y}'' + a_1(x)\, \bar{y}' + b_1(x)\, \bar{y} = 0, \tag{3.3.20}$$

are equivalent by function, i.e., can be mapped one into another by an appropriate linear substitution (3.3.9) if and only if their invariants

$$J(x) = b(x) - \frac{1}{4}\, a^2(x) - \frac{1}{2}\, a'(x)$$

and

$$J_1(x) = b_1(x) - \frac{1}{4}\, a_1^2(x) - \frac{1}{2}\, a_1'(x)$$

are identical, i.e., $J(x) = J_1(x)$. In particular, Eq. (3.3.6) is equivalent by function with Eq. (3.3.10), $\bar{y}'' = 0$, if and only if its invariant $J(x)$ vanishes, i.e., Eq. (3.3.6) has the form

$$y'' + a(x)\, y' + \left[\frac{1}{4}\, a^2(x) + \frac{1}{2}\, a'(x)\right] y = 0. \tag{3.3.21}$$

Example 3.3.1. The equation

$$x^2 y'' + xy' + \left(x^2 - \frac{1}{4}\right) y = 0 \tag{3.3.22}$$

is not of form (3.3.21) and hence is not equivalent by function with the equation $\bar{y}'' = 0$. The invariant of Eq. (3.3.22) is $J = 1$. On the other hand, the invariant J_1 of the equation

$$\bar{y}'' + \bar{y} = 0 \tag{3.3.23}$$

has the same value, hence, $J_1 = J = 1$. In consequence, Eq. (3.3.22) is connected with Eq. (3.3.23) by the linear substitution

$$\bar{y} = \sqrt{x}\, y$$

obtained by employing (3.3.13) with $a(x) = 1/x$. Since the general solution of Eq. (3.3.23) has the form (see Example 3.3.2 in the next section)

$$\bar{y} = C_1 \sin x + C_2 \cos x,$$

the general solution to Eq. (3.3.22) is given by

$$y = \frac{1}{\sqrt{x}}\, (C_1 \sin x + C_2 \cos x).$$

3.3.3 Homogeneous equation: Constant coefficients

Consider the homogeneous linear equation (3.3.6) with constant coefficients:

$$y'' + Ay' + By = 0, \quad A, B = \text{const.} \tag{3.3.24}$$

Its solution was given by Leonard Euler in 1743. He looked for particular solutions of the form (*Euler's ansatz*)

$$y = e^{\lambda x}, \quad \lambda = \text{const.} \tag{3.3.25}$$

Then the differential equation (3.3.24) reduces to the algebraic equation

$$\lambda^2 + A\lambda + B = 0 \tag{3.3.26}$$

called the *characteristic equation*. Accordingly,

$$P_2[\lambda] = \lambda^2 + A\lambda + B$$

is called the *characteristic polynomial* for Eq. (3.3.24). There are the following three possibilities.

(i) The characteristic equation (3.3.26) has two distinct real solutions, λ_1 and λ_2. Then one has two linearly independent particular solutions:

$$y_1(x) = e^{\lambda_1 x}, \quad y_2(x) = e^{\lambda_2 x},$$

and the general solution of Eq. (3.3.24) is given by

$$y = C_1 e^{\lambda_1 x} + C_2 e^{\lambda_2 x}. \tag{3.3.27}$$

(ii) The solutions of the characteristic equation (3.3.26) are complex, namely $\lambda_1 = \alpha + i\beta$ and its complex conjugate $\lambda_2 = \alpha - i\beta$. The corresponding complex solutions can be written, using Euler's formula (1.2.41), in the form

$$y = e^{\alpha x}(\cos \beta x + i \sin \beta x), \quad \bar{y} = e^{\alpha x}(\cos \beta x - i \sin \beta x).$$

Since their linear combinations with arbitrary complex coefficients are again solutions, we replace the complex solutions by the real ones by setting

$$y_1 = \frac{1}{2}(y + \bar{y}), \quad y_2 = \frac{1}{2i}(y - \bar{y}).$$

Thus, the pair of conjugate complex roots provide two distinct real solutions:

$$y_1(x) = e^{\alpha x} \cos \beta x, \quad y_2(x) = e^{\alpha x} \sin \beta x. \tag{3.3.28}$$

Hence, the general solution to Eq. (3.3.24) is given in this case by

$$y = C_1 e^{\alpha x} \cos \beta x + C_2 e^{\alpha x} \sin \beta x. \tag{3.3.29}$$

Example 3.3.2. Consider Eq. (2.3.14) of free harmonic oscillations

$$y'' + \omega^2 y = 0, \qquad (3.3.30)$$

where $\omega \neq 0$ is a real number. The characteristic polynomial $\lambda^2 + \omega^2 = 0$ has the complex roots, $\lambda_1 = i\omega$, $\lambda_2 = -i\omega$, and solution (3.3.29) has the form

$$y = C_1 \cos \omega x + C_2 \sin \omega x. \qquad (3.3.31)$$

(iii) The characteristic equation (3.3.26) has repeated roots, $\lambda_1 = \lambda_2$. Then formula (3.3.25) provides only one solution,

$$y = e^{\lambda_1 x}.$$

However, in this case we can employ Theorem 3.3.2. Indeed, since the characteristic polynomial has repeated roots, its discriminant vanishes:

$$A^2 - 4B = 0.$$

But then the invariant

$$J = B - \frac{1}{4} A^2$$

also vanishes, and hence Eq. (3.3.24) reduces to $\bar{y}'' = 0$. The root of the characteristic polynomial is $\lambda_1 = -A/2$, and transformation (3.3.13) is written:

$$y = \bar{y} e^{-\frac{1}{2} \int A \, dx} = \bar{y} e^{\int \lambda_1 \, dx} = \bar{y} e^{\lambda_1 x}.$$

Whence, substituting the solution $\bar{y} = (C_1 + C_2 x)$ of the equation $\bar{y}'' = 0$, we obtain the following general solution to Eq. (3.3.24):

$$y = (C_1 + C_2 x) e^{\lambda_1 x}. \qquad (3.3.32)$$

Example 3.3.3. Consider the equation

$$y'' + 2y' + y = 0.$$

Its characteristic polynomial

$$P_2[\lambda] = \lambda^2 + 2\lambda + 1 \equiv (\lambda + 1)^2$$

has the repeated real root $\lambda = -1$. Consequently the general solution of the differential equation is given by

$$y = (C_1 + C_2 x) e^{-x}.$$

3.3.4 Non-homogeneous equation: Variation of parameters

Provided that fundamental system of solutions for the homogeneous equation (3.3.6) is known, the non-homogeneous equation

$$y'' + a(x)\,y' + b(x)\,y = f(x) \qquad (3.3.33)$$

can be solved by quadratures using the following method of *variation of the parameters.*

Let us assume that we know a fundamental system of solutions $y_1(x), y_2(x)$ for the homogeneous equation, i.e., let

$$y_1'' + a(x)\,y_1' + b(x)\,y_1 = 0, \quad y_2'' + a(x)\,y_2' + b(x)\,y_2 = 0. \qquad (3.3.34)$$

Then we can obtain the general solution to the non-homogeneous equation (3.3.33) by the following *method of variation of parameters.* Just as in the case of the first-order equation, we replace the constants C_1 and C_2 in solution (3.3.7)

$$y = C_1 y_1(x) + C_2 y_2(x)$$

of the homogenous equation by functions $u_1(x)$ and $u_2(x)$, respectively. Thus, we set

$$y = u_1(x)y_1(x) + u_2(x)y_2(x). \qquad (3.3.35)$$

It follows:

$$y' = u_1(x)y_1'(x) + u_2(x)y_2'(x) + y_1(x)u_1'(x) + y_2(x)u_2'(x). \qquad (3.3.36)$$

If we substitute (3.3.35) into Eq. (3.3.33), we obtain only one equation for two unknown functions $u_1(x)$ and $u_2(x)$. Therefore, we can subject these function to one more condition. We will take this condition in the form

$$y_1(x)u_1'(x) + y_2(x)u_2'(x) = 0. \qquad (3.3.37)$$

Then, invoking (3.3.35) and (3.3.36), we have

$$y = u_1(x)\,y_1(x) + u_2(x)\,y_2(x),$$
$$y' = u_1(x)\,y_1'(x) + u_2(x)\,y_2'(x),$$
$$y'' = u_1(x)\,y_1''(x) + u_2(x)\,y_2''(x) + y_1'(x)\,u_1'(x) + y_2'(x)\,u_2'(x).$$

Thus,

$$y'' + a(x)\,y' + b(x)\,y$$
$$= u_1(x)[y_1'' + a(x)\,y_1' + b(x)\,y_1] + u_2(x)[y_2'' + a(x)\,y_2' + b(x)\,y_2]$$
$$+ y_1'(x)\,u_1'(x) + y_2'(x)\,u_2'(x),$$

and hence, invoking Eqs. (3.3.34), we obtain from Eq. (3.3.33):

$$y_1'(x)\, u_1'(x) + y_2'(x)\, u_2'(x) = f(x). \tag{3.3.38}$$

Since $y_1(x), y_2(x)$ are known functions we obtain for determining u_1, u_2 the system of two equations (3.3.37) and (3.3.38):

$$\begin{aligned} y_1(x)u_1'(x) + y_2(x)u_2'(x) &= 0, \\ y_1'(x)\, u_1'(x) + y_2'(x)\, u_2'(x) &= f(x). \end{aligned} \tag{3.3.39}$$

Since $y_1(x), y_2(x)$ are linearly independent, the determinant of Eqs. (3.3.39)

$$W(x) = y_1(x)\, y_2'(x) - y_2(x)\, y_1'(x), \tag{3.3.40}$$

known as the Wronskian, does not vanish. Hence, system (3.3.39) can be solved with respect to the derivatives of the unknown functions:

$$u_1' = -\frac{y_2(x)f(x)}{W(x)}, \quad u_2' = \frac{y_1(x)f(x)}{W(x)},$$

and integration yields

$$u_1 = -\int \frac{y_2(x)f(x)}{W(x)}\,dx + C_1, \quad u_2 = \int \frac{y_1(x)f(x)}{W(x)}\,dx + C_2. \tag{3.3.41}$$

Substituting (3.3.41) into (3.3.35), we arrive at the following result.

Theorem 3.3.3. Let $y_1(x)$, $y_2(x)$ be a fundamental system of solutions for the homogeneous equation (3.3.6),

$$y'' + a(x)\, y' + b(x)\, y = 0.$$

Then the general solution to the non-homogeneous equation (3.3.33),

$$y'' + a(x)\, y' + b(x)\, y = f(x),$$

is given by quadratures and has the form:

$$y = C_1 y_1(x) + C_2 y_2(x) - y_1(x) \int \frac{y_2(x)f(x)}{W(x)}\,dx + y_2(x) \int \frac{y_1(x)f(x)}{W(x)}\,dx,$$

where $W(x)$ is Wronskian (3.3.40).

Example 3.3.4. Let us solve the non-homogeneous equation

$$y'' + y = \sin x.$$

We have the fundamental system of solutions

$$y_1(x) = \cos x, \quad y_2(x) = \sin x,$$

with the Wronskian $W[y_1(x), y_2(x)] = 1$. Formulae (3.3.41) yield

$$u_1(x) = -\int \sin^2 x dx = -\frac{x}{2} + \frac{1}{4} \sin 2x + C_1,$$

$$u_2(x) = \int \cos x \sin x dx = -\frac{1}{2} \cos^2 x + C_2.$$

Hence, we obtain, after elementary simplification the following solution:

$$y = -\frac{x}{2} \cos x + C_1 \cos x + C_2 \sin x.$$

In this example, we could express the general solution of the differential equation in question in elementary functions. However, this fact has no significance and a representation of a general solution by quadratures is equally useful, e.g. for solving initial value problems. The following two examples illustrate the statement.

Example 3.3.5. Let us integrate the equation

$$y'' + 2y' - 8y = xe^{4x} \tag{3.3.42}$$

and solve the Cauchy problem with the initial conditions

$$y\big|_{x=0} = 0, \quad y'\big|_{x=0} = 1. \tag{3.3.43}$$

The characteristic equation $\lambda^2 + 2\lambda - 8 = 0$ for (3.3.42) has the roots $\lambda_1 = 2$, $\lambda_2 = -4$, and hence the fundamental system of solutions is provided by

$$y_1(x) = e^{2x}, \quad y_2(x) = e^{-4x}.$$

Wronskian (3.3.40) is $W[y_1(x), y_2(x)] = -6e^{-2x}$. Using Theorem 3.3.3, we will write the general solution to Eq. (3.3.42) in the following form convenient for satisfying the initial conditions at $x = 0$:

$$y = C_1 e^{2x} + C_2 e^{-4x} + \frac{1}{6}\left[e^{2x}\int_0^x \tau e^{2\tau} d\tau - e^{-4x}\int_0^x \tau e^{8\tau} d\tau\right]. \tag{3.3.44}$$

Differentiating (3.3.44) we obtain

$$y' = 2C_1 e^{2x} - 4C_2 e^{-4x} + \frac{1}{3}\left[e^{2x}\int_0^x \tau e^{2\tau} d\tau + 2e^{-4x}\int_0^x \tau e^{8\tau} d\tau\right].$$

The initial conditions (3.3.43) are written $C_1 + C_2 = 0$, $2C_1 - 4C_2 = 1$ and yield $C_1 = 1/6$, $C_2 = -1/6$. Substituting into (3.3.44), we obtain the following solution to the Cauchy problem (3.3.42)~(3.3.43):

$$y = \frac{1}{6}\left[e^{2x}\left(1 + \int_0^x \tau e^{2\tau} d\tau\right) - e^{-4x}\left(1 + \int_0^x \tau e^{8\tau} d\tau\right)\right]. \tag{3.3.45}$$

Remark 3.3.1. One can work out the integrals in (3.3.44) and rewrite the general solution (3.3.44), and hence (3.3.45), in terms of elementary functions. Indeed, integration by parts yields

$$\int_0^x \tau e^{2\tau} d\tau = \frac{\tau}{2} e^{2\tau} \Big|_0^x - \frac{1}{2} \int_0^x e^{2\tau} d\tau = \frac{x}{2} e^{2x} - \frac{1}{4} e^{2x} + \frac{1}{4},$$

$$\int_0^x \tau e^{8\tau} d\tau = \frac{\tau}{8} e^{8\tau} \Big|_0^x - \frac{1}{64} \int_0^x e^{8\tau} d\tau = \frac{x}{8} e^{8x} - \frac{1}{64} e^{8x} + \frac{1}{64}.$$

Substitution into (3.3.44) yields

$$y = C_1 e^{2x} + C_2 e^{-4x} + \frac{1}{6} \left[\frac{1}{4} e^{2x} - \frac{1}{64} e^{-4x} - \frac{15}{64} e^{4x} + \frac{3}{8} x e^{4x} \right] \quad (3.3.46)$$

and solution (3.3.45) to the Cauchy problem (3.3.42)~(3.3.43) is written

$$y = \frac{5}{24} e^{2x} - \frac{65}{384} e^{-4x} - \frac{5}{128} e^{4x} + \frac{1}{16} x e^{4x}. \quad (3.3.47)$$

Example 3.3.6. Let us integrate the equation

$$y'' + 2y' - 8y = \frac{1}{x+1} e^{3x+1} \quad (3.3.48)$$

and solve the Cauchy problem with the initial conditions

$$y\big|_{x=0} = 0, \quad y'\big|_{x=0} = 1. \quad (3.3.49)$$

Proceeding as in Example 3.3.5 one obtains the solution to Eq. (3.3.48):

$$y = C_1 e^{2x} + C_2 e^{-4x} + \frac{1}{6} \left[e^{2x} \int_0^x \frac{e^{\tau+1}}{\tau+1} d\tau - e^{-4x} \int_0^x \frac{e^{7\tau+1}}{\tau+1} d\tau \right]. \quad (3.3.50)$$

Differentiating (3.3.50) one obtains

$$y' = 2C_1 e^{2x} - 4C_2 e^{-4x} + \frac{1}{3} \left[e^{2x} \int_0^x \frac{e^{\tau+1}}{\tau+1} d\tau + 2e^{-4x} \int_0^x \frac{e^{7\tau+1}}{\tau+1} d\tau \right].$$

The initial conditions (3.3.43) yield $C_1 = 1/6$, $C_2 = -1/6$. Hence, the solution to the Cauchy problem (3.3.48)~(3.3.49) is

$$y = \frac{1}{6} \left[e^{2x} \left(1 + \int_0^x \frac{e^{\tau+1}}{\tau+1} d\tau \right) - e^{-4x} \left(1 + \int_0^x \frac{e^{7\tau+1}}{\tau+1} d\tau \right) \right]. \quad (3.3.51)$$

Both integrals in (3.3.51) cannot be worked out in elementary functions.

3.3.5 Bessel's equation and the Bessel functions

Bessel's equation is the following homogeneous linear second-order equation with variable coefficients:

$$x^2 y'' + x y' + (x^2 - n^2)y = 0. \tag{3.3.52}$$

The solutions of this equation are termed the *Bessel functions* and play an important part in mathematical physics. One of the solutions is denoted by $J_n(x)$ and is known as the *Bessel function of nth-order*. The power series expansions, e.g. for $n = 0$ and $n = 1$ are as follows:

$$J_0(x) = 1 - \left(\frac{x}{2}\right)^2 + \frac{1}{(2!)^2}\left(\frac{x}{2}\right)^4 - \frac{1}{(3!)^2}\left(\frac{x}{2}\right)^6 + \cdots ,$$

$$J_1(x) = \frac{x}{2} - \frac{1}{2!}\left(\frac{x}{2}\right)^3 + \frac{1}{2!3!}\left(\frac{x}{2}\right)^5 - \cdots . \tag{3.3.53}$$

3.3.6 Hypergeometric equation

The second-order linear differential equation

$$x(1 - x)\, y'' + [\gamma - (\alpha + \beta + 1)x]\, y' - \alpha\beta\, y = 0 \tag{3.3.54}$$

with arbitrary parameters α, β, and γ are known as the *hypergeometric equation*. It has singularities at $x = 0$, $x = 1$ and $x = \infty$.

Furthermore, any homogeneous linear second-order differential equation

$$(x^2 + Ax + B)\, y'' + (Cx + D)\, y' + E\, y = 0 \tag{3.3.55}$$

is transformable to the hypergeometric equation (3.3.54), provided that the equation $x^2 + Ax + B = 0$ has two distinct roots x_1 and x_2. Indeed, rewriting Eq. (3.3.54) in the new independent variable t defined by

$$x = x_1 + (x_2 - x_1)t \tag{3.3.56}$$

one obtains

$$t(1 - t)\frac{\mathrm{d}^2 y}{\mathrm{d}t^2} + \left[\frac{Cx_1 + D}{x_1 - x_2} - Ct\right]\frac{\mathrm{d}y}{\mathrm{d}t} - Ey = 0.$$

Whence setting

$$\frac{Cx_1 + D}{x_1 - x_2} = \gamma, \quad C = \alpha + \beta + 1, \quad E = \alpha\beta$$

and denoting the new independent variable t again by x, one arrives at equation (3.3.54).

If $\alpha\beta = 0$, then the hypergeometric equation (3.3.54) is integrable by two quadratures. Indeed, letting, e.g. $\beta = 0$ and integrating the equation

$$\frac{\mathrm{d}y'}{y'} = \frac{(\alpha + 1)x - \gamma}{x(1 - x)}\,\mathrm{d}x,$$

one obtains $y' = C_1 e^{q(x)}$, where

$$q(x) = \int \frac{(\alpha + 1)x - \gamma}{x(1 - x)}\, dx.$$

Now the second integration yields

$$y = C_1 \int e^{q(x)}\, dx + C_2, \quad C_1,\ C_2 = \text{const}.$$

In the theory of hypergeometric functions, the main emphasis is on asymptotics of the hypergeometric equation and its series solutions near the singular points (see, e.g. the classical book [40]). However, in practice one often needs analytic expressions for the general solutions of certain types of the hypergeometric equation. Therefore, the following theorem[3] determines a class of hypergeometric equations integrable by elementary functions or by quadrature. Numerous particular cases of this class can be found in various books on special functions.

Theorem 3.3.4. The general solution of the hypergeometric equation (3.3.54) with $\beta = -1$ and two arbitrary parameters α and γ :

$$x(1 - x)\, y'' + (\gamma - \alpha\, x)\, y' + \alpha\, y = 0 \tag{3.3.57}$$

is given by quadrature and has the form

$$y = C_1 \left(x - \frac{\gamma}{\alpha}\right) \int \left(|x|^{-\gamma}\, |x - 1|^{\gamma - \alpha}\, [x - (\gamma/\alpha)]^{-2}\right) dx + C_2 \left(x - \frac{\gamma}{\alpha}\right), \tag{3.3.58}$$

where C_1 and C_2 are arbitrary constants.

Remark 3.3.2. If γ and $\gamma - \alpha$ are rational numbers, one can reduce the integral in (3.3.58) to integration of a rational function by standard substitutions and represent solution (3.3.58) in terms of elementary functions.

3.4   Higher-order linear equations

The general nth-order linear equation with variable coefficients has the form

$$L_n[y] \equiv y^{(n)} + a_1(x)y^{(n-1)} + \cdots + a_{n-1}(x)y' + a_n(x)y = f(x). \tag{3.4.1}$$

The term *linear* refers to the fundamental property

$$L_n[C_1 y_1 + C_2 y_2] = C_1 L_n[y_1] + C_2 L_n[y_2] \tag{3.4.2}$$

of the nth-order differential operator

$$L_n = D^n + a_1 D^{n-1} + \cdots + a_{n-1}D + a_n,$$

where $D = d/dx$. Accordingly, L_n is termed a *linear differential operator*. Eq. (3.4.1) is said to be *homogeneous* if $f(x) = 0$, and *non-homogeneous* otherwise (cf. Sections 3.2.6 and 3.3).

---

[3]N.H. Ibragimov, 'Invariant Lagrangians and a new method of integration of nonlinear equations', *J. Mathematical Analysis and Applications*, V. 304, No. 1, 2005, pp. 212-235.

3.4.1   Homogeneous equations. Fundamental system

The *linear superposition principle* follows from property (3.4.2) and states that the general solution of the homogeneous linear equation

$$y^{(n)} + a_1(x)y^{(n-1)} + \cdots + a_{n-1}(x)y' + a_n(x)y = 0 \qquad (3.4.3)$$

is a *linear superposition* of n linearly independent particular solutions:

$$y = C_1 y_1(x) + \cdots + C_n y_n(x), \qquad (3.4.4)$$

where $C_1, \ldots, C_n$ are arbitrary constants. Any set $y_1(x), \ldots, y_n(x)$ of n linearly independent solutions is termed a *fundamental system* for (3.4.3).

3.4.2   Non-homogeneous equations. Variation of parameters

Theorem 3.4.1. Let a fundamental system of solutions for the homogeneous equation (3.4.3) be known. Then the general solution to the non-homogeneous equation (3.4.1) can be obtained by quadratures.

Proof. The solution can be obtained by the general method of *variation of parameters* due to Lagrange (1774). Namely, we replace the constants C_i in (3.4.4) by functions $u_i(x)$ (cf. Section 3.3):

$$y = u_1(x)y_1(x) + \cdots + u_n(x)y_n(x).$$

Lagrange's method provides the following relations for determining the unknown functions $u_i(x)$:

$$y_1 \frac{du_1}{dx} + \cdots + y_n \frac{du_n}{dx} = 0,$$

$$y_1' \frac{du_1}{dx} + \cdots + y_n' \frac{du_n}{dx} = 0,$$

$$\cdots \cdots$$

$$y_1^{(n-2)} \frac{du_1}{dx} + \cdots + y_n^{(n-2)} \frac{du_n}{dx} = 0,$$

$$y_1^{(n-1)} \frac{du_1}{dx} + \cdots + y_n^{(n-1)} \frac{du_n}{dx} = f(x). \qquad (3.4.5)$$

Since $y_1, \ldots, y_n$ are known and they are linearly independent, equations (3.4.5) can be solved with respect to the derivatives of the unknown functions and written in the form integrable by quadrature:

$$\frac{du_k}{dx} = \psi_k(x), \quad k = 1, \ldots, n.$$

3.4.3 Equations with constant coefficients

Euler's ansatz discussed in the case of second-order equations applies also to higher-order homogeneous linear equations with constant coefficients,

$$y^{(n)} + a_1 y^{(n-1)} + \cdots + a_{n-1} y' + a_n y = 0, \quad a_1, \ldots, a_n = \text{const.} \qquad (3.4.6)$$

Looking for a particular solutions of form (3.3.25),

$$y = e^{\lambda x}, \quad \lambda = \text{const.},$$

one reduces the nth-order differential equation (3.4.6) to the algebraic equation of the nth degree,

$$P_n[\lambda] \equiv \lambda^n + a_1 \lambda^{n-1} + \cdots + a_{n-1}\lambda + a_n = 0 \qquad (3.4.7)$$

called the *characteristic equation*.

The polynomial $P_n(\lambda)$ is known as the *characteristic polynomial* for (3.4.6). Let $\lambda_1, \ldots, \lambda_n$ be real distinct roots of the characteristic equation (3.4.7). Then the particular solutions $e^{\lambda_1 x}, \ldots, e^{\lambda_n x}$ provide a fundamental system. According to the superposition principle (3.4.4), the general solution of Eq. (3.4.6) with constant coefficients is given by the linear combination

$$y = C_1 e^{\lambda_1 x} + \cdots + C_n e^{\lambda_n x}. \qquad (3.4.8)$$

The cases of complex as well as repeated roots are treated as in the case of second-order equations discussed in Section 3.3. For example, let λ_1 be repeated s times. Then the corresponding solution is given by

$$y_1 = \left(C_1 + C_2 x + \cdots + C_s x^{s-1} \right) e^{\lambda_1 x}, \qquad (3.4.9)$$

with arbitrary constants C_i. Taking into account all multiple roots, one obtains the following modification of formula (3.4.8) for the general solution:

$$y = q_1(x) e^{\lambda_1 x} + \ldots + q_r(x) e^{\lambda_r x}. \qquad (3.4.10)$$

Here $q_s(x)$ is the polynomial with arbitrary coefficients of degree $s - 1$, where s is the order of multiplicity of the corresponding root λ_s $(1 \le s \le r)$.

In the case of complex $\lambda_1 = \alpha_1 + \beta_1 i$, the right-hand side of expression (3.4.9) (and hence the first term of (3.4.10)) should be replaced by

$$\left(C_1 + C_2 x + \cdots + C_s x^{s-1} \right) e^{\alpha_1 x} \cos(\beta_1 x)$$
$$+ \left(C_{s+1} + C_{s+2} x + \cdots + C_{2s} x^{s-1} \right) e^{\alpha_1 x} \sin(\beta_1 x).$$

Example 3.4.1. Consider Eq. (2.3.20),

$$\frac{d^4 u}{dx^4} = \alpha^4 u, \quad \alpha = \text{const.}$$

The characteristic equation $\lambda^4 - \alpha^4 = 0$ has four distinct roots:

$$\lambda_1 = \alpha, \quad \lambda_2 = -\alpha, \quad \lambda_3 = \alpha i, \quad \lambda_4 = -\alpha i.$$

Thus, one arrives at formula (2.3.21) for the general solution

$$u = C_1 e^{\alpha x} + C_2 e^{-\alpha x} + C_3 \cos(\alpha x) + C_4 \sin(\alpha x).$$

Example 3.4.2. Consider the equation

$$\frac{d^4 y}{dx^4} + 2 \frac{d^2 y}{dx^2} + y = 0.$$

The characteristic equation

$$\lambda^4 + 2\lambda^2 + 1 = 0$$

has the repeated imaginary roots

$$\lambda_{1,2} = i, \quad \lambda_{3,4} = \overline{\lambda}_{1,2} = -i.$$

Hence, the general solution

$$y = (C_1 + C_2 x) \cos x + (C_3 + C_4 x) \sin x.$$

3.4.4 Euler's equation

The equation

$$x^n \frac{d^n y}{dx^n} + a_1 x^{n-1} \frac{d^{n-1} y}{dx^{n-1}} + \cdots + a_{n-1} x \frac{dy}{dx} + a_n y = 0, \qquad (3.4.11)$$

where $a_1, \ldots, a_n = $ const., is known as *Euler's equation*. This equation with variable coefficients is invariant under the dilation, i.e., it does not alter after replacing x by $k x$ with a parameter $k \neq 0$. Therefore, the transformation

$$t = \ln |x| \qquad (3.4.12)$$

converting the dilation $k x$ into the translation $t + \ln |k|$, maps Eq. (3.4.11) to an equation with constant coefficients for the function $y(t)$.

Example 3.4.3. Consider the second-order Euler equation

$$x^2 \frac{d^2 y}{dx^2} + 3x \frac{dy}{dx} + y = 0.$$

Introducing the new independent variable $t = \ln |x|$, we obtain

$$\frac{d^2 y}{dt^2} + 2 \frac{dy}{dt} + y = 0.$$

For the latter equation, we have the characteristic equation

$$\lambda^2 + 2\lambda + 1 = 0,$$

which has the repeated root $\lambda_1 = \lambda_2 = -1$. Hence,

$$y = (C_1 + C_2 t)e^{-t}.$$

Returning to the original variable x, we obtain the following general solution of the equation in question:

$$y = \frac{1}{x}(C_1 + C_2 \ln|x|).$$

3.5   Systems of first-order equations

3.5.1   General properties of systems

Consider the general system of first-order ordinary differential equations

$$\frac{dy^i}{dx} = f^i(x, y^1, y^2, \ldots, y^n), \quad i = 1, 2, \ldots, n. \tag{3.5.1}$$

Let the functions f^i be continuous in a neighborhood of $x_0, y_0^1, \ldots, y_0^n$.

We will use the vector notation

$$\boldsymbol{y} = (y^1, \ldots, y^n), \quad \boldsymbol{f} = (f^1, \ldots, f^n)$$

and write an initial value problem for (3.5.1) in the compact form:

$$\frac{d\boldsymbol{y}}{dx} = \boldsymbol{f}(x, \boldsymbol{y}), \quad \boldsymbol{y}\big|_{x=x_0} = \boldsymbol{y}_0, \tag{3.5.2}$$

where

$$\boldsymbol{y}_0 = (y_0^1, \ldots, y_0^n).$$

Thus, $\boldsymbol{y}$ is an n-tuple of dependent variables, the ith one of which is denoted by y^i and called the ith coordinate of the vector $\boldsymbol{y}$.

The definition of classical solutions applies to systems of differential equations as well with the natural replacement of the single variable y by the vector $\boldsymbol{y}$. We will use the following simple version of the existence and uniqueness theorem for systems.

Theorem 3.5.1. Let the function $\boldsymbol{f}(x, \boldsymbol{y})$ be continuously differentiable in a neighborhood of the point $(x_0, \boldsymbol{y}_0)$. Then problem (3.5.2) has one and only one solution defined in a neighborhood of x_0. It follows that the general solution of a system of n first-order differential equations (3.5.1) depends precisely on n arbitrary constants $C_1, \ldots, C_n$, e.g. on arbitrarily chosen initial values $y_0^1, \ldots, y_0^n$ of the dependent variables at $x = x_0$. Accordingly, the general solution to (3.5.1) is written

$$y^i = \phi^i(x, C_1, \ldots, C_n), \quad i = 1, 2, \ldots, n. \tag{3.5.3}$$

3.5.2 First integrals

Consider a system of ordinary differential equations of the first order with $n-1$ dependent variables:

$$\frac{dy^i}{dx} = f^i(x, y^1, y^2, \ldots, y^{n-1}), \quad i = 1, \ldots, n-1. \tag{3.5.4}$$

According to Theorem 3.5.1, its general solution has the form

$$y^i(x) = \phi^i(x, C_1, \ldots, C_{n-1}), \quad i = 1, \ldots, n-1,$$

whence, upon solving with respect to the constants of integration C_i,

$$\psi_i(x, y^1, y^2, \ldots, y^{n-1}) = C_i, \quad i = 1, \ldots, n-1. \tag{3.5.5}$$

The system of relations (3.5.5) is called the *general integral* of Eqs. (3.5.4). The left-hand side of each Eq. (3.5.5) reduces to a constant when $y^1, y^2, \ldots, y^{n-1}$ are replaced by the coordinates $y^1(x), y^2(x), \ldots, y^{n-1}(x)$ of any solution of system (3.5.4). For this reason every single relation in (3.5.5) is known as a *first integral* of the system of equations (3.5.4).

Example 3.5.1. Consider the system

$$\frac{dx}{dt} = y, \quad \frac{dy}{dt} = -x. \tag{3.5.6}$$

One can integrate this system as follows. Differentiating the first equation of (3.5.6) and substituting dy/dt from the second equation, one reduces the problem to integration of the single second-order equation

$$\frac{d^2x}{dt^2} + x = 0.$$

Its fundamental system is provided by

$$x_{(1)} = \cos t, \quad x_{(2)} = \sin t$$

and hence,

$$x = C_1 \cos t + C_2 \sin t.$$

The first equation of (3.5.6), $y = dx/dt$, yields

$$y = C_2 \cos t - C_1 \sin t.$$

Hence, the general solution to system (3.5.6) is given by

$$x = C_1 \cos t + C_2 \sin t, \quad y = C_2 \cos t - C_1 \sin t. \tag{3.5.7}$$

Solving Eqs. (3.5.7) for C_1, C_2, one obtains the following first integrals:

$$x \cos t - y \sin t = C_1, \quad x \sin t + y \cos t = C_2. \tag{3.5.8}$$

Hence, the functions ψ in Eqs. (3.5.5) are

$$\psi_1(t, x, y) = x\,\cos t - y\,\sin t, \quad \psi_2(t, x, y) = x\,\sin t + y\,\cos t. \qquad (3.5.9)$$

The first integrals can be obtained also by rewriting system (3.5.6) in the following form (see further Eqs. (3.5.13)):

$$\frac{dx}{y} = -\frac{dy}{x} = dt.$$

Integration of the first equation written in the form $x\,dx + y\,dy = 0$ yields

$$x^2 + y^2 = a^2, \quad a = \text{const.} \qquad (3.5.10)$$

Now the second equation is written

$$dt + \frac{dy}{\sqrt{a^2 - y^2}} = 0$$

and yields

$$t + \arcsin(y/a) = C.$$

Invoking (3.5.10) and the elementary formula (1.1.7), we have

$$\arcsin \frac{y}{a} = \arctan \frac{y}{\sqrt{a^2 - y^2}} = \arctan \frac{y}{x}, \qquad (3.5.11)$$

and hence,

$$t + \arctan(y/x) = C.$$

Finally, we arrive at the first integrals (3.5.5) with the functions

$$\tilde{\psi}_1(t, x, y) = x^2 + y^2, \quad \tilde{\psi}_2(t, x, y) = t + \arctan(y/x). \qquad (3.5.12)$$

instead of (3.5.9).

The set of the first integrals (3.5.5) is not the only possible representation of the general solution. Indeed, any relation $\tilde{\psi}(\psi_1, \ldots, \psi_{n-1}) = C$ is a first integral, and hence one can replace the functions ψ by any $n-1$ functionally independent functions $\tilde{\psi}_i(\psi_1, \ldots, \psi_{n-1})$, $i = 1, \ldots, n-1$. Therefore, it is useful to have the definition of the first integrals independent on the general integral (3.5.5).

Definition 3.5.1. Given a system (3.5.4), its *first integral* is a relation

$$\psi(x, y^1, y^2, \ldots, y^{n-1}) = C$$

satisfied for any solution $y^i = y^i(x)$, $i = 1, \ldots, n-1$, where the function ψ is not identically constant. In other words, the function ψ, which is also called a *first integral* for the sake of brevity, holds a constant value along each solution with the constant C depending on the solution.

System (3.5.4) can be rewritten in the form

$$\frac{dx}{1} = \frac{dy^1}{f^1} = \frac{dy^2}{f^2} = \cdots = \frac{dy^{n-1}}{f^{n-1}}.$$

Since the denominators can be multiplied by any function distinct from zero, one can rewrite these equations (using the notation $x = (x^1, x^2, \ldots, x^n)$ for the variables $x, y^1, \ldots, y^{n-1}$) in the *symmetric form*:

$$\frac{dx^1}{\xi^1(x)} = \frac{dx^2}{\xi^2(x)} = \cdots = \frac{dx^n}{\xi^n(x)}. \tag{3.5.13}$$

The term *symmetric* is due to the fact that form (3.5.13) of $n - 1$ first-order ordinary differential equations does not specify the independent variable, which may be now any of the n variables $x^1, x^2, \ldots, x^n$. A *first integral* of system (3.5.13) is given by Definition 3.5.1 and is written

$$\psi(x) = C. \tag{3.5.14}$$

The first integral (3.5.14) is often identified with the function $\psi(x)$.

Lemma 3.5.1. A function $\psi(x) = \psi(x^1, \ldots, x^n)$ is a first integral of system (3.5.13) if and only if $u = \psi(x)$ solves the partial differential equation

$$\xi^1(x)\frac{\partial u}{\partial x^1} + \cdots + \xi^n(x)\frac{\partial u}{\partial x^n} = 0. \tag{3.5.15}$$

Proof. Let a function $\psi(x)$ provide a first integral. Since $\psi(x) = \text{const.}$ for any solution $x = (x^1, \ldots, x^n)$ of system (3.5.13), the differential $d\psi$ taken along any integral curve of Eqs. (3.5.13) vanishes:

$$d\psi \equiv \frac{\partial \psi}{\partial x^1}dx^1 + \cdots + \frac{\partial \psi}{\partial x^n}dx^n = 0. \tag{3.5.16}$$

In other words, Eq. (3.5.16) holds whenever $dx = (dx^1, \ldots, dx^n)$ is proportional to the vector $\xi = (\xi^1, \ldots, \xi^n)$, i.e., $dx = \lambda\xi$, $\lambda \neq 0$. Substituting $dx^i = \lambda\xi^i$ into (3.5.16), one arrives at Eq. (3.5.15). Thus, we have proved that Eq. (3.5.15) is satisfied at points x belonging to the integral curves of system (3.5.13). But, according to Theorem 3.5.1, integral curves pass through any point. Hence, Eq. (3.5.15) is satisfied identically in a neighborhood of any point x. The above steps are reversible. This completes the proof.

Definition 3.5.2. A set of $n - 1$ first integrals

$$\psi_k(x) = C_k, \quad k = 1, \ldots, n-1, \tag{3.5.17}$$

is said to be *independent* if the functions $\psi_k(x)$ are *functionally independent*, i.e., if there is no relation of the form $F(\psi_1, \ldots, \psi_{n-1}) = 0$.

Example 3.5.2. Consider the system

$$\frac{dx}{x} = \frac{dy}{y} = \frac{dz}{z}.$$

Integrating the equations

$$\frac{dx}{x} = \frac{dy}{y} \quad \text{and} \quad \frac{dx}{x} = \frac{dz}{z},$$

we obtain $y/x = C_1$ and $z/x = C_2$, respectively. Hence, we have the following two independent first integrals:

$$\psi_1(x, y, z) = \frac{y}{x}, \quad \psi_2(x, y, z) = \frac{z}{x}.$$

Equation (3.5.15) has the form

$$x\frac{\partial u}{\partial x} + y\frac{\partial u}{\partial y} + z\frac{\partial u}{\partial z} = 0.$$

One can easily verify that this equation is satisfied by $u = \psi_1(x, y, z) = y/x$ and $u = \psi_2(x, y, z) = z/x$.

Any set of $n - 1$ independent first integrals represents the general solution of system (3.5.13). Since the general solution of a system of $n - 1$ first order equations depends precisely on $n - 1$ arbitrary constants (see Theorem 3.5.1), one arrives at the following statement.

Theorem 3.5.2. A system of $n - 1$ first-order ordinary differential equations (3.5.13) has $n - 1$ independent first integrals (3.5.17). Any other first integral (3.5.14) of system (3.5.13) is expressible in terms of (3.5.17):

$$\psi = F(\psi_1, \ldots, \psi_{n-1}). \tag{3.5.18}$$

Example 3.5.3. Consider the system

$$\frac{dx}{yz} = \frac{dy}{xz} = \frac{dz}{xy}.$$

It is equivalently rewritten as

$$\frac{dx}{yz} = \frac{dy}{xz}, \quad \frac{dy}{xz} = \frac{dz}{xy},$$

or, multiplying through by z and x, respectively,

$$\frac{dx}{y} = \frac{dy}{x}, \quad \frac{dy}{z} = \frac{dz}{y}.$$

Rewriting them in the form $ydy - xdx = 0$ and $ydy - zdz = 0$ and integrating, one arrives at the following two independent first integrals:

$$\psi_1 \equiv x^2 - y^2 = C_1, \quad \psi_2 \equiv z^2 - y^2 = C_2.$$

Alternatively, the system in question can be written in the form

$$\frac{dx}{y} = \frac{dy}{x}, \quad \frac{dx}{z} = \frac{dz}{x}.$$

Then one arrives at the first integrals

$$\psi_1 \equiv x^2 - y^2 = C_1, \quad \psi_3 \equiv x^2 - z^2 = C_3,$$

and hence one obtains three different first integrals, $\psi_1 = C_1, \psi_2 = C_2$, and $\psi_3 = C_3$. However, they are not independent. Indeed, e.g. $\psi_3 = \psi_1 - \psi_2$.

Thus, in this example, representation (3.5.18) of an arbitrary first integral can be taken in the form $\psi = F(x^2 - y^2, z^2 - y^2)$.

3.5.3   Linear systems with constant coefficients

Euler's method discussed above applies also to the general system of linear homogeneous equations of the first order with constant coefficients:

$$\frac{dy^i}{dx} + \sum_{j=1}^{n} a_{ij}y^j = 0, \quad i = 1, \ldots, n, \quad \text{or} \quad y' + Ay = 0. \tag{3.5.19}$$

Here $y = (y^1, \ldots, y^n)$ denotes the dependent variables, $y' = dy/dx$ and $A = (a_{ij})$ is a constant $n \times n$ matrix so that $(Ay)^i = \sum_{j=1}^{n} a_{ij}y^j$.

Euler's formula for particular solutions is now written

$$y = e^{\lambda x} l, \quad \lambda = \text{const.} \tag{3.5.20}$$

Here $l = (l^1, \ldots, l^n)$ is an unknown constant vector to be determined from Eq. (3.5.19). Substitution of (3.5.20) into (3.5.19) yields

$$(A + \lambda E)l = 0, \tag{3.5.21}$$

where E is the unit $n \times n$ matrix. The system of linear equations (3.5.21) has a solution $l \neq 0$ if and only if the determinant $|A + \lambda E| = \det(A + \lambda E)$ vanishes. Subsequently, the *characteristic polynomial* P_n and the *characteristic equation* for system (3.5.19) are defined by

$$P_n(\lambda) \equiv |A + \lambda E| = 0. \tag{3.5.22}$$

Let the characteristic equation (3.5.22) have distinct roots, $\lambda_1, \ldots, \lambda_n$. Then one obtains precisely n linearly independent solutions $l_{(1)}, \ldots, l_{(n)}$ of (3.5.21), and hence the following *fundamental system* of solutions:

$$y_{(1)} = e^{\lambda_1 x} l_{(1)}, \quad \ldots, \quad y_{(n)} = e^{\lambda_n x} l_{(n)}. \tag{3.5.23}$$

The general solution to system (3.5.19) is given by

$$y = C_1 e^{\lambda_1 x} l_{(1)} + \cdots + C_n e^{\lambda_n x} l_{(n)}. \tag{3.5.24}$$

If Eq. (3.5.22) has a complex root, $\lambda = \alpha + i\beta$ (and hence its complex conjugate), then Eq. (3.5.21) has a complex solution $l = p + iq$. The corresponding solution (3.5.20) splits into two real solutions:

$$\begin{aligned}
y_{(1)} &= e^{\alpha x}(p \cos \beta x - q \sin \beta x), \\
y_{(2)} &= e^{\alpha x}(p \sin \beta x + q \cos \beta x).
\end{aligned} \tag{3.5.25}$$

Example 3.5.4. Let us solve the system (cf. Example 3.5.1)

$$\frac{dx}{dt} = y, \quad \frac{dy}{dt} = -x.$$

Here $y = (x, y)$, $a_{11} = a_{22} = 0$, $a_{12} = -1$, $a_{21} = 1$. The characteristic equation has the complex roots $\lambda = i$, $\overline{\lambda} = -i$ and Eq. (3.5.21) yields $l = p + iq$ with $p = (1, 0)$, $q = (0, 1)$. Eqs. (3.5.25) provide the fundamental system of solutions

$$y_{(1)} = (\cos t, -\sin t), \quad y_{(2)} = (\sin t, \cos t). \tag{3.5.26}$$

Writing $y = C_1 y_{(1)} + C_2 y_{(2)}$, one arrives at solution (3.5.7):

$$x = C_1 \cos t + C_2 \sin t, \quad y = C_2 \cos t - C_1 \sin t.$$

3.5.4 Variation of parameters for systems

Consider now systems of non-homogeneous linear equations:

$$\frac{dy^i}{dx} + \sum_{j=1}^{n} a_{ij}(x) y^j = f_i(x), \quad i = 1, \ldots, n. \tag{3.5.27}$$

It is called *homogeneous* if $f_i = 0$, $i = 1, \ldots, n$, and *non-homogeneous* otherwise (cf. Section 3.2.6).

Let the general solution of the homogeneous system

$$\frac{dy^i}{dx} + \sum_{j=1}^{n} a_{ij}(x) y^j = 0, \quad i = 1, \ldots, n,$$

be known. Then the non-homogeneous system (3.5.27) can be solved by the method of variation of parameters discussed for a single equation. Let us illustrate the method by the following example.

Example 3.5.5. Consider the system

$$\frac{dx}{dt} - y = \cos t, \quad \frac{dy}{dt} + x = 1. \tag{3.5.28}$$

The solution of the homogeneous system,

$$\frac{dx}{dt} - y = 0, \quad \frac{dy}{dt} + x = 0,$$

is given by (3.5.7):

$$x = C_1 \cos t + C_2 \sin t, \quad y = C_2 \cos t - C_1 \sin t.$$

One can proceed as in the case of a single equation. Namely, replacing the constants C_1, C_2 by unknown functions $u(t), v(t)$ and substituting the expressions

$$x = u(t) \cos t + v(t) \sin t, \quad y = v(t) \cos t - u(t) \sin t \qquad (3.5.29)$$

into Eq. (3.5.28), one obtains

$$\frac{du}{dt} \cos t + \frac{dv}{dt} \sin t = \cos t, \quad \frac{dv}{dt} \cos t - \frac{du}{dt} \sin t = 1,$$

whence:

$$\frac{du}{dt} = \cos^2 t - \sin t, \quad \frac{dv}{dt} = \cos t \sin t + \cos t. \qquad (3.5.30)$$

Integration of Eqs. (3.5.30) yields

$$u = \frac{t}{2} + \frac{1}{2} \sin t \cos t + \cos t + K_1, \quad v = -\frac{1}{2} \cos^2 t + \sin t + K_2. \qquad (3.5.31)$$

Substituting (3.5.31) into (3.5.29) and denoting the arbitrary constants K_1 and K_2 by C_1 and C_2, respectively, one obtains the following solution of system (3.5.28):

$$x = 1 + \frac{t}{2} \cos t + C_1 \cos t + C_2 \sin t,$$

$$y = -\frac{t}{2} \sin t - \frac{1}{2} \cos t + C_2 \cos t - C_1 \sin t.$$

It is convenient to represent the method of variation of parameters in the vector form as follows. System (3.5.27) is written:

$$y' + A(x)y = f(x). \qquad (3.5.32)$$

Furthermore, let us write the general solution (3.2.35) to the non-homogeneous single linear equation $y' + P(x)y = Q(x)$ in the form

$$y = C y_1(x) + y_1(x) \int y_1^{-1}(x)Q(x)\, dx,$$

where

$$y_1(x) = e^{-\int P(x)\, dx}$$

is a particular solution to the homogeneous equation $y' + P(x)y = 0$. Proceeding as in Section 3.2.7, one arrives at the following result.

Theorem 3.5.3. Let $y_1, \ldots, y_n$ be a fundamental system of solutions for the homogeneous system

$$y' + A(x)y = 0.$$

Then the general solution to the non-homogeneous system (3.5.32) is given by quadrature by the following formula:

$$y = C_1 y_1(x) + \cdots + C_n y_n(x) + Y(x) \int Y^{-1}(x)\, f(x)\, dx, \qquad (3.5.33)$$

where $Y(x)$ is the $n \times n$ matrix defined by

$$Y(x) = (y_1(x), \cdots, y_n(x)) \qquad (3.5.34)$$

and known as the fundamental matrix, and $Y^{-1}(x)$ is the inverse matrix.

Example 3.5.6. Consider again system (3.5.28) from Example 3.5.5. It is written in the vector form (3.5.32) as follows:

$$\begin{pmatrix} x' \\ y' \end{pmatrix} + \begin{pmatrix} 0 & -1 \\ 1 & 0 \end{pmatrix} \begin{pmatrix} x \\ y \end{pmatrix} = \begin{pmatrix} \cos t \\ 1 \end{pmatrix}.$$

Thus, we have a two-dimensional non-homogeneous vector equation of form (3.5.32) with the independent variable t and

$$y = \begin{pmatrix} x \\ y \end{pmatrix}, \quad A = \begin{pmatrix} 0 & -1 \\ 1 & 0 \end{pmatrix}, \quad f(t) = \begin{pmatrix} \cos t \\ 1 \end{pmatrix}.$$

The fundamental system of solutions of the homogeneous system is given by (3.5.26). We write it in the form

$$y_{(1)} = \begin{pmatrix} \cos t \\ -\sin t \end{pmatrix}, \quad y_{(2)} = \begin{pmatrix} \sin t \\ \cos t \end{pmatrix}$$

and obtain the following fundamental matrix (3.5.34) and its inverse:

$$Y = \begin{pmatrix} \cos t & \sin t \\ -\sin t & \cos t \end{pmatrix}, \quad Y^{-1} = \begin{pmatrix} \cos t & -\sin t \\ \sin t & \cos t \end{pmatrix}.$$

Thus, formula (3.5.33) for the solution is written:

$$y = C_1 \begin{pmatrix} \cos t \\ -\sin t \end{pmatrix} + C_2 \begin{pmatrix} \sin t \\ \cos t \end{pmatrix}$$

$$+ \begin{pmatrix} \cos t & \sin t \\ -\sin t & \cos t \end{pmatrix} \int \begin{pmatrix} \cos t & -\sin t \\ \sin t & \cos t \end{pmatrix} \begin{pmatrix} \cos t \\ 1 \end{pmatrix} dt.$$

We have

$$\int \begin{pmatrix} \cos t & -\sin t \\ \sin t & \cos t \end{pmatrix} \begin{pmatrix} \cos t \\ 1 \end{pmatrix} dt = \int \begin{pmatrix} \cos^2 t - \sin t \\ \sin t \cos t + \cos t \end{pmatrix} dt.$$

The integral in the right-hand side is written (see Eqs. (3.5.30)):

$$\begin{pmatrix} \int (\cos^2 t - \sin t) dt \\ \int (\sin t \cos t + \cos t) dt \end{pmatrix} = \begin{pmatrix} \dfrac{t}{2} + \dfrac{1}{2} \sin t \cos t + \cos t \\ -\dfrac{1}{2} \cos^2 t + \sin t \end{pmatrix}.$$

Substituting this into the above expression for y and changing the arbitrary constants C_1, C_2, we arrive at the solution given in Example 3.5.5. Namely,

$$y = C_1 \begin{pmatrix} \cos t \\ -\sin t \end{pmatrix} + C_2 \begin{pmatrix} \sin t \\ \cos t \end{pmatrix} + \frac{1}{2} \begin{pmatrix} 2 + t \cos t \\ -t \sin t - \cos t \end{pmatrix}.$$

Problems to Chapter 3

3.1. Integrate the following first-order equations:

(i) $y' = 0$,   (ii) $y' = 2xy$,   (iii) $y' = \dfrac{y}{1 + x^2}$,

(iv) $y' = y + x^2$,   (v) $y' + C_1 y + x + x^2 = 0$,

3.2. Integrate the following second-order equations:

(i) $y'' = 0$,   (ii) $y'' = 2y$,   (iii) $y'' = -2y$,   (iv) $y'' = 2y'$,

(v) $y'' = y + x^2$,   (vi) $y'' = [(x + x^2)e^y]'$,

3.3. Integrate the following third-order equations:

(i) $y''' = 0$,   (ii) $y''' = y$,   (iii) $y''' + y = 0$,

(iv) $y''' = y + x^2$,

3.4. Single out the homogeneous equations of the form $y' + y^2 = C x^s$, where C and s are any constants.

3.5. Verify the uniform homogeneity of the equations from Example 3.1.5.

3.6. Give detailed calculations of the proof of Theorem 3.3.1. Specifically, verify that transformation (3.3.11) maps the general linear homogeneous second-order equation (3.3.6) to the simplest form (3.3.10).

3.7. Test the linearization for the following Riccati equations:

(i) $y' = 1 + y^2$,   (ii) $y' = 1 - y^2$,   (iii) $y' = x + 2xy + xy^2$,

(iv) $y' = x + y^2$,   (v) $y' = P(x) + Q(x)y + [Q(x) - P(x)]y^2$,

(vi) $y' = x + xy^2$,   (vii) $y' = P(x) + Q(x)y + [Q(x) - 2P(x)]y^2$,

(viii) $y' = x - xy^2$,   (ix) $y' = P(x) + Q(x)y + 2[Q(x) - 2P(x)]y^2$,

(x) $y' = P(x) + [1 + P(x)]y + y^2$,   (xi) $y' = P(x) + [1 + 2P(x)]y + y^2$,

(xii) $y' = \dfrac{2}{x^2} - y^2$,   (xiii) $y' = \dfrac{2}{x^2} + \dfrac{2 - x^2}{x^2}y - y^2$,

(xiv) $y' = x + (1 + x)^2 y + (1 + x + x^2) y^2$

by checking both properties (A) and (B) of Theorem 3.2.2.

3.8. Solve the following system with initial conditions:

$$\frac{dy^1}{dt} = y^2, \ \frac{dy^2}{dt} = y^1, \quad y^1\big|_{t=0} = x^1, \ y^2\big|_{t=0} = x^2.$$

3.9. Show that the following equation is exact and integrate it:

$$\left(\frac{1}{x} - \frac{y^2}{x^2}\right) dx + \frac{2y}{x} \, dy = 0.$$

3.10. Integrate the following equation describing free oscillations of a damped mechanical system with a small damping force:

$$\frac{d^2 y}{dt^2} + 2b\frac{dy}{dt} + cy = 0,$$

where b, c are positive constants such that $b^2 < c$.

3.11. Solve the following equations:

(i) $y'' + y = \tan x$,   (ii) $y'' + y = \dfrac{1}{\cos x}$,   (iii) $y'' + y = \dfrac{1}{\sin x}$.

3.12. Solve the exact equation $(y\,e^{xy} + \cos x)dx + x\,e^{xy}dy = 0$ from Example 3.2.1 by using Eqs. (3.2.7) and (3.2.8).

3.13. Find all second-order equations $f(x, y, y', y'') = 0$ that are reducible to the form $g(x, y, y')y''' = 0$ through differentiation.

3.14. Integrate Euler's equation $x^2y'' + 2xy' + 4y = 0$, $x > 0$.

3.15. Integrate the equation

$$y'' - 3\frac{y'}{x} + 3\frac{y}{x^2} = 0.$$

Chapter 4

First-order partial differential equations

Partial differential equations of the first order with one dependent variable pertain to the theory of ordinary differential equations, the link between these two, seemingly distinctly different, classes of equations being provided by characteristics. Furthermore, an acquaintance with the theory of first-order partial differential equations is a prerequisite for Lie's theory.

Additional reading: E. Goursat [9], V.I. Smirnov [36], N.H. Ibragimov [21].

4.1 Introduction

Let $x = (x^1, \ldots, x^n)$ be $n \geq 2$ independent variables and u a dependent variable. We denote by $p = (p_1, \ldots, p_n)$ the partial derivatives $p_i = \partial u / \partial x^i$.

Recall that equations in which the number of independent variables is greater than one are termed partial differential equations. An equation is said to be of the first order if the partial derivatives of highest order that occur are of order one. A single partial differential equation of the first order with one dependent variable is written

$$F(x^1, \ldots, x^n, u, p_1, \ldots, p_n) = 0. \tag{4.1.1}$$

If $n = 2$, the independent variables are denoted by x, y, the derivatives by $p = \partial u / \partial x$, $q = \partial u / \partial y$. Then Eq. (4.1.1) is written $F(x, y, u, p, q) = 0$. Its solution $u = \phi(x, y)$ defines a surface in the three-dimensional space x, y, u and therefore it is often termed an *integral surface*.

The general *linear* first-order partial differential equation is written

$$\xi^1(x)p_1 + \cdots + \xi^n(x)p_n + c(x)u = f(x), \tag{4.1.2}$$

or

$$\xi^1(x)\frac{\partial u}{\partial x^1} + \cdots + \xi^n(x)\frac{\partial u}{\partial x^n} + c(x)u = f(x). \tag{4.1.3}$$

If $f(x) = 0$, Eq. (4.1.2) is homogeneous by function (see Sections 3.1.4 and 3.2.6). However, the term *homogeneous* applies in the literature to Eq. (4.1.2) with $c(x) = 0$ and $f(x) = 0$, i.e., to the equation

$$\xi^1(x)p_1 + \cdots + \xi^n(x)p_n = 0. \tag{4.1.4}$$

The general *quasi-linear* equation of the first order has the form

$$\xi^1(x, u)p_1 + \cdots + \xi^n(x, u)p_n = g(x, u). \tag{4.1.5}$$

4.2 Homogeneous linear equation

Let us introduce the linear partial differential operator of the first order:

$$X = \xi^1(x)\frac{\partial}{\partial x^1} + \cdots + \xi^n(x)\frac{\partial}{\partial x^n}. \tag{4.2.1}$$

Lemma 4.2.1. Let $\tilde{x}^i$ be new independent variables defined by

$$\tilde{x}^i = \varphi^i(x), \quad i = 1, \ldots, n. \tag{4.2.2}$$

Then operator (4.2.1) is written in the new variables in the form

$$\tilde{X} = X(\varphi^1)\frac{\partial}{\partial \tilde{x}^1} + \cdots + X(\varphi^n)\frac{\partial}{\partial \tilde{x}^n}, \tag{4.2.3}$$

where $X(\varphi^i) = \xi^1 \partial \varphi^i / \partial x^1 + \cdots + \xi^n(x)\partial \varphi^i / \partial x^n$.

Proof. The chain rule for the partial derivatives yields

$$\frac{\partial}{\partial x^i} = \sum_{k=1}^n \frac{\partial \varphi^k}{\partial x^i} \frac{\partial}{\partial \tilde{x}^k}.$$

One can easily verify that the substitution of the above expressions into operator (4.2.1) transforms it to form (4.2.3).

In terms of this operator, the homogeneous linear partial differential equation (4.1.4) is written as follows:

$$X(u) \equiv \xi^1(x)\frac{\partial u}{\partial x^1} + \cdots + \xi^n(x)\frac{\partial u}{\partial x^n} = 0. \tag{4.2.4}$$

Theorem 4.2.1. The general solution to Eq. (4.2.4) has the form

$$u = F(\psi_1(x), \ldots, \psi_{n-1}(x)), \tag{4.2.5}$$

where F is an arbitrary function of $n - 1$ variables and

$$\psi_1(x) = C_1, \quad \ldots, \quad \psi_{n-1}(x) = C_{n-1}$$

are independent first integrals of the following system of $n-1$ ordinary differential equations called the *characteristic system* for Eq. (4.2.4):

$$\frac{dx^1}{\xi^1(x)} = \frac{dx^2}{\xi^2(x)} = \cdots = \frac{dx^n}{\xi^n(x)}. \tag{4.2.6}$$

Proof. The function u defined by (4.2.5) solves Eq. (4.2.4). Indeed, we have $X(\psi_1) = 0, \ldots, X(\psi_{n-1}) = 0$ by Lemma 3.5.1. Therefore, Eq. (4.2.4) follows from the chain rule:

$$X\left(F(\psi_1, \ldots, \psi_{n-1})\right) = \frac{\partial F}{\partial \psi_1} X(\psi_1) + \cdots + \frac{\partial F}{\partial \psi_{n-1}} X(\psi_{n-1}) = 0.$$

Let us verify now that any solution to Eq. (4.2.4) has the form (4.2.5). Let us introduce new independent variables

$$x'^1 = \psi_1(x), \ldots, \quad x'^{n-1} = \psi_{n-1}(x), \quad x'^n = \phi(x), \tag{4.2.7}$$

where $\psi_1(x), \ldots, \psi_{n-1}(x)$ are the left-hand sides of $n-1$ independent first integrals of the characteristic system (4.2.6), and $\phi(x)$ is any function which is functionally independent of $\psi_1(x), \ldots, \psi_{n-1}(x)$. Lemma 3.5.1 yields $X(\psi_1) = \cdots = X(\psi_{n-1}) = 0$, whereas $X(\phi) \neq 0$. According to Lemma 4.2.1, Eq. (4.2.4) takes the form

$$X(u) = X(\phi)\frac{\partial u}{\partial x'^n} = 0,$$

whence $\partial u/\partial x'^n = 0$. Therefore, the general solution is an arbitrary function of $x'^1, \ldots, x'^{n-1}$, i.e., $u = F(x'^1, \ldots, x'^{n-1})$. Using Eqs. (4.2.7), we obtain representation (4.2.5) of the general solution.

Example 4.2.1. Let us solve the following equation:

$$x\frac{\partial u}{\partial x} + y\frac{\partial u}{\partial y} = 0.$$

The characteristic system (4.2.6) has the form $dx/x = dy/y$ and provides the first integral $y/x = C$. Consequently, the general solution (4.2.5) is written $u = F(y/x)$.

Example 4.2.2. Consider the equation

$$y\frac{\partial u}{\partial x} - x\frac{\partial u}{\partial y} = 0.$$

The characteristic equation (4.2.6) has the form $dx/y = -dy/x$, or $xdx + ydy = 0$. Integration yields the first integral $x^2 + y^2 = C$. Hence, the general solution (4.2.5) is written $u = F(x^2 + y^2)$.

4.3 Particular solutions of non-homogeneous equations

Let us begin with the non-homogeneous linear equation (4.1.3) of the form

$$\xi^1(x)\frac{\partial u}{\partial x^1} + \cdots + \xi^n(x)\frac{\partial u}{\partial x^n} = f(x). \tag{4.3.1}$$

Using operator (4.2.1), Eq. (4.3.1) is written

$$X(u) = f(x). \tag{4.3.2}$$

Note that knowledge of a single solution $u = \varphi(x)$ of the non-homogeneous equation $X(u) = f(x)$ provides the general solution. Namely, the general solution u of Eq. (4.3.1) is given by

$$u = \varphi(x) + v, \tag{4.3.3}$$

where v is the general solution of the homogeneous equation, $X(v) = 0$. Indeed, let $X(\varphi(x)) = f(x)$. By setting $u = v + \varphi(x)$, one obtains

$$X(u) = X(v) + X(\varphi(x)) = X(v) + f(x).$$

It follows that $X(u) = f(x)$ if and only if $X(v) = 0$.

Thus, knowledge of a particular solution $\varphi(x)$ of (4.3.1) allows one to reduce the integration of a non-homogeneous linear partial differential equation (4.3.1) to integration of the homogeneous equation or, equivalently, to determination of $n-1$ independent first integrals of the characteristic system (4.2.6). In general, it is not a simple matter to find a solution $\varphi(x)$. However, one can easily arrive at a desired particular solution in special cases, e.g. in the following case.

Example 4.3.1. Let us solve Eq. (4.3.1) where one of the ξ^i's and the function f depend upon the single variable x^i, e.g.

$$\xi^1(x^1)\frac{\partial u}{\partial x^1} + \xi^2(x^1,\ldots,x^n)\frac{\partial u}{\partial x^2} + \cdots + \xi^n(x^1,\ldots,x^n)\frac{\partial u}{\partial x^n} = f(x^1). \tag{4.3.4}$$

One readily obtains a particular solution to Eq. (4.3.4) by letting $u = \varphi(x^1)$. Substitution into Eq. (4.3.4) yields the ordinary differential equation

$$\xi^1(x^1)\frac{d\varphi}{dx^1} = f(x^1),$$

whence the solution is obtained by quadrature:

$$\varphi(x^1) = \int \frac{f(x^1)}{\xi^1(x^1)}dx^1.$$

The general solution is provided now by formula (4.3.3).

Example 4.3.2. The equation with the independent variables x and y,

$$x^2\frac{\partial u}{\partial x} + xy\frac{\partial u}{\partial y} = 1,$$

has form (4.3.4) with $\xi^1 = x^2$ and $f = 1$. Consequently, one can look for a particular solution of the form $u = \varphi(x)$. Then the equation in question reduces to the ordinary differential equation $x^2 d\varphi/dx = 1$, whence one obtains (ignoring the additive constant of integration) the particular solution $\varphi = -1/x$. The associated system (4.2.6),

$$\frac{dx}{x^2} = \frac{dy}{xy},$$

has the first integral

$$\frac{y}{x} = C.$$

Consequently, the general solution (4.3.3) is written

$$u = -\frac{1}{x} + F\left(\frac{y}{x}\right).$$

Example 4.3.3. Consider the equation

$$y\frac{\partial u}{\partial x} - x\frac{\partial u}{\partial y} = y.$$

Upon dividing by y, it takes form (4.3.4) with $\xi^1 = 1$ and $f = 1$. Consequently, assuming $u = \varphi(x)$, one obtains from $d\varphi/dx = 1$ a particular solution $\varphi = x$. The general solution of the corresponding homogeneous equation (see Example 4.2.2) is given by $v = F(x^2 + y^2)$ and therefore the general solution of the non-homogeneous equation has the form:

$$u = x + F(x^2 + y^2).$$

Example 4.3.4. The general solution to the non-homogeneous equation

$$x^1\frac{\partial u}{\partial x^1} + \cdots + x^n\frac{\partial u}{\partial x^n} = 1$$

has the form

$$u = \ln|x^n| + F(x^1/x^n, x^2/x^n, \ldots, x^{n-1}/x^n).$$

4.4 Quasi-linear equations

Arbitrary non-homogeneous equations (4.1.3) can be solved by the general method for the quasi-linear equations discussed in this section. We will show that the general quasi-linear equation (4.1.5) with n independent variables,

$$\xi^1(x, u)\frac{\partial u}{\partial x^1} + \cdots + \xi^n(x, u)\frac{\partial u}{\partial x^n} = g(x, u), \tag{4.4.1}$$

in particular, an arbitrary non-homogeneous linear equation (4.1.3), can be reduced to a homogeneous linear equation with $n + 1$ variables as follows.

Let us define u as an implicit function of $x = (x^1, \ldots, x^n)$ by the equation

$$V(x^1, \ldots, x^n, u) = 0 \qquad (4.4.2)$$

and treat V as an unknown function of $n + 1$ variables, $x^1, \ldots, x^n$ and u. Differentiating Eq. (4.4.2) by means of the *total differentiation*

$$D_i = \frac{\partial}{\partial x^i} + p_i \frac{\partial}{\partial u} \qquad (4.4.3)$$

one obtains

$$D_i V \equiv \frac{\partial V}{\partial x^i} + p_i \frac{\partial V}{\partial u} = 0, \quad i = 1, \ldots, n,$$

whence

$$p_i = -\frac{\partial V/\partial x^i}{\partial V/\partial u}, \quad i = 1, \ldots, n. \qquad (4.4.4)$$

Inserting expressions (4.4.4) for p_i into Eq. (4.4.1) one obtains the homogeneous linear equation

$$\xi^1(x, u)\frac{\partial V}{\partial x^1} + \cdots + \xi^n(x, u)\frac{\partial V}{\partial x^n} + g(x, u)\frac{\partial V}{\partial u} = 0 \qquad (4.4.5)$$

for an unknown function V of $n + 1$ variables $x^1, \ldots, x^n$ and u. Now we apply Theorem 4.2.1 to the linear equation (4.4.5) and obtain the following.

Theorem 4.4.1. The general solution of the quasi-linear equation (4.4.1) is defined implicitly by the equation

$$V(x, u) = \Phi\left(\psi_1(x, u), \ldots, \psi_n(x, u)\right), \qquad (4.4.6)$$

with an arbitrary function Φ of n variables such that $\partial V/\partial u \neq 0$. Here,

$$\psi_1(x, u) = C_1, \quad \ldots, \quad \psi_n(x, u) = C_n$$

are independent first integrals of the system of equations

$$\frac{dx^1}{\xi^1(x, u)} = \frac{dx^2}{\xi^2(x, u)} = \cdots = \frac{dx^n}{\xi^n(x, u)} = \frac{du}{g(x, u)} \qquad (4.4.7)$$

called the *characteristic system for the quasi-linear equation* (4.4.1).

Example 4.4.1. Let us apply the method of this section to equation

$$y\frac{\partial u}{\partial x} - x\frac{\partial u}{\partial y} = 1. \qquad (4.4.8)$$

Here $g(x, y, u) = 1$ and hence the characteristic system (4.4.7) is written

$$\frac{dx}{y} = -\frac{dy}{x} = \frac{du}{1}.$$

We have to find two independent first integrals of this system. The first equation, $x dx + y dy = 0$, yields $x^2 + y^2 = a^2 = $ const. By virtue of this relation, the second equation is rewritten

$$du + \frac{dy}{\sqrt{a^2 - y^2}} = 0,$$

whence, upon integration, $u + \arcsin(y/a) = C$. Using formula (3.5.11), we have $u + \arctan(y/x) = C$. Hence, the two independent first integrals have the form

$$\psi_1 \equiv x^2 + y^2 = C_1, \quad \psi_2 \equiv u + \arctan(y/x) = C_2.$$

Therefore, the general solution of the corresponding equation (4.4.5),

$$y \frac{\partial V}{\partial x} - x \frac{\partial V}{\partial y} + \frac{\partial V}{\partial u} = 0,$$

is given by formula (4.4.6):

$$V = \Phi(\psi_1, \psi_2) \equiv \Phi\left(x^2 + y^2, u + \arctan(y/x)\right).$$

Hence, Eq. (4.4.2) is written

$$\Phi\left(x^2 + y^2, u + \arctan(y/x)\right) = 0.$$

If $\partial \Phi / \partial \psi_2 \neq 0$, one can solve the latter equation with respect to u and obtain the solution in the explicit form

$$u = -\arctan(y/x) + F\left(x^2 + y^2\right). \tag{4.4.9}$$

Remark 4.4.1. In polar coordinates defined by

$$x = r \cos\theta, \quad y = r \sin\theta, \tag{4.4.10}$$

or

$$r = \sqrt{x^2 + y^2}, \quad \theta = \arctan(y/x),$$

solution (4.4.9) is written $u = -\theta + f(r)$, thus suggesting to use the polar coordinates. We have

$$\frac{\partial u}{\partial x} = \frac{\partial r}{\partial x} \frac{\partial u}{\partial r} + \frac{\partial \theta}{\partial x} \frac{\partial u}{\partial \theta}, \quad \frac{\partial u}{\partial y} = \frac{\partial r}{\partial y} \frac{\partial u}{\partial r} + \frac{\partial \theta}{\partial y} \frac{\partial u}{\partial \theta},$$

whence, substituting

$$\frac{\partial r}{\partial x} = \frac{x}{r}, \quad \frac{\partial r}{\partial y} = \frac{y}{r}, \quad \frac{\partial \theta}{\partial x} = -\frac{y}{r^2}, \quad \frac{\partial \theta}{\partial y} = \frac{x}{r^2},$$

we obtain

$$\frac{\partial u}{\partial x} = \frac{x}{r} \frac{\partial u}{\partial r} - \frac{y}{r^2} \frac{\partial u}{\partial \theta}, \quad \frac{\partial u}{\partial y} = \frac{y}{r} \frac{\partial u}{\partial r} + \frac{x}{r^2} \frac{\partial u}{\partial \theta}.$$

Thus, Eq. (4.4.8) reduces to $\partial u / \partial \theta = -1$ and yields $u = -\theta + f(r)$.

Example 4.4.2. Consider the transfer equation known also as the Hopf equation $u_t + uu_x = 0$. The characteristic system (4.4.7) can be written formally as

$$\frac{dt}{1} = \frac{dx}{u} = \frac{du}{0},$$

where the last term simply means that the system has the first integral $u = C_1$. By virtue of this first integral, the characteristic system reduces to $dx - C_1 dt = 0$, whence $x - C_1 t = C_2$. We have two first integrals:

$$u = C_1 \quad \text{and} \quad x - tu = C_2.$$

Therefore, $V = \Phi(u, x - tu)$, and the solution to the Hopf equation is given implicitly by (4.4.2):

$$\Phi(u, x - tu) = 0, \quad \text{or} \quad u = F(x - tu).$$

4.5   Systems of homogeneous equations

When several equations of form (4.1.2) for one dependent variable u are given instead of a single one, they furnish a *system* (known also as a *simultaneous system*) of linear partial differential equations of the first order. Since we have several equations for one dependent variable, we deal here with what is called *over-determined systems*.

We will consider here systems of homogeneous equations when the equations composing the simultaneous system have form (4.1.4). After introducing r differential operators of form (4.2.1),

$$X_\alpha = \xi_\alpha^1(x)\frac{\partial}{\partial x^1} + \cdots + \xi_\alpha^n(x)\frac{\partial}{\partial x^n}, \quad \alpha = 1, \ldots, r, \qquad (4.5.1)$$

a system of r homogeneous linear equations is written in the compact form

$$X_1(u) = 0, \quad \ldots, \quad X_r(u) = 0. \qquad (4.5.2)$$

Equations (4.5.2) have a *trivial solution* $u = $ const., which is of no interest to us. Furthermore, it is apparent that any equation of the system can be multiplied by a function of x. Therefore, if a function $u = u(x)$ solves s equations

$$X_\alpha(u) = 0, \quad \alpha = 1, \ldots, s \leq r,$$

it also satisfies their linear combination with any variable coefficients $\lambda^\alpha(x)$:

$$\sum_{\alpha=1}^{s} \lambda^\alpha(x) X_\alpha(u) = 0.$$

This motivates the following definitions.

Definition 4.5.1. Differential operators $X_1, \ldots, X_s$ are said to be *connected* if there exist functions $\lambda^\alpha(x)$, not all zero, such that

$$\lambda^1(x)X_1 + \cdots + \lambda^s(x)X_s = 0, \tag{4.5.3}$$

this being satisfied as an operator identity in a neighborhood of a generic x. If relation (4.5.3) implies $\lambda^1 = \cdots = \lambda^s = 0$, we say that the operators $X_1, \ldots, X_s$ are *unconnected*. In the latter case, the corresponding differential equations $X_1(u) = 0, \ldots, X_s(u) = 0$ are said to be *independent*.

Let Z_α be linear combinations of operators (4.5.1):

$$Z_\alpha = \sum_{\beta=1}^{r} h_\alpha^\beta(x)X_\beta, \quad \alpha = 1, \ldots, r,$$

with variable coefficients $h_\alpha^\beta(x)$ whose determinant $\left| h_\alpha^\beta(x) \right|$ is not zero. The system of linear homogeneous equations

$$Z_1(u) = 0, \ldots, Z_r(u) = 0 \tag{4.5.4}$$

has the same set of solutions as the original system (4.5.2).

Definition 4.5.2. Systems (4.5.2) and (4.5.4), as well as the corresponding operators X_α and Z_α, are said to be *equivalent*.

Lemma 4.5.1. The number r_* of unconnected operators among operators (4.5.1) is equal to the rank of the $r \times n$ matrix of their coefficients:

$$r_* = \text{rank}(\xi_\alpha^i(x)), \tag{4.5.5}$$

where α and i denote rows and columns, respectively. The number r_* is the same for equivalent operators Z_α.

According to Lemma 4.5.1, any system of r homogeneous linear equations can be replaced by a system of r_* independent equations. It is clear that more than n equations cannot be independent. Furthermore, if $r = n$, and if operators (4.5.1) are unconnected (i.e., $r_* = r = n$), then the determinant of the coefficients $\xi_\alpha^i(x)$ is not zero. In this case the solution of system (4.5.2) is trivial, $u = \text{const}$. Thus, a necessary condition for the existence of non-trivial solutions is $r_* < n$. The condition $r_* < n$ alone is *not sufficient*, however for the existence of non-trivial solutions.

Example 4.5.1. The system

$$X_1(u) \equiv z\frac{\partial u}{\partial y} - y\frac{\partial u}{\partial z} = 0, \quad X_2(u) \equiv y\frac{\partial u}{\partial x} + z\frac{\partial u}{\partial y} = 0$$

is composed of two independent equations since the operators X_1 and X_2 involve differentiations in different variables. Integration of the first equation of the system yields $u = v(x, \rho)$, where $\rho = \sqrt{y^2 + z^2}$. Upon substituting this expression into the second equation, one has

$$\rho \frac{\partial v}{\partial x} + z \frac{\partial v}{\partial \rho} = 0.$$

Since v does not involve the variable z *explicitly*, it follows that

$$\frac{\partial v}{\partial \rho} = 0, \quad \text{and hence} \quad \frac{\partial v}{\partial x} = 0.$$

Consequently, $u = v = $ const. Thus, the system in question does not have a non-trivial solution even though $r_* = r = 2$ is less than the number $n = 3$ of the independent variables x, y, z.

To discern the true nature of the situation, one needs the following notion of *complete systems*. Note that if $u = u(x)$ solves system (4.5.2), then it solves also the equations $X_\alpha(X_\beta(u)) = 0$ for any values of the indices α and β. Hence, u solves the following first-order equations:

$$X_\alpha(X_\beta(u)) - X_\beta(X_\alpha(u)) \equiv \sum_{i=1}^{n} \left(X_\alpha(\xi_\beta^i) - X_\beta(\xi_\alpha^i) \right) \frac{\partial u}{\partial x^i} = 0.$$

In other words, u annuls, together with operators (4.5.1), all their *commutators* defined as follows.

Definition 4.5.3. The commutator of any two operators X_α and X_β of form (4.5.1) is the first-order differential operator $[X_\alpha, X_\beta]$ defined by

$$[X_\alpha, X_\beta] = X_\alpha X_\beta - X_\beta X_\alpha,$$

or in the following equivalent form exhibiting the coefficients explicitly:

$$[X_\alpha, X_\beta] = \sum_{i=1}^{n} \left(X_\alpha(\xi_\beta^i) - X_\beta(\xi_\alpha^i) \right) \frac{\partial}{\partial x^i}. \tag{4.5.6}$$

Thus, any solution of Eqs. (4.5.2) solves the equations $[X_\alpha, X_\beta](u) = 0$ as well. In consequence, one has the following alternatives: either *some of commutators* (4.5.6) *are independent of the original operators* (4.5.1), or *commutators* (4.5.6) *are linear combinations with variable coefficients of operators* (4.5.1). The latter case means that the combined set of operators (4.5.1) and (4.5.6) is connected.

In the first case, one should consider an extended system of differential equations of the first order obtained by combining (4.5.1) with all independent commutators. Then one can apply the above operations to this new system. Proceeding in this manner, one ultimately reaches the second case and hence arrives at what is called a complete system.

Definition 4.5.4. Let (4.5.2) be a system of independent equations. It is called a *complete system* if all commutators (4.5.6) are *dependent* on operators (4.5.1):

$$[X_\alpha, X_\beta] = \sum_{\gamma=1}^{r} h_{\alpha\beta}^\gamma(x) X_\gamma. \tag{4.5.7}$$

If $h_{\alpha\beta}^\gamma(x) = 0$, i.e., if all commutators of operators (4.5.1) vanish, we have a particular case of a complete system known as a *Jacobian system*.

The system of equations $X_1(u) = 0, X_2(u) = 0$ in Example 4.5.1 is not complete. Consequently, the process of solution gave rise to a new equation, and the corresponding complete system was self-generated.

If system (4.5.2) is complete, then any equivalent system (4.5.4) is also complete. Furthermore, any complete system is equivalent to a Jacobian system. To illustrate the integration procedure for complete systems, consider one more example.

Example 4.5.2. Consider the system of equations $X_1(u) = 0, X_2(u) = 0$ with the operators

$$X_1 = z\frac{\partial}{\partial y} - y\frac{\partial}{\partial z}, \quad X_2 = \frac{\partial}{\partial x} + t\frac{\partial}{\partial y} + y\frac{\partial}{\partial t}.$$

The commutator of these operators has the form $[X_1, X_2] = X_3$, where

$$X_3 = t\frac{\partial}{\partial z} + z\frac{\partial}{\partial t}.$$

The three equations $X_1(u) = 0, X_2(u) = 0$, and $X_3(u) = 0$ form a complete system since

$$[X_1, X_2] = X_3, \quad [X_1, X_3] = -\left(\frac{t}{z}X_1 + \frac{y}{z}X_3\right), \quad [X_2, X_3] = -X_1.$$

The equation $X_1(u) = 0$ yields $u = v(x, t, \rho)$, where $\rho = \sqrt{y^2 + z^2}$. Then $X_3(u) = 0$ reduces to

$$t\frac{\partial v}{\partial \rho} + \rho\frac{\partial v}{\partial t} = 0,$$

whence $v = w(x, \lambda)$, where $\lambda = \rho^2 - t^2 = y^2 + z^2 - t^2$. Now the last equation, $X_2(u) = 0$, reduces to $\partial w/\partial x = 0$. Thus,

$$u = \phi(y^2 + z^2 - t^2).$$

Problems to Chapter 4

4.1. Find the first integrals and the general solution of the system

$$\frac{dx}{dt} = x^2, \quad \frac{dy}{dt} = xy.$$

4.2. Find a first integral $\psi(x, y) = C$ for

(i) the equation $\dfrac{dx}{2y} = \dfrac{dy}{3x^2}$,

(ii) the system of the Lotka–Volterra equations (2.2.4).

4.3. Solve the homogeneous linear equations

(i) $x^1 \dfrac{\partial u}{\partial x^1} + \cdots + x^n \dfrac{\partial u}{\partial x^n} = 0;$ (ii) $y \dfrac{\partial u}{\partial x} - x \dfrac{\partial u}{\partial y} = 0;$

(iii) $y \dfrac{\partial u}{\partial x} + x \dfrac{\partial u}{\partial y} = 0;$ (iv) $2y \dfrac{\partial u}{\partial x} + 3x^2 \dfrac{\partial u}{\partial y} = 0.$

4.4. Solve the non-homogeneous linear equation

$$x \frac{\partial u}{\partial x} + 2y \frac{\partial u}{\partial y} - 2z \frac{\partial u}{\partial z} = 1.$$

4.5. Solve the equations

(i) $y \dfrac{\partial u}{\partial x} - x \dfrac{\partial u}{\partial y} = x,$ (ii) $y \dfrac{\partial u}{\partial x} - x \dfrac{\partial u}{\partial y} = yg(x),$ (iii) $y \dfrac{\partial u}{\partial x} - x \dfrac{\partial u}{\partial y} = xh(y),$

where $g(x)$ and $h(y)$ are arbitrary functions.

4.6. Show that $u + \arctan(y/x) = C$ is a first integral of the system

$$\frac{dx}{y} = -\frac{dy}{x} = du.$$

4.7. Solve the equation

$$y \frac{\partial u}{\partial x} - x \frac{\partial u}{\partial y} = x^2.$$

4.8. Solve the following linear equation:

$$x^1 \frac{\partial u}{\partial x^1} + \cdots + x^n \frac{\partial u}{\partial x^n} = \sigma u, \quad \sigma = \text{const.} \neq 0.$$

4.9. Investigate the completeness for the following system:

$$X_1(u) \equiv z \frac{\partial u}{\partial y} - y \frac{\partial u}{\partial z} = 0, \quad X_2(u) \equiv y \frac{\partial u}{\partial x} + z \frac{\partial u}{\partial y} = 0.$$

4.10. Solve the following system of three equations with three independent variables x, y, z :

$$X_1(u) \equiv z \frac{\partial u}{\partial y} - y \frac{\partial u}{\partial z} = 0,$$

$$X_2(u) \equiv x\frac{\partial u}{\partial z} - z\frac{\partial u}{\partial x} = 0,$$

$$X_3(u) \equiv y\frac{\partial u}{\partial x} - x\frac{\partial u}{\partial y} = 0.$$

4.11. Consider the following system of two linear equations with four independent variables, t, x, y, and z :

$$t\frac{\partial u}{\partial t} - \frac{\partial u}{\partial z} = 0, \quad \sin x\frac{\partial u}{\partial x} - \cos x\frac{\partial u}{\partial y} + 2\cos x\frac{\partial u}{\partial z} = 0.$$

Is this system complete? Solve the system.

4.12. Check that $F(u, x - tu) = 0$ defines implicitly the solution to the equation $u_t + uu_x = 0$.

4.13. Solve the system

$$\frac{\partial u}{\partial x} + y\frac{\partial u}{\partial z} = 0, \quad \frac{\partial u}{\partial y} + x\frac{\partial u}{\partial z} = 0.$$

Chapter 5

Linear partial differential equations of the second order

This chapter contains mainly the second-order partial differential equations in two independent variables. The emphasis is on the classification and methods of integration. Before beginning this chapter, it's recommended to read Section 1.1.4.

Additional reading: R. Courant and D. Hilbert [4], A. Sommerfeld [38], G.F.D. Duff [6], S.L. Sobolev [37], J. Hadamard [12], A.N. Tikhonov and A.A. Samarskii [39], I.G. Petrovsky [33].

5.1 Equations with several variables

5.1.1 Classification at a fixed point

The general linear second-order partial differential equation (PDE) with one dependent variable u and n independent variables $x = (x^1, \ldots, x^n)$ is

$$a^{ij}(x)u_{ij} + b^i(x)u_i + c(x)u = f(x), \qquad (5.1.1)$$

where the usual notation $u_i = \partial u/\partial x^i$, $u_{ij} = \partial^2 u/\partial x^i \partial x^j$ is used for the partial derivatives. According to the summation convention (Section 1.2.3), the summation is assumed over $i, j = 1, \ldots, n$. The coefficients $a^{ij}(x)$ are assumed to be symmetric, i.e., $a^{ij}(x) = a^{ji}(x)$.

Equation (5.1.1) is homogeneous by function if and only if $f(x) = 0$ (see Section 3.1.4). Consequently, Eq. (5.1.1) is called *homogeneous* if $f(x) = 0$, and *non-homogeneous* otherwise (cf. Section 3.2.6). Thus, the homogeneous linear second-order partial differential equation has the form

$$a^{ij}(x)u_{ij} + b^i(x)u_i + c(x)u = 0. \qquad (5.1.2)$$

Let us write the left-hand side of Eq. (5.1.1) in the form

$$L[u] = a^{ij}(x)u_{ij} + b^i(x)u_i + c(x)u, \tag{5.1.3}$$

where L is the linear second-order differential operator defined by

$$L = a^{ij}(x)D_iD_j + b^i(x)D_i + c(x). \tag{5.1.4}$$

We will simplify the *principal part* of L, i.e., the terms with second-order derivatives, by means of a change of variables $\bar{x}^i = \bar{x}^i(x)$. The differentiations D_i and $\overline{D}_k$ with respect to x and $\bar{x}$, respectively, are related by

$$D_i = \frac{\partial \bar{x}^k}{\partial x^i}\overline{D}_k.$$

Hence,

$$L = \bar{a}^{kl}\overline{D}_k\overline{D}_l + \cdots,$$

where

$$\bar{a}^{kl} = \frac{\partial \bar{x}^k}{\partial x^i}\frac{\partial \bar{x}^l}{\partial x^j}a^{ij}. \tag{5.1.5}$$

Let us fix on a definite point $x_0 = (x_0^1, \ldots, x_0^n)$ and introduce, at this point, the quadratic form in $\mu = (\mu_1, \ldots, \mu_n)$:

$$K(\mu) = a_0^{ij}\mu_i\mu_j, \tag{5.1.6}$$

with constant coefficients $a_0^{ij} = a^{ij}(x_0)$. Consider a linear transformation

$$\mu_i = \alpha_i^k \bar{\mu}_k, \quad i = 1, \ldots, n. \tag{5.1.7}$$

We assume that it is invertible, i.e., the determinant $|\alpha_i^k|$ does not vanish. Transformation (5.1.7) changes the quadratic form (5.1.6) as follows:

$$K(\bar{\mu}) = a_0^{ij}\alpha_i^k\alpha_j^l\bar{\mu}_k\bar{\mu}_l. \tag{5.1.8}$$

It is known from linear algebra that there exists a transformation (5.1.7) such that the quadratic form (5.1.6) becomes a sum of squares, i.e.,

$$K(\bar{\mu}) = \sum_{i=1}^n \varepsilon_i\bar{\mu}_i^2, \tag{5.1.9}$$

where the ε_i are either ± 1 or 0. Now we compare (5.1.5) and (5.1.8) and set

$$\frac{\partial \bar{x}^k}{\partial x^i} = \alpha_i^k.$$

Integration of these equations furnishes us with the linear change of variables

$$\bar{x}^k = \alpha_i^k x^i, \quad k = 1, \ldots, n, \tag{5.1.10}$$

leading to the following statement.

Theorem 5.1.1. At any fixed point x_0, Eq. (5.1.2) can be mapped by the linear change of variables (5.1.10) to the following form:

$$\sum_{i=1}^{n} \varepsilon_i \frac{\partial^2 u}{\partial (\bar{x}^i)^2} + \cdots = 0. \tag{5.1.11}$$

Equation (5.1.11) is called a *canonical* or *standard* form of Eq. (5.1.2).

If all of the ε_i have the same sign, e.g. $+1$ (if all of them are -1 we multiply Eq. (5.1.2) by -1), we say that Eq. (5.1.2) has the *elliptic type*.

If none of the ε_i is zero and some of the ε_i are $+1$, some -1, we have Eq. (5.1.2) of the *hyperbolic type*. Of these, equations of the *normal hyperbolic type* are most frequent in applications. They are defined by the conditions that one only of the ε_i is positive (or one only negative).

If some of the ε_i are 0, (5.1.2) is an equation of the *parabolic type*.

The above nomenclature applies to the non-homogeneous equation (5.1.1) as well.

Equations with variable coefficients a^{ij} may have different types at different points x. Moreover, it is impossible, in general, to find a change of variables, defined not only at a fixed point but also in a certain domain, such that it maps Eq. (5.1.2) to a canonical form in the whole domain. This is possible only if Eq. (5.1.2) with several variables has constant coefficients a^{ij} or if there exists some coordinate system in which the a^{ij} are all constants. The only exception is provided by Eq. (5.1.2) with *two independent variables* (see Section 5.2).

5.1.2   Adjoint linear differential operators

The concept of the adjoint operator defined below plays an important part in the theory and applications of linear differential equations.

Definition 5.1.1. Let L be a linear differential operator of any order. A linear differential operator L^* is called an *adjoint operator* to L if

$$vL[u] - uL^*[v] = D_i(p^i) \equiv \mathrm{div} P \tag{5.1.12}$$

for all functions u and v, where $P = (p^1, \ldots, p^n)$ is a vector field with components $p^i(x)$. Equation $L^*[v] = 0$ is termed the *adjoint equation* to Eq. $L[u] = 0$.

One can prove that the adjoint operator L^* is uniquely determined by Eq. (5.1.12). We will demonstrate this statement in the case of the second-order operator (5.1.4):

$$L = a^{ij}(x)D_iD_j + b^i(x)D_i + c(x).$$

Theorem 5.1.2. The adjoint operator L^* to L is uniquely determined and has the form

$$L^*[v] = D_iD_j(a^{ij}v) - D_i(b^iv) + cv. \tag{5.1.13}$$

Proof. The crucial idea is to consider the expression $vL[u]$ and to transfer differentiation from u to v. We have

$$vL[u] = va^{ij}D_iD_ju + vb^iD_iu + cuv =$$
$$D_i(va^{ij}D_ju) - D_i(va^{ij})D_ju + D_i(vb^iu) - uD_i(vb^i) + ucv.$$

Furthermore, we write the term $-D_i(va^{ij})D_ju$ in the form

$$-D_i(va^{ij})D_ju = -D_j(uD_i(va^{ij})) + uD_iD_j(a^{ij}v),$$

whence, interchanging i and j in the first term of the right-hand side and invoking that $a^{ij} = a^{ji}$, we obtain

$$-D_i(va^{ij})D_ju = -D_i(uD_j(va^{ij})) + uD_iD_j(a^{ij}v).$$

Finally, invoking that $D_ju = u_j$ we arrive at the equation

$$vL[u] = u\{D_iD_j(a^{ij}v) - D_i(b^iv) + cv\} + D_i\{a^{ij}vu_j + b^iuv - uD_j(a^{ij}v)\}.$$

It follows that the adjoint operator is defined by (5.1.13) and satisfies Eq. (5.1.12) with

$$p^i = a^{ij}vu_j + b^iuv - uD_j(a^{ij}v). \tag{5.1.14}$$

Remark 5.1.1. The definition of the adjoint operator is the same for systems of differential equations, e.g. in the case of second-order equations when the function u in Eq. (5.1.3) is an m-dimensional vector and the coefficients $a^{ij}(x), b^i(x)$ and $c(x)$ of operator (5.1.4) are $m \times m$ matrices.

Definition 5.1.2. An operator L is said to be *self-adjoint* if

$$L[u] = L^*[u] \tag{5.1.15}$$

for any function $u(x)$. Then Eq. $L[u] = 0$ is also termed self-adjoint.

Theorem 5.1.3. Operator (5.1.4), $L = a^{ij}(x)D_iD_j + b^i(x)D_i + c(x)$, is self-adjoint if and only if

$$b^i(x) = D_j(a^{ij}(x)), \quad i = 1,\ldots,n. \tag{5.1.16}$$

Proof. Expression (5.1.13) of the adjoint operator L^* can be written in the form

$$L^*[v] = a^{ij}v_{ij} + (2D_j(a^{ij}) - b^i)v_i + (c - D_i(b^i) + D_iD_j(a^{ij}))v. \tag{5.1.17}$$

Substituting (5.1.17) in Eq. (5.1.15) and using the symmetry $a^{ji} = a^{ij}$ one obtains

$$2D_j(a^{ij}) - b^i = b^i, \quad c - D_i(b^i) + D_iD_j(a^{ij}) = c. \tag{5.1.18}$$

The first equation of (5.1.18) yields (5.1.16) whereas the second equation of (5.1.18) is a consequence of (5.1.16). Note, that the proof is the same when (5.1.3) is a system of second-order linear equations.

Remark 5.1.2. It is stated in [4] (Appendix 1 to Chapter III, §2.2) that a self-adjoint operator must obey, along with (5.1.16), one more condition:

$$D_i(b^i) = 0. \tag{5.1.19}$$

However, condition (5.1.19) is needless. Indeed, let us consider, e.g. the following scalar equation in two independent variables x, y:

$$L[u] \equiv x^2 u_{xx} + y^2 u_{yy} + 2x u_x + 2y u_y = 0.$$

The operator L is self-adjoint (see Problem 5.7) but does not meet condition (5.1.19) since $D_i(b^i) = D_x(b^1) + D_y(b^2) = 4$.

A simple example of a self-adjoint system that does not meet condition (5.1.19) is

$$x^2 u_{xx} + u_{yy} + 2x u_x + w = 0, \quad w_{xx} + y^2 w_{yy} + 2y w_y + u = 0.$$

The coefficients of this system with two independent variables x, y and two dependent variables u, w satisfy equations (5.1.16) but do not meet condition (5.1.19). Indeed,

$$a^{11} = \begin{vmatrix} x^2 & 0 \\ 0 & 1 \end{vmatrix}, \quad a^{22} = \begin{vmatrix} 1 & 0 \\ 0 & y^2 \end{vmatrix}, \quad a^{12} = a^{21} = 0, \quad c = \begin{vmatrix} 0 & 1 \\ 1 & 0 \end{vmatrix},$$

$$b^1 = \begin{vmatrix} 2x & 0 \\ 0 & 0 \end{vmatrix}, \quad b^2 = \begin{vmatrix} 0 & 0 \\ 0 & 2y \end{vmatrix}, \quad D_x(b^1) + D_y(b^2) = \begin{vmatrix} 2 & 0 \\ 0 & 2 \end{vmatrix} \neq 0.$$

5.2 Classification of equations in two independent variables

5.2.1 Characteristics. Three types of equations

The general form of the homogeneous linear second-order partial differential equations with two independent variables, x and y, is

$$A u_{xx} + 2B u_{xy} + C u_{yy} + a u_x + b u_y + c u = 0, \tag{5.2.1}$$

where $A = A(x, y), \ldots, c = c(x, y)$ are prescribed functions. The terms with the second derivatives,

$$A u_{xx} + 2B u_{xy} + C u_{yy}, \tag{5.2.2}$$

compose the *principal part* of Eq. (5.2.1).

The crucial step in studying Eq. (5.2.1) is the reduction of its principal part (5.2.2) to so-called *standard forms* by a change of variables

$$\xi = \varphi(x, y), \quad \eta = \psi(x, y). \tag{5.2.3}$$

Let us obtain the standard forms of the principal parts for all equations of form (5.2.1). The change of variables (5.2.3) leads to the following transformation of the derivatives (see Section 1.4.5):

$$u_x = \varphi_x u_\xi + \psi_x u_\eta, \qquad u_y = \varphi_y u_\xi + \psi_y u_\eta,$$

$$u_{xx} = \varphi_x^2 u_{\xi\xi} + 2\varphi_x\psi_x u_{\xi\eta} + \psi_x^2 u_{\eta\eta} + \varphi_{xx} u_\xi + \psi_{xx} u_\eta, \qquad (5.2.4)$$

$$u_{yy} = \varphi_y^2 u_{\xi\xi} + 2\varphi_y\psi_y u_{\xi\eta} + \psi_y^2 u_{\eta\eta} + \varphi_{yy} u_\xi + \psi_{yy} u_\eta,$$

$$u_{xy} = \varphi_x\varphi_y u_{\xi\xi} + (\varphi_x\psi_y + \varphi_y\psi_x)u_{\xi\eta} + \psi_x\psi_y u_{\eta\eta} + \varphi_{xy} u_\xi + \psi_{xy} u_\eta.$$

Substituting expressions (5.2.4) in (5.2.2) and keeping only the terms with the second-order derivatives $u_{\xi\xi}, u_{\xi\eta}, u_{\eta\eta}$, we obtain the following principal part of Eq. (5.2.1) in the new variables:

$$\widetilde{A}\, u_{\xi\xi} + 2\widetilde{B}\, u_{\xi\eta} + \widetilde{C}\, u_{\eta\eta}, \qquad (5.2.5)$$

where

$$\widetilde{A} = A\varphi_x^2 + 2B\varphi_x\varphi_y + C\varphi_y^2,$$

$$\widetilde{B} = A\varphi_x\psi_x + B(\varphi_x\psi_y + \varphi_y\psi_x) + C\varphi_y\psi_y, \qquad (5.2.6)$$

$$\widetilde{C} = A\psi_x^2 + 2B\psi_x\psi_y + C\psi_y^2.$$

It is manifest from (5.2.6) that the principal part (5.2.5) will have only one term, $2\widetilde{B}$, if we choose for $\varphi(x,y)$ and $\psi(x,y)$ two solutions of the equation

$$A\omega_x^2 + 2B\omega_x\omega_y + C\omega_y^2 = 0$$

provided that the latter has two functionally independent solutions, $\omega_1 = \varphi(x,y)$ and $\omega_2 = \psi(x,y)$. However, the equation under consideration may have only one solution or even no solutions at all since it is nonlinear. Therefore, we will dwell upon this problem.

Definition 5.2.1. The first-order nonlinear partial differential equation

$$A\omega_x^2 + 2B\omega_x\omega_y + C\omega_y^2 = 0 \qquad (5.2.7)$$

is called the *characteristic equation* for Eq. (5.2.1). If $\omega(x,y)$ is a solution of Eq. (5.2.1), the curves

$$\omega(x,y) = \text{const.} \qquad (5.2.8)$$

are referred to as *characteristics* curves of Eq. (5.2.1).

The characteristics are important for integrating and/or understanding the behaviour of the solutions of Eq. (5.2.1). In order to find the characteristics, we set

$$\frac{\omega_x}{\omega_y} = \lambda \qquad (5.2.9)$$

and rewrite the characteristic equation (5.2.7) in the form

$$A(x,y)\lambda^2 + 2B(x,y)\lambda + C(x,y) = 0. \tag{5.2.10}$$

Equations (5.2.1) are classified into three types in accordance with the number of their characteristics, i.e., with the number of real roots to the quadratic equation (5.2.10).

Definition 5.2.2. Equation (5.2.1) is said to be *hyperbolic* if the quadratic equation (5.2.10) has two distinct real roots, $\lambda_1(x,y)$ and $\lambda_2(x,y)$, i.e., if

$$B^2 - AC > 0, \tag{5.2.11}$$

parabolic if (5.2.10) has repeated roots, $\lambda_1(x,y) = \lambda_2(x,y)$, i.e., if

$$B^2 - AC = 0, \tag{5.2.12}$$

and *elliptic* if the roots $\lambda_1(x,y)$, $\lambda_2(x,y)$ are complex, i.e., if

$$B^2 - AC < 0. \tag{5.2.13}$$

5.2.2   The standard form of the hyperbolic equations

Consider the hyperbolic type. Equation (5.2.10) has two distinct real roots

$$\lambda_1(x,y) = \frac{-B + \sqrt{B^2 - AC}}{A}, \quad \lambda_2(x,y) = \frac{-B - \sqrt{B^2 - AC}}{A}. \tag{5.2.14}$$

Substituting them into (5.2.9), we see that the characteristic equation (5.2.7) splits into two different linear first-order partial differential equations:

$$\frac{\partial \omega}{\partial x} - \lambda_1 \frac{\partial \omega}{\partial y} = 0, \quad \frac{\partial \omega}{\partial x} - \lambda_2 \frac{\partial \omega}{\partial y} = 0. \tag{5.2.15}$$

The characteristic systems (4.2.6) for Eqs. (5.2.15) are

$$\frac{dx}{1} + \frac{dy}{\lambda_1(x,y)} = 0, \quad \frac{dx}{1} + \frac{dy}{\lambda_2(x,y)} = 0. \tag{5.2.16}$$

Each equation of (5.2.16) has one independent first integral, $\varphi(x,y) = $ const. and $\psi(x,y) = $ const. for the first and the second equation of (5.2.16), respectively. Accordingly, the functions $\varphi(x,y)$ and $\psi(x,y)$ satisfy the first and the second equation of (5.2.15), respectively:

$$\frac{\partial \varphi}{\partial x} - \lambda_1 \frac{\partial \varphi}{\partial y} = 0, \quad \frac{\partial \psi}{\partial x} - \lambda_2 \frac{\partial \psi}{\partial y} = 0, \tag{5.2.17}$$

and hence they are functionally independent. Thus, they provide two functionally independent solutions of the characteristic equation (5.2.10) and therefore one can take them as the right-hand sides in the change of variables (5.2.3) thus reducing the principal part (5.2.5) to one term, $2\widetilde{B}u_{\xi\eta}$. The new variables ξ and η are termed the *characteristic variables*. Finally, we divide by $2\widetilde{B}$ the equation obtained from (5.2.1) after the change of variables and arrive at what is called the *standard form* of the hyperbolic equations. We summarize (cf. Eq. (1.1.56) in Theorem 1.1.3):

Theorem 5.2.1. The hyperbolic equations (5.2.1) are written in characteristic variables in the following *standard form*

$$u_{\xi\eta} + \tilde{a}(\xi,\eta)u_\xi + \tilde{b}(\xi,\eta)u_\eta + \tilde{c}(\xi,\eta)u = 0. \qquad (5.2.18)$$

Exercise 5.2.1. Prove that the nontrivial (i.e., not identically constant) solutions $\varphi(x,y)$ and $\psi(x,y)$ of the first and second equation of (5.2.17), respectively, are functionally independent.

Example 5.2.1. A typical representative of the hyperbolic equations is the wave equation (2.6.5)

$$u_{tt} - k^2 u_{xx} = 0.$$

Setting $t = y$, we have $A = -k^2$, $B = 0$, $C = 1$, hence $B^2 - AC = k^2 > 0$. We will continue the discussion of this example in Section 5.3.1.

5.2.3   The standard form of the parabolic equations

For the parabolic type, Eq. (5.2.10) has the repeated real root,

$$\lambda = -\frac{B}{A},$$

and the two equations (5.2.17) collapse into one equation

$$A\frac{\partial\varphi}{\partial x} + B\frac{\partial\varphi}{\partial y} = 0. \qquad (5.2.19)$$

Now we take the change of variables (5.2.3) in the form

$$\xi = \varphi(x,y), \quad \eta = x, \qquad (5.2.20)$$

where $\varphi(x,y)$ is a solution of Eq. (5.2.19). One can readily verify that Eqs. (5.2.6) yield $\widetilde{A} = 0$, $\widetilde{B} = 0$. Dividing by $\widetilde{C} \neq 0$, we obtain the following *standard form* of parabolic equations (cf. Eq. (1.1.57) in Theorem 1.1.3).

Theorem 5.2.2. The parabolic equations (5.2.1) are written in the variables (5.2.20) in the following *standard form*:

$$u_{\eta\eta} + \tilde{a}(\xi,\eta)u_\xi + \tilde{b}(\xi,\eta)u_\eta + \tilde{c}(\xi,\eta)u = 0. \qquad (5.2.21)$$

We assumed above that $A \neq 0$. But if $A = 0$, then condition (5.2.12), $B^2 - AC = 0$, yields that $B = 0$, and hence Eq. (5.2.1) is already in the standard form (5.2.21).

Example 5.2.2. A typical representative of the parabolic equations is the heat equation (2.4.7)

$$u_t - a^2 u_{xx} = 0.$$

Here $A = -a^2$, $B = 0$, $C = 0$, and hence $B^2 - AC = 0$.

5.2.4 The standard form of the elliptic equations

For the elliptic type, the condition $B^2 - AC < 0$ yields that Eq. (5.2.10) has no real roots but it has the complex root

$$\lambda_1 = \frac{-B + \sqrt{B^2 - AC}}{A},$$

and the complex conjugate root $\lambda_2 = \bar{\lambda}_1$. We take the first equation of (5.2.15) with the complex root λ_1:

$$\frac{\partial \omega}{\partial x} - \lambda_1 \frac{\partial \omega}{\partial y} = 0, \qquad (5.2.22)$$

and write its (complex) solution in the form

$$\omega = \varphi(x, y) + i\psi(x, y). \qquad (5.2.23)$$

The second equation of (5.2.15) is merely the complex conjugate to Eq. (5.2.22):

$$\frac{\partial \bar{\omega}}{\partial x} - \bar{\lambda}_1 \frac{\partial \bar{\omega}}{\partial y} = 0,$$

where $\bar{\omega} = \varphi(x, y) - i\psi(x, y)$. Therefore, we consider only function (5.2.23). It solves the characteristic equation (5.2.7):

$$A (\varphi_x + i\psi_x)^2 + 2B (\varphi_x + i\psi_x)(\varphi_y + i\psi_y) + C (\varphi_y + i\psi_y)^2 = 0.$$

Substituting $(\varphi_x + i\psi_x)^2 = \varphi_x^2 - \psi_x^2 + 2i\varphi_x\psi_x, \ldots$ into the left-hand side of the above equation and annulling the real and imaginary parts, we obtain

$$A \varphi_x^2 + 2B \varphi_x\varphi_y + C \varphi_y^2 = A \psi_x^2 + 2B \psi_x\psi_y + C \psi_y^2,$$

$$A \varphi_x\psi_x + B (\varphi_y\psi_x + \varphi_x\psi_y) + C \varphi_y\psi_y = 0.$$

Hence, using the change of variables (5.2.3),

$$\xi = \varphi(x, y), \quad \eta = \psi(x, y),$$

with $\varphi(x, y)$ and $\psi(x, y)$ taken from (5.2.23), and invoking (5.2.6), we have $\tilde{A} = \tilde{C} \neq 0$ and $\tilde{B} = 0$. Now we divide by $\tilde{A}$ Eq. (5.2.1) written in the new variables and arrive at the standard form of elliptic equations given in the following theorem (cf. Eq. (1.1.58) in Theorem 1.1.3).

Theorem 5.2.3. Let Eq. (5.2.1) be an elliptic equation. We introduce new variables

$$\xi = \varphi(x, y), \quad \eta = \psi(x, y),$$

where $\varphi(x, y)$ and $\psi(x, y)$ are the real and imaginary parts of the solution (5.2.23) to the complex characteristic equation (5.2.22). This change of variables transforms Eq. (5.2.1) to the following *standard form:*

$$u_{\xi\xi} + u_{\eta\eta} + \tilde{a}(\xi, \eta)u_\xi + \tilde{b}(\xi, \eta)u_\eta + \tilde{c}(\xi, \eta)u = 0. \tag{5.2.24}$$

Example 5.2.3. A typical representative of the elliptic equations is the Laplace equation in two variables,

$$u_{xx} + u_{yy} = 0. \tag{5.2.25}$$

It is manifest that the characteristic equation (5.2.7), $\omega_x^2 + \omega_y^2 = 0$, has no real-valued solutions $\omega(x, y)$.

Remark 5.2.1. It is manifest from Eqs. (5.2.18), (5.2.24) that hyperbolic and elliptic equations are connected by complex transformations (cf. Remark 1.1.7).

5.2.5   Equations of a mixed type

An example of second-order equations of a mixed type is provided by the Tricomi equation (2.6.37)

$$xu_{yy} + u_{xx} = 0. \tag{5.2.26}$$

It is hyperbolic when $x < 0$ and elliptic when $x > 0$. The Tricomi equation is used in gas dynamics as an approximate model in studying transonic flows. Specifically, Eq. (5.2.26) corresponds to the subsonic gas flow in the elliptic domain, and to the supersonic flow in the hyperbolic domain.

5.2.6   The type of nonlinear equations

The type of a nonlinear partial differential equation

$$A(x, y, u, u_x, u_y)\, u_{xx} + 2B(x, y, u, u_x, u_y)\, u_{xy}$$

$$+\, C(x, y, u, u_x, u_y)\, u_{yy} + \Phi(x, y, u, u_x, u_y) = 0 \tag{5.2.27}$$

depends on its solutions and is defined as follows.

Definition 5.2.3. Let

$$u^* = h(x, y)$$

be any particular solution of Eq. (5.2.27). The type of the *nonlinear* equation (5.2.27) on the solution u^* is identified with the type of the *linear equation*

$$A^*(x, y)\, u_{xx} + 2B^*(x, y)\, u_{xy} + C^*(x, y)\, u_{yy} = 0, \tag{5.2.28}$$

where the coefficients A^*, B^*, C^* are obtained by replacing the dependent variable u and its first derivatives by the function $h(x,y)$ and its derivatives, respectively, e.g.

$$A^*(x,y) = A\left(x,y,h(x,y),h_x(x,y),h_y(x,y)\right).$$

The terms containing the derivatives of lower order are omitted in the linearized equation (5.2.28) since they do not affect the type of the equation.

Example 5.2.4. Consider the nonlinear wave equation (2.6.31):

$$u_{tt} = \phi(u)\, u_{xx} + \frac{1}{2}\, \phi'(u)u_x^2. \qquad (5.2.29)$$

If, e.g. $\phi(u) = u^2$, then Eq. (5.2.29),

$$u_{tt} = u^2\, u_{xx} + u\, u_x^2,$$

is hyperbolic for any solution $u^* = h(x,y)$. Likewise, if $\phi(u) = -u^2$, then Eq. (5.2.29),

$$u_{tt} = -u^2\, u_{xx} - u\, u_x^2,$$

is elliptic for any solution $u^* = h(x,y)$. On the other hand, letting, e.g. $\phi(u) = u$ we obtain the nonlinear equation of the mixed type:

$$u_{tt} = u\, u_{xx} + \frac{1}{2}\, u_x^2.$$

It is hyperbolic for any positive solution $u^* = h(x,y), h(x,y) > 0$, and elliptic for any negative solution $u^* = h(x,y), h(x,y) < 0$.

5.3 Integration of hyperbolic equations in two variables

We present here the simple and efficient integration methods due to d'Alembert (1747), Euler (1770) and Laplace (1773). The Laplace invariants are vital to this section.

5.3.1 d'Alembert's solution

The first partial differential equation, the wave equation for vibrating strings,

$$u_{tt} = k^2 u_{xx}, \quad k = \text{const.}, \qquad (5.3.1)$$

was formulated and solved by d'Alembert in 1747.

Let us solve Eq. (5.3.1) by transforming it to the standard form. The characteristic equation (5.2.7) is written

$$\omega_t^2 - k^2\omega_x^2 = (\omega_t - k\omega_x)(\omega_t + k\omega_x) = 0.$$

It splits into two equations of the first order (cf. (5.2.15)):

$$\frac{\partial \omega}{\partial t} + k \frac{\partial \omega}{\partial x} = 0$$

and

$$\frac{\partial \omega}{\partial t} - k \frac{\partial \omega}{\partial x} = 0.$$

The associated equations (5.2.16),

$$dt = \frac{dx}{k}$$

and

$$dt = -\frac{dx}{k},$$

have the first integrals $x - kt = $ const. and $x + kt = $ const., respectively. Hence, the characteristic variables for the wave equation are

$$\xi = x - kt, \quad \eta = x + kt. \tag{5.3.2}$$

In the characteristic variables (5.3.2), Eq. (5.3.1) is written in the standard form

$$u_{\xi\eta} = 0. \tag{5.3.3}$$

Integrating first with respect to η we have

$$u_\xi = f(\xi)$$

whence, integrating now with respect to ξ and denoting $F(\xi) = \int f(\xi)d\xi$, we obtain the general solution to Eq. (5.3.3):

$$u = F(\xi) + H(\eta). \tag{5.3.4}$$

Now we return to the original variables by substituting in (5.3.4) the expressions (5.3.2) of the characteristic variables and ultimately arrive at the following general solution of the wave equation (5.3.1):

$$u = F(x - kt) + H(x + kt) \tag{5.3.5}$$

known as *d'Alembert's solution*.

5.3.2   Equations reducible to the wave equation

Consider hyperbolic equations in two independent variables written in the standard form (5.2.18):

$$u_{\xi\eta} + a(\xi, \eta)u_\xi + b(\xi, \eta)u_\eta + c(\xi, \eta)u = 0. \tag{5.3.6}$$

Some of equations of form (5.3.6) can be reduced to the wave equation (5.3.3) by a change of variables, and hence, solved by d'Alembert's method. Let us single out all these equations. First of all, we identify the most general form of changes of variables that can be utilized without loss of linearity and homogeneity of Eqs. (5.3.6) as well as their standard form. These changes of variables are termed *equivalence transformations* and have the following form:

$$\tilde{\xi} = f(\xi), \quad \tilde{\eta} = g(\eta), \quad v = \sigma(\xi, \eta)\, u, \tag{5.3.7}$$

where $f'(\xi) \neq 0$, $g'(\eta) \neq 0$, and $\sigma(\xi, \eta) \neq 0$. Here u and v are regarded as functions of ξ, η and $\tilde{\xi}, \tilde{\eta}$, respectively. Eqs. (5.3.6) related by an equivalence transformation (5.3.7) are said to be *equivalent*.

Let us begin with the restricted equivalence transformations (5.3.7) by setting $\tilde{\xi} = \xi$, $\tilde{\eta} = \eta$ and find Eq. (5.3.6) reducible to the wave equation by the linear transformation of the dependent variable written in the form

$$v = u\, e^{\varphi(\xi, \eta)}.$$

We substitute the expressions

$$u = v\, e^{-\varphi(\xi, \eta)},$$

$$u_\xi = (v_\xi - v\varphi_\xi)\, e^{-\varphi(\xi, \eta)}, \quad u_\eta = (v_\eta - v\varphi_\eta)\, e^{-\varphi(\xi, \eta)},$$

$$u_{\xi\eta} = (v_{\xi\eta} - v_\xi\varphi_\eta - v_\eta\varphi_\xi - v\,\varphi_{\xi\eta} + v\,\varphi_\xi\,\varphi_\eta)\, e^{-\varphi(\xi, \eta)}$$

into the left-hand side of Eq. (5.3.6) and obtain

$$
\begin{aligned}
u_{\xi\eta} &+ a\, u_\xi + b\, u_\eta + c\, u \\
&= \big[v_{\xi\eta} + (a - \varphi_\eta)\, v_\xi + (b - \varphi_\xi)\, v_\eta \\
&\quad + (-\varphi_{\xi\eta} + \varphi_\xi\, \varphi_\eta - a\varphi_\xi - b\varphi_\eta + c)\, v\big]e^{-\varphi}.
\end{aligned}
\tag{5.3.8}
$$

Therefore, Eq. (5.3.6) reduces to the wave equation $v_{\xi\eta} = 0$ if

$$a - \varphi_\eta = 0, \quad b - \varphi_\xi = 0 \tag{5.3.9}$$

and

$$\varphi_{\xi\eta} - \varphi_\xi\, \varphi_\eta + a\varphi_\xi + b\varphi_\eta - c = 0. \tag{5.3.10}$$

Equations (5.3.9) provide a system of *two* equations for *one* unknown function $\varphi(\xi, \eta)$ of two variables. Recall that a system of equations is called an *over-determined system* if it contains more equations than unknown functions to be determined by the system in question. Over-determined systems have solutions only if they satisfy certain *compatibility conditions*.

Thus, the system of equations (5.3.9) is over-determined. Its compatibility condition is obtained from the equation $\varphi_{\xi\eta} = \varphi_{\eta\xi}$ (independence of successive partial differentiation on the order of differentiation) and has the form

$$a_\xi = b_\eta. \tag{5.3.11}$$

Equation (5.3.10), upon using Eqs. (5.3.9) and (5.3.11), is written as

$$a_\xi + ab - c = 0. \tag{5.3.12}$$

Upon introducing the quantities h and k are defined by[1]

$$h = a_\xi + ab - c, \quad k = b_\eta + ab - c, \tag{5.3.13}$$

conditions (5.3.11) and (5.3.12) are written in the following symmetric form:

$$h = 0, \quad k = 0. \tag{5.3.14}$$

Remark 5.3.1. Equations (5.3.14) are invariant under the general equivalence transformation (5.3.7). In consequence, the change of the independent variables does not provide new equations reducible to the wave equation.

Summing up the above calculations and taking into account Remark 5.3.1, we arrive at the following result.

Theorem 5.3.1. Equation (5.3.6) is equivalent to the wave equation if and only if its Laplace invariants (5.3.13) vanish, $h = k = 0$. Any equation (5.3.6) with $h = k = 0$ can be reduced to the wave equation $v_{\xi\eta} = 0$ by the linear transformation of the dependent variable:

$$u = v\, e^{-\varphi(\xi,\eta)} \tag{5.3.15}$$

without changing the independent variables ξ and η. The function φ in (5.3.15) is obtained by solving the following compatible system:

$$\frac{\partial \varphi}{\partial \xi} = b(\xi, \eta), \quad \frac{\partial \varphi}{\partial \eta} = a(\xi, \eta). \tag{5.3.16}$$

Theorem 5.3.1 furnishes us with a practical method for solving a wide class of equations (5.2.1)

$$A\,u_{xx} + 2B\,u_{xy} + C\,u_{yy} + a\,u_x + b\,u_y + cu = 0$$

of the hyperbolic type by reducing them to the wave equation. The method requires the following two steps.

First step. Check if Eq. (5.2.1) is hyperbolic, i.e., if $B^2 - AC > 0$. Provided that this condition is satisfied, reduce the equation in question to its standard form (5.3.6) by introducing the characteristic variables (Section 5.2.2)

$$\xi = \omega_1(x, y), \quad \eta = \omega_2(x, y). \tag{5.3.17}$$

---

[1]Quantities (5.3.13) were introduced by Euler [7]. Then they were rediscovered by Laplace [23] and became known in the literature as the *Laplace invariants*. See further Sections 5.3.3 and 5.3.4.

Second step. Find the Laplace invariants (5.3.13). If $h = k = 0$, find $\varphi(\xi, \eta)$ by solving Eqs. (5.3.16) and reduce your equation to the wave equation $v_{\xi\eta} = 0$ by transformation (5.3.15). Finally, substituting $v = f(\xi) + g(\eta)$ into (5.3.15) you will obtain the solution to your equation in characteristic variables:

$$u = [f(\xi) + g(\eta)] e^{-\varphi(\xi,\eta)}. \tag{5.3.18}$$

Substitute here expressions (5.3.17) for ξ and η to obtain the solution in the original variables x, y.

Example 5.3.1. Let us illustrate the method by the equation

$$\frac{u_{xx}}{x^2} - \frac{u_{yy}}{y^2} + 3\left(\frac{u_x}{x^3} - \frac{u_y}{y^3}\right) = 0. \tag{5.3.19}$$

First step. Here $A = x^{-2}, B = 0, C = -y^{-2}$, and hence $B^2 - AC = (xy)^{-2} > 0$. Equation (5.2.7) for the characteristics has the form

$$\left(\frac{\omega_x}{x}\right)^2 - \left(\frac{\omega_y}{y}\right)^2 = \left(\frac{\omega_x}{x} - \frac{\omega_y}{y}\right)\left(\frac{\omega_x}{x} + \frac{\omega_y}{y}\right) = 0.$$

It splits into two equations:

$$\frac{\omega_x}{x} + \frac{\omega_y}{y} = 0, \quad \frac{\omega_x}{x} - \frac{\omega_y}{y} = 0.$$

They have the following first integrals:

$$x^2 - y^2 = \text{const.}, \quad x^2 + y^2 = \text{const.}$$

Hence, the characteristic variables (5.3.17) are defined by

$$\xi = x^2 - y^2, \quad \eta = x^2 + y^2.$$

We have

$$u_x = u_\xi \cdot \xi_x + u_\eta \cdot \eta_x = 2x(u_\xi + u_\eta),$$
$$u_y = u_\xi \cdot \xi_y + u_\eta \cdot \eta_y = 2y(u_\eta - u_\xi),$$
$$u_{xx} = 2(u_\xi + u_\eta) + 4x^2[(u_\xi + u_\eta)_\xi + (u_\xi + u_\eta)_\eta]$$
$$= 2(u_\xi + u_\eta) + 4x^2(u_{\xi\xi} + 2u_{\xi\eta} + u_{\eta\eta}),$$
$$u_{yy} = 2(u_\eta - u_\xi) + 4y^2[(u_\eta - u_\xi)_\eta - (u_\eta - u_\xi)_\xi]$$
$$= 2(u_\eta - u_\xi) + 4y^2(u_{\xi\xi} - 2u_{\xi\eta} + u_{\eta\eta}).$$

Therefore, Eq. (5.3.19) takes the following form:

$$u_{\xi\eta} + \frac{x^2 + y^2}{2x^2y^2} u_\xi - \frac{x^2 - y^2}{2x^2y^2} u_\eta = 0.$$

Invoking that $x^2 - y^2 = \xi$, $x^2 + y^2 = \eta$ and noting that

$$2x^2 y^2 = \frac{\eta^2 - \xi^2}{2},$$

we ultimately arrive at the following standard form of Eq. (5.3.19):

$$u_{\xi\eta} + \frac{2\eta}{\eta^2 - \xi^2} u_\xi - \frac{2\xi}{\eta^2 - \xi^2} u_\eta = 0. \qquad (5.3.20)$$

Second step. The coefficients of Eq. (5.3.20) are

$$a = \frac{2\eta}{\eta^2 - \xi^2}, \qquad b = -\frac{2\xi}{\eta^2 - \xi^2}, \qquad c = 0.$$

Substituting into (5.3.13) the expressions for a, b, c and their derivatives

$$a_\xi = b_\eta = \frac{4\xi\eta}{(\eta^2 - \xi^2)^2},$$

we see that $h = k = 0$. Now we solve Eqs. (5.3.16):

$$\frac{\partial\varphi}{\partial\eta} = \frac{2\eta}{\eta^2 - \xi^2}, \qquad \frac{\partial\varphi}{\partial\xi} = -\frac{2\xi}{\eta^2 - \xi^2}$$

and obtain

$$\varphi = \ln(\eta^2 - \xi^2).$$

Consequently, substitution (5.3.15)

$$v = u e^{\ln(\eta^2 - \xi^2)} = (\eta^2 - \xi^2) u \qquad (5.3.21)$$

maps Eq. (5.3.20) to the wave equation

$$v_{\xi\eta} = 0.$$

Therefore,

$$v(\xi, \eta) = f(\xi) + g(\eta),$$

and (5.3.21) yields

$$u(\xi, \eta) = \frac{f(\xi) + g(\eta)}{\eta^2 - \xi^2}.$$

Returning to the original variables by substituting $\xi = x^2 - y^2$, $\eta = x^2 + y^2$ and denoting $F = f/4$, $H = g/4$ we finally obtain the following general solution to Eq. (5.3.19):

$$u(x, y) = \frac{F(x^2 - y^2) + H(x^2 + y^2)}{x^2 y^2}. \qquad (5.3.22)$$

5.3.3 Euler's method

We owe to Leonard Euler [7] the first significant results in integration theory of hyperbolic equations not necessarily equivalent to the wave equation. He generalized d'Alembert's solution to a wide class of equations (5.3.6). Namely, he introduced quantities (5.3.13) and showed that Eq. (5.3.6) is factorable if and only if at least one of the quantities h and k vanishes. The solution of the factorized equation (5.3.6) reduces to consecutive integration of two first-order ordinary differential equations.

Euler's method consists in the following. Consider Eq. (5.3.6),

$$u_{\xi\eta} + a(\xi,\eta)u_\xi + b(\xi,\eta)u_\eta + c(\xi,\eta)u = 0,$$

with $h = 0$. Then this equation is factorable in the form

$$\left(\frac{\partial}{\partial\xi} + b\right)\left(\frac{\partial u}{\partial\eta} + au\right) = 0. \tag{5.3.23}$$

Setting

$$v = u_\eta + a\,u \tag{5.3.24}$$

one can rewrite Eq. (5.3.23) as a first-order equation

$$v_\xi + b\,v = 0$$

and integrate it to obtain

$$v = Q(\eta)e^{-\int b(\xi,\eta)d\xi}. \tag{5.3.25}$$

Now we substitute (5.3.25) into (5.3.24), integrate the resulting non-homogeneous linear equation

$$u_\eta + au = Q(\eta)e^{-\int b(\xi,\eta)d\xi} \tag{5.3.26}$$

with respect to η and obtain the following result.

Theorem 5.3.2. The general solution of Eq. (5.3.6),

$$u_{\xi\eta} + a(\xi,\eta)u_\xi + b(\xi,\eta)u_\eta + c(\xi,\eta)u = 0,$$

with $h = 0$ is given by the formula

$$u = \left[P(\xi) + \int Q(\eta)e^{\int a d\eta - b d\xi}d\eta\right]e^{-\int a d\eta} \tag{5.3.27}$$

containing two arbitrary functions, $P(\xi)$ and $Q(\eta)$.

Likewise, if $k = 0$, Eq. (5.3.6) is factorable in the form

$$\left(\frac{\partial}{\partial\eta} + a\right)\left(\frac{\partial u}{\partial\xi} + bu\right) = 0. \tag{5.3.28}$$

In this case, we replace substitution (5.3.24) by

$$w = u_\xi + bu. \tag{5.3.29}$$

Now we repeat the calculations made in the case $h = 0$ and obtain the following result.

Theorem 5.3.3. The general solution of (5.3.6),

$$u_{\xi\eta} + a(\xi, \eta)u_\xi + b(\xi, \eta)u_\eta + c(\xi, \eta)u = 0,$$

with $k = 0$ is given by the formula

$$u = \left[Q(\eta) + \int P(\xi)e^{\int bd\xi - ad\eta}d\xi \right] e^{-\int bd\xi}. \tag{5.3.30}$$

One can apply the above method when Eq. (5.3.6) is replaced by a non-homogeneous equation

$$u_{\xi\eta} + a(\xi, \eta)u_\xi + b(\xi, \eta)u_\eta + c(\xi, \eta)u = f(\xi, \eta). \tag{5.3.31}$$

Then, for example, when $h = 0$ one obtains the following result.

Theorem 5.3.4. The solution of Eq. (5.3.31) with $h = 0$ has the form

$$u = \left[P(\xi) + \int \left(Q(\eta) + \int f(\xi, \eta)e^{\int bd\xi}d\xi \right) e^{\int ad\eta - bd\xi}d\eta \right] e^{-\int ad\eta}. \tag{5.3.32}$$

Example 5.3.2. Let us consider the following equation known as Darboux's equation:

$$u_{xy} + \frac{\beta u_y}{x - y} = 0, \quad \beta = \text{const.} \tag{5.3.33}$$

Here,

$$a = 0, \quad b = \frac{\beta}{x - y}, \quad c = 0.$$

The Laplace invariants have the form

$$h = a_x + ab - c = 0, \quad k = b_y = \frac{\beta}{(x - y)^2} \neq 0.$$

Formula (5.3.27) yields the general solution

$$u(x, y) = P(x) + \int Q(y)(x - y)^{-\beta}dy.$$

Let us employ the general solution in the Cauchy problem with the initial data given on the non-characteristic line $x - y = 1$:

$$u|_{x-y=1} = u_0(x), \quad u_y|_{x-y=1} = u_1(x).$$

Then the solution is written in the form

$$u = P(x) + \int_{-1}^{y} Q(\tau)(x - \tau)^{-\beta} d\tau.$$

The initial conditions yield

$$u|_{x-y=1} = P(x) + \int_{-1}^{x-1} Q(\tau)(x - \tau)^{-\beta} d\tau = u_0(x),$$

$$u_y|_{x-y=1} = Q(x - y) = u_1(x).$$

Hence,

$$Q(y) = u_1(y + 1), \quad P(x) = u_0(x) - \int_{-1}^{x-1} u_1(\tau + 1)(x - \tau)^{-\beta} d\tau.$$

Thus, the solution to the Cauchy problem is

$$u(x, y) = u_0(x) - \int_{-1}^{x-1} u_1(\tau + 1)(x - \tau)^{-\beta} d\tau + \int_{-1}^{y} u_1(\tau + 1)(x - \tau)^{-\beta} d\tau.$$

5.3.4   Laplace's cascade method

In 1773, Laplace [23] developed a more general method than that of Euler. In Laplace's method, known also as the *cascade method*, the quantities h, k play the central part. Laplace introduced two transformations. Laplace's first transformation has form (5.3.24):

$$v = u_\eta + a\,u, \tag{5.3.34}$$

and the second transformation has form (5.3.29):

$$w = u_\xi + b\,u. \tag{5.3.35}$$

Laplace's transformations allow one to solve certain equations when both Laplace invariants are different from zero. Thus, we let $h \neq 0, k \neq 0$ and consider transformation (5.3.34). It maps Eq. (5.3.6) to an equation of the same form, namely,

$$v_{\xi\eta} + a_1 v_\xi + b_1 v_\eta + c_1 v = 0, \tag{5.3.36}$$

with the coefficients

$$a_1 = a - \frac{\partial \ln|h|}{\partial \eta}, \quad b_1 = b, \quad c_1 = c + b_\eta - a_\xi - b\frac{\partial \ln|h|}{\partial \eta}. \tag{5.3.37}$$

The Laplace invariants for Eq. (5.3.36) have the form:

$$h_1 = 2h - k - \frac{\partial^2 \ln|h|}{\partial \xi \partial \eta}, \quad k_1 = h. \tag{5.3.38}$$

Likewise, one can utilize the second transformation (5.3.35) and arrive to a linear equation for w with the Laplace invariants

$$h_2 = k, \quad k_2 = 2k - h - \frac{\partial^2 \ln |k|}{\partial \xi \partial \eta}. \tag{5.3.39}$$

If $h_1 = 0$, one can solve Eq. (5.3.36) using Euler's method described above. Then it remains to substitute the solution $v = v(x, y)$ in (5.3.34) and to integrate the non-homogeneous first-order linear equation (5.3.34) for u. If $h_1 \neq 0$ but $k_2 = 0$, we find in a similar way the function $w = w(x, y)$ and solve the non-homogeneous first-order linear equation (5.3.35) for u. If $h_1 \neq 0$ and $k_2 \neq 0$, one can iterate the Laplace transformations by applying transformations (5.3.34) and (5.3.35) to equations for v and w, etc. This is the essence of Laplace's *cascade method.*

Example 5.3.3. Let us apply Laplace's cascade method to the following Darboux equation:

$$u_{xy} - \alpha \frac{u_x}{x - y} + \beta \frac{u_y}{x - y} = 0, \qquad \alpha, \beta = \text{const.} \tag{5.3.40}$$

We already considered its particular case (5.3.33) obtained from (5.3.40) when $\alpha = 0$. The coefficients of Eq. (5.3.40) are

$$a = -\frac{\alpha}{x - y}, \quad b = \frac{\beta}{x - y},$$

and hence,

$$a_x + ab - c = \frac{\alpha}{(x - y)^2} - \frac{\alpha\beta}{(x - y)^2} = \frac{\alpha(1 - \beta)}{(x - y)^2},$$

$$b_y + ab - c = \frac{\beta}{(x - y)^2} - \frac{\alpha\beta}{(x - y)^2} = \frac{\beta(1 - \alpha)}{(x - y)^2}.$$

Thus, the Laplace invariants have the form

$$h = \frac{\alpha(1 - \beta)}{(x - y)^2}, \quad k = \frac{\beta(1 - \alpha)}{(x - y)^2}. \tag{5.3.41}$$

It follows that $h = 0$ if $\alpha = 0$ or $\beta = 1$, and $k = 0$ if $\beta = 0$ or $\alpha = 1$. Thus, Eq. (5.3.40) can be solved by Euler's method in the cases

$$\alpha = 0, \quad \beta = 1; \qquad \beta = 0, \quad \alpha = 1. \tag{5.3.42}$$

Let us consider the general case and apply Laplace's first transformation (5.3.34). We have

$$\frac{\partial^2 \ln |h|}{\partial x \partial y} = -\frac{2}{(x - y)^2},$$

and Eq. (5.3.38),

$$h_1 = 2h - k - \frac{\partial^2 \ln |h|}{\partial x \partial y},$$

yields

$$h_1 = \frac{(\alpha + 1)(2 - \beta)}{(x - y)^2}.$$

Hence, $h_1 = 0$ if and only if $\alpha = -1$ or $\beta = 2$. Thus, invoking (5.3.41), we see that we can solve now Eq. (5.3.40) in the cases

$$\alpha = 0, \quad \alpha = \pm 1; \quad \beta = 0, \quad \beta = 1, \quad \beta = 2. \tag{5.3.43}$$

Applying Laplaces' second transformation (5.3.35), we can further extend (5.3.43) and obtain the following integrable cases:

$$\alpha = 0, \quad \alpha = \pm 1 \quad \alpha = 2; \quad \beta = 0, \quad \beta = \pm 1, \quad \beta = 2. \tag{5.3.44}$$

Continuing Laplace's cascade method, one can solve the Darboux equation (5.3.40) with any integers α and β.

Remark 5.3.2. We summarize. Eq. (5.3.40),

$$u_{xy} - \alpha \frac{u_x}{x - y} + \beta \frac{u_y}{x - y} = 0$$

can be solved in the following cases.
(i) By d'Alembert's method when

$$\alpha = \beta = 0.$$

(ii) By Euler's method when

$$\alpha = 0, \quad \beta = 1 \quad \text{and} \quad \beta = 0, \quad \alpha = 1.$$

(iii) By Laplace's cascade method for any integers α and β.

5.4 The initial value problem

5.4.1 The wave equation

Let us use d'Alembert's formula (5.3.5) for solving the Cauchy problem for the wave equation (5.3.1), $u_{tt} = k^2 u_{xx}$, with the initial data

$$u\big|_{t=0} = u_0(x), \quad u_t\big|_{t=0} = u_1(x). \tag{5.4.1}$$

Equations (5.3.5) and (5.4.1) yield

$$F(x) + H(x) = u_0(x), \quad H'(x) - F'(x) = \frac{1}{k} u_1(x). \tag{5.4.2}$$

Differentiating the first of Eq. (5.4.2) and adding to the second of Eq. (5.4.2) we have

$$H'(x) = \frac{1}{2}\left[u_0'(x) + \frac{u_1(x)}{k}\right],$$

and hence,

$$H(x) = \frac{u_0(x)}{2} + \frac{1}{2k}\int_{x_0}^{x} u_1(s)ds. \tag{5.4.3}$$

Now we get from (5.4.2):

$$F(x) = \frac{u_0(x)}{2} - \frac{1}{2k}\int_{x_0}^{x} u_1(s)ds. \tag{5.4.4}$$

Substitution of Eqs. (5.4.3) and (5.4.4) into d'Alembert's formula (5.3.5) leads to Theorem 5.4.1.

Theorem 5.4.1. The solution to the Cauchy problem for the wave equation with the initial data (5.4.1) is given by

$$u(x,t) = \frac{u_0(x - kt) + u_0(x + kt)}{2} + \frac{1}{2k}\int_{x-kt}^{x+kt} u_1(s)ds. \tag{5.4.5}$$

Solution (5.4.5) discloses a physical significance of characteristics, namely, that waves propagate along characteristics. The following two examples provide a good illustration. For more details, see [11].

Example 5.4.1. Let a plucked guitar string have the initial configuration $u_0(x)$ in the form of a pulse at a point x_0. We release the string from the rest, i.e., we let $u_1(x) = 0$. Then solution (5.4.5) has the form

$$u(x,t) = \frac{u_0(x - kt) + u_0(x + kt)}{2} \tag{5.4.6}$$

and describes the propagation of the initial configuration.

Example 5.4.2. Let us imitate the vibration of a piano string by the solution of the Cauchy problem, where the initial displacement of the piano wire is zero, $u_0(x) = 0$, but the string is subjected to a localized initial velocity $u_1(x)$. Then solution (5.4.5) has the form

$$u(x,t) = \frac{1}{2k}\int_{x-kt}^{x+kt} u_1(s)ds. \tag{5.4.7}$$

5.4.2 Non-homogeneous wave equation

Let us consider the Cauchy problem with the initial conditions (5.4.1) for the non-homogeneous one-dimensional wave equation

$$u_{tt} - k^2 u_{xx} = f(x,t), \quad k = \text{const.} \tag{5.4.8}$$

Lemma 5.4.1. The function $v(x,t)$ defined by

$$v(x,t) = \frac{1}{2k} \int_0^t d\tau \int_{x-k(t-\tau)}^{x+k(t-\tau)} f(s,\tau)ds \qquad (5.4.9)$$

solves the non-homogeneous wave equation (5.4.8),

$$v_{tt} - k^2 v_{xx} = f(x,t),$$

and satisfies the following initial conditions:

$$v\big|_{t=0} = 0, \quad v_t\big|_{t=0} = 0.$$

Proof. Using rules (1.2.13) for differentiation of definite integrals, one obtains the following (see Problem 5.11):

$$v_t = \frac{1}{2} \int_0^t [f(x+k(t-\tau),\tau) + f(x-k(t-\tau),\tau)]d\tau,$$

$$v_{tt} = \frac{k}{2} \int_0^t [f_x(x+k(t-\tau),\tau) - f_x(x-k(t-\tau),\tau)]d\tau + f(x,t),$$

$$v_x = \frac{1}{2k} \int_0^t [f(x+k(t-\tau),\tau) - f(x-k(t-\tau),\tau)]d\tau, \qquad (5.4.10)$$

$$v_{xx} = \frac{1}{2k} \int_0^t [f_x(x+k(t-\tau),\tau) - f_x(x-k(t-\tau),\tau)]d\tau.$$

The statements of the lemma follow from Eqs. (5.4.9) and (5.4.10).

Theorem 5.4.2. The solution of the non-homogeneous equation (5.4.8),

$$u_{tt} - k^2 u_{xx} = f(x,t),$$

satisfying the initial conditions

$$u\big|_{t=0} = u_0(x), \quad u_t\big|_{t=0} = u_1(x)$$

is unique and is given by

$$u(t,x) = \frac{u_0(x-kt) + u_0(x+kt)}{2} + \frac{1}{2k} \int_{x-kt}^{x+kt} u_1(\xi)d\xi$$

$$+ \frac{1}{2k} \int_0^t d\tau \int_{x-k(t-\tau)}^{x+k(t-\tau)} f(\xi,\tau)d\xi. \qquad (5.4.11)$$

Proof. Use Theorem 5.4.1 and Lemma 5.4.1.

5.5   Mixed problem. Separation of variables

The method of *separation of variables* (Fourier's method) is used, e.g. for solving *mixed problems* when the solution of a differential equation under consideration should satisfy prescribed *initial* and *boundary* values. In what follows, we will use the following integrals (see Problem 1.5) .

Lemma 5.5.1. Let k and m be any integers, $k \neq 0$. Then

$$\int_0^\pi \sin(kx)\,\sin(mx)\mathrm{d}x = 0 \ (m \neq k), \quad \int_0^\pi \sin^2(kx)\mathrm{d}x = \frac{\pi}{2}\,.$$

It is convenient to use these equations in the compact form

$$\int_0^\pi \sin(kx)\,\sin(mx)\mathrm{d}x = \frac{\pi}{2}\,\delta_{km}, \qquad (5.5.1)$$

where δ_{km} are the *Kronecker symbols*: $\delta_{km} = 0$ if $m \neq k$ and $\delta_{km} = 1$ if $m = k$ (see Section 1.4.3).

Proof. Using the equations

$$\sin\alpha\,\sin\beta = \frac{1}{2}[\cos(\alpha - \beta) - \cos(\alpha + \beta)], \quad \sin^2\alpha = \frac{1}{2}[1 - \cos(2\alpha)]$$

we obtain

$$\int_0^\pi \sin(kx)\,\sin(mx)\mathrm{d}x = \frac{1}{2}\int_0^\pi [\cos[(k-m)x] - \cos[(k+m)x]]\mathrm{d}x$$

$$= \left[\frac{\sin[(k-m)x]}{2(k-m)} - \frac{\sin[(k+m)x]}{2(k+m)}\right]_0^\pi = 0$$

and

$$\int_0^\pi \sin^2(kx)\,\mathrm{d}x = \frac{1}{2}\int_0^\pi [1 - \cos(2kx)]\mathrm{d}x = \frac{\pi}{2} - \frac{1}{4k}\sin(2kx)\Big|_0^\pi = \frac{\pi}{2}\,.$$

5.5.1   Vibration of a string tied at its ends

Consider a mixed problem for describing vibrations of a string tied at its ends $x = 0$ and $x = l$ with given initial configuration and velocity. Thus, the problem is to find the solution of the wave equation

$$u_{tt} = u_{xx}, \qquad (5.5.2)$$

with prescribed *initial values*:

$$u\big|_{t=0} = u_0(x), \quad u_t\big|_{t=0} = u_1(x), \qquad (5.5.3)$$

and the *boundary values*:

$$u\big|_{x=0} = 0, \quad u\big|_{x=l} = 0. \tag{5.5.4}$$

Consistency of conditions (5.5.3) and (5.5.4) requires that

$$u_0(0) = u_0(l) = 0, \quad u_1(0) = u_1(l) = 0. \tag{5.5.5}$$

The method of separation of variables consists in seeking particular solutions in the product form

$$u(t, x) = T(t) X(x) \tag{5.5.6}$$

such that none of factors vanishes identically. Substituting (5.5.6) into Eq. (5.5.2) gives

$$T''X = TX''.$$

Whence, separating the functions depending on t and x, respectively:

$$\frac{T''}{T} = \frac{X''}{X} = -\lambda$$

where λ is a positive constant. The above equation is equivalent to the following two equations:

$$T'' + \lambda T = 0 \tag{5.5.7}$$

and

$$X'' + \lambda X = 0. \tag{5.5.8}$$

The boundary conditions (5.5.4) yield

$$X(0) = 0, \quad X(l) = 0. \tag{5.5.9}$$

The general solution of Eq. (5.5.8) is

$$X = C_1 \sin \sqrt{\lambda}\, x + C_2 \cos \sqrt{\lambda}\, x \qquad (\lambda > 0).$$

The first condition in (5.5.9), $X(0) = 0$, yields $C_2 = 0$. Therefore, the solution takes the form

$$X = C_1 \sin \sqrt{\lambda}\, x. \tag{5.5.10}$$

Then the second condition in (5.5.9), $X(l) = 0$, is written

$$C_1 \sin \sqrt{\lambda}\, l = 0.$$

Since the function $X(x)$ should not be identical to zero, we require that $C_1 \neq 0$, and the above equation yields

$$\sin \sqrt{\lambda}\, l = 0.$$

Hence, λ assumes the following values:

$$\lambda_k = \left(\frac{k\pi}{l}\right)^2, \quad k = \pm 1, \pm 2, \pm 3, \dots . \tag{5.5.11}$$

Substituting (5.5.11) into (5.5.10) we obtain the infinite sequence of solutions to the boundary value problem (5.5.8)—(5.5.9):

$$X_k(x) = C_1 \sin \frac{k\pi x}{l}.$$

It is convenient to choose the constant C_1 from the *normalization condition*:

$$\int_0^l [X_k(x)]^2 dx = 1.$$

Using Lemma 5.5.1, we have

$$\int_0^l [X_k(x)]^2 dx = C_1^2 \int_0^l \sin^2 \frac{k\pi x}{l} dx = C_1^2 \frac{l}{\pi} \int_0^\pi \sin^2(ky) dy = \frac{l}{2} C_1^2.$$

Thus, by setting $C_1 = \sqrt{2/l}$, we obtain the normalized functions

$$X_k(x) = \sqrt{\frac{2}{l}} \sin \frac{k\pi x}{l}. \qquad (5.5.12)$$

The constants λ_k given by (5.5.11) and the functions $X_k(x)$ given by (5.5.12) are termed the *eigenvalues* and *eigenfunctions*, respectively, for the boundary value problem (5.5.8)—(5.5.9). Note that it suffices to take only positive integers $k = 1, 2, \ldots$ since eigenfunctions (5.5.12) have the property $X_{-k} = -X_k$.

The general solution of Eq. (5.5.7) with $\lambda = \lambda_k$ has the form

$$T_k(t) = a_k \cos\left(\sqrt{\lambda_k}\, t\right) + b_k \sin\left(\sqrt{\lambda_k}\, t\right), \quad a_k, b_k = \text{const}.$$

We obtain a *formal solution* $u(t, x)$ to the wave equation by taking the series

$$u(t, x) = \sum_{k=1}^\infty T_k(t) X_k(x)$$

$$= \sqrt{\frac{2}{l}} \sum_{k=1}^\infty \left[a_k \cos\left(\sqrt{\lambda_k}\, t\right) + b_k \sin\left(\sqrt{\lambda_k}\, t\right) \right] \sin\left(\sqrt{\lambda_k}\, x\right),$$

or, upon substituting eigenvalues (5.5.11),

$$u(t, x) = \sqrt{\frac{2}{l}} \sum_{k=1}^\infty \left(a_k \cos \frac{k\pi t}{l} + b_k \sin \frac{k\pi t}{l} \right) \sin \frac{k\pi x}{l}. \qquad (5.5.13)$$

Subjecting (5.5.13) into the first initial condition in (5.5.3), $u(0, x) = u_0(x)$, we obtain

$$u_0(x) = \sqrt{\frac{2}{l}} \sum_{k=1}^\infty a_k \sin \frac{k\pi x}{l}. \qquad (5.5.14)$$

Equation (5.5.14) allows one to determine the coefficients a_k. Indeed, multiplying (5.5.14) by $\sin(m\pi x/l)$ and integrating from 0 to l, one obtains

$$\int_0^l u_0(x) \sin \frac{m\pi x}{l} dx = \sqrt{\frac{2}{l}} \sum_{k=1}^\infty a_k \int_0^l \sin \frac{k\pi x}{l} \sin \frac{m\pi x}{l} dx. \qquad (5.5.15)$$

Rewriting the integral in the right-hand side of Eq. (5.5.15) in the new variable $y = \pi x/l$ and invoking Eq. (5.5.1), one has

$$\int_0^l \sin \frac{k\pi x}{l} \sin \frac{m\pi x}{l} dx = \frac{l}{\pi} \int_0^\pi \sin(ky) \sin(my) dy = \frac{l}{\pi} \frac{\pi}{2} \delta_{km} = \frac{l}{2} \delta_{km}.$$

Hence, Eq. (5.5.15) yields

$$\int_0^l u_0(x) \sin \frac{m\pi x}{l} dx = \sqrt{\frac{l}{2}} \sum_{k=1}^\infty a_k \delta_{km} = \sqrt{\frac{l}{2}} a_m.$$

Ultimately, we arrive at the following expression for the coefficients:

$$a_m = \sqrt{\frac{2}{l}} \int_0^l u_0(x) \sin \frac{m\pi x}{l} dx, \quad m = 1, 2, \ldots. \qquad (5.5.16)$$

The second initial condition in (5.5.3), $u_t(0, x) = u_1(x)$, is written

$$\sqrt{\frac{2}{l}} \sum_{k=1}^\infty \frac{k\pi}{l} b_k \sin \frac{k\pi x}{l} = u_1(x).$$

Whence, multiplying by $\sin(m\pi x/l)$, integrating from 0 to l and proceeding as above, one obtains

$$\frac{m\pi}{\sqrt{2l}} b_m = \int_0^l u_1(x) \sin \frac{m\pi x}{l} dx.$$

Hence,

$$b_m = \frac{\sqrt{2l}}{m\pi} \int_0^l u_1(x) \sin \frac{m\pi x}{l} dx, \quad m = 1, 2, \ldots. \qquad (5.5.17)$$

Now we substitute into (5.5.13) expressions (5.5.16) and (5.5.17) for the coefficients a_k and b_k, respectively, and obtain the solution $u(x, t)$ of the mixed problem (5.5.2)—(5.5.4). It is still a formal solution because the function $u(x, t)$ is represented by the formal series (5.5.13) called the Fourier series. To verify the solution, one has to prove that the Fourier series (5.5.13) converges and twice continuously differentiable.

5.5.2 Mixed problem for the heat equation

Let us use the method of separation of variables to solve the following mixed problem:

$$u_t = u_{xx}, \qquad (5.5.18)$$

$$u\big|_{t=0} = u_0(x), \qquad\qquad (5.5.19)$$

$$u\big|_{x=0} = 0, \quad u\big|_{x=l} = 0. \qquad\qquad (5.5.20)$$

Consistency of conditions (5.5.19) and (5.5.20) requires that

$$u_0(0) = u_0(l) = 0. \qquad\qquad (5.5.21)$$

We begin by seeking the solutions of the product form

$$u(t, x) = T(t)X(x).$$

Substitution into (5.5.18) yields

$$XT' = TX'',$$

and hence,

$$\frac{T'}{T} = \frac{X''}{X} = -\lambda.$$

It follows that

$$T' + \lambda T = 0 \qquad\qquad (5.5.22)$$

and

$$X'' + \lambda X = 0, \quad X(0) = X(l) = 0. \qquad\qquad (5.5.23)$$

The boundary value problem (5.5.23) is readily solved and gives the following eigenvalues and eigenfunctions:

$$\lambda_k = \left(\frac{k\pi}{l}\right)^2, \quad X_k(x) = \sqrt{\frac{2}{l}} \sin \frac{k\pi x}{l}. \qquad\qquad (5.5.24)$$

Equation (5.5.22) gives the following solutions:

$$T_k(t) = c_k e^{-(k\pi/l)^2 t}.$$

Thus, the function

$$u(t, x) = \sqrt{\frac{2}{l}} \sum_{k=1}^{\infty} c_k\, e^{-(k\pi/l)^2 t} \sin \frac{k\pi x}{l} \qquad\qquad (5.5.25)$$

solves (formally) the heat equation (5.5.18) and satisfies the boundary conditions (5.5.20). The initial condition (5.5.19) yields

$$\sqrt{\frac{2}{l}} \sum_{k=1}^{\infty} c_k \sin \frac{k\pi x}{l} = u_0(x),$$

whence

$$c_k = \sqrt{\frac{2}{l}} \int_0^l u_0(x) \sin \frac{k\pi x}{l} dx. \qquad\qquad (5.5.26)$$

Substituting (5.5.26) into (5.5.25), we obtain the solution of the mixed problem (5.5.18)—(5.5.20).

Problems to Chapter 5

5.1. Indicate all points (t, x, y) where the operator

$$L[u] = u_{xx} + u_{yy} + u_{zz} + \left(x^2 + y^2 + z^2 - t\right) u_{tt} - u_t$$

is elliptic, hyperbolic and parabolic.

5.2. Find the adjoint operator for the following operator of order zero: $L[u] = c(x, y)u$.

5.3. Find the adjoint operator for the following operator of the first order: $L[u] = a(x, y)u_x + b(x, y)u_y + c(x, y)u$.

5.4. Find the adjoint equation for the general hyperbolic equation in two variables written in the standard form (5.2.18):

$$u_{xy} + a(x, y)u_x + b(x, y)u_y + c(x, y)u = 0. \qquad \text{(P5.1)}$$

5.5. Find the adjoint equation for each of the following equations of the second order:

(i) Laplace equation: $u_{xx} + u_{yy} + u_{zz} = 0$,

(ii) Wave equation: $u_{tt} - k^2(u_{xx} + u_{yy} + u_{zz}) = 0$,

(iii) Heat equation: $u_t - k^2(u_{xx} + u_{yy} + u_{zz}) = 0$,

(iv) $x^2 u_{xx} + y^2 u_{yy} + 2xu_x + 2yu_y = 0$,

(v) Telegraph equation: $u_{tt} - c^2 u_{xx} - k^2 u = 0$ $(c, k = \text{const.})$,

(vi) Equation: $u_{tt} - \mu^2(x)u_{xx} - k^2 u = 0$ $(k = \text{const.})$,

(vii) Black-Scholes model: $u_t + \dfrac{1}{2}A^2 x^2 u_{xx} + Bxu_x - Cu = 0$,

(viii) $u_{xy} + \dfrac{u_x + u_y}{x + y} = 0$.

Which of the given equations are self-adjoint?

5.6. Find the adjoint equation and the adjoint operator for the following system of the first order:

$$u_x^1 + a(x, y)u_x^2 + b(x, y)u_y^2 = 0, \quad u_y^1 + c(x, y)u_x^2 + d(x, y)u_y^2 = 0.$$

5.7. Rewrite the telegraph equation (2.3.26), $u_{tt} - c^2 u_{xx} - k^2 u = 0$, in the characteristic variables, i.e., in the standard form (5.2.18).

5.8. Consider the following wave equations with a variable coefficient:

$$u_{tt} - \mu^2(x)u_{xx} = 0. \qquad (P5.2)$$

Find all equations (P5.2) reducible to the wave equation $v_{\xi\eta} = 0$.

5.9. The following statement is valid.

Theorem 5.5.1. Equation (P5.1) is equivalent to the telegraph equation, i.e., can be reduced to the telegraph equation $v_{\xi\eta} + kv = 0$ ($k =$ const.) by an appropriate equivalence transformation (5.3.7),

$$\xi = f(x), \quad \eta = g(y), \quad v = \sigma(x,y)\,u,$$

if and only if the Laplace invariants of Eq. (P5.1) obey the conditions

$$h = k \neq 0, \quad (\ln|h|)_{xy} = 0.$$

Using this theorem, show that the equation

$$u_{tt} - x^2 u_{xx} = 0$$

is equivalent to the telegraph equation $v_{\xi\eta}+v = 0$ and find the appropriate change of variables.

5.10. Single out Eqs. (P5.2) equivalent to the telegraph equation $v_{\xi\eta} + kv = 0$, $k =$ const.

5.11. Calculate the derivatives of the function $v(x,t)$ defined by (5.4.9):

$$v(x,t) = \frac{1}{2k}\int_0^t d\tau \int_{x-k(t-\tau)}^{x+k(t-\tau)} f(s,\tau)ds,$$

and show that its first and second derivatives v_t, v_{tt} and v_x, v_{xx} are given by Eqs. (5.4.10).

5.12. Let L be any linear differential operator and L^* its adjoint operator. Show that $(L^*)^* = L$, i.e., the adjoint operator to L^* coincides with the original operator.

5.13. Solve the mixed problem

$$u_{tt} = u_{xx}, \quad u\big|_{x=0} = u\big|_{x=2\pi} = 0, \quad u\big|_{t=0} = \sin x, \quad u_t\big|_{t=0} = 0.$$

5.14. Solve the mixed problem

$$u_{tt} = u_{xx}, \quad u\big|_{x=0} = u\big|_{x=2\pi} = 0, \quad u\big|_{t=0} = 0, \quad u_t\big|_{t=0} = \sin x.$$

5.15. Solve the mixed problem

$$u_t = u_{xx}, \quad u\big|_{x=0} = u\big|_{x=2\pi} = 0, \quad u\big|_{t=0} = \sin x.$$

5.16. Discuss the mixed problem

$$u_t = u_{xx}, \quad u\big|_{x=0} = u\big|_{x=2\pi} = 0, \quad u\big|_{t=0} = \cos x.$$

Chapter 6

Nonlinear ordinary differential equations

Mathematical models of fundamental natural laws and of technological problems are formulated frequently, even prevalently, in terms of *nonlinear differential equations*. Many of them are based on Newton's second law, and therefore they involve differential equations of the second order.

Thus, mathematical models of real world problems provide many nonlinear differential equations of the second order. The only general method for solving these equations analytically is provided by Lie group analysis which is particularly simple and efficient in the case of second-order equations.

Therefore, this chapter focuses on group analysis of nonlinear ordinary differential equations with emphasis on integration of first and second order equations. Applications of Lie group methods to linear and nonlinear partial differential equations are discussed in two last chapters.

Additional reading: S. Lie [26], L.V. Ovsyannikov [32], N.H. Ibragimov [21], P.J. Olver [31], G.W. Bluman and S. Kumei [2].

6.1   Introduction

The idea of symmetry permeates all mathematical models formulated in terms of differential equations. Mathematical tools for revealing and using the symmetry of differential equations are provided by the theory of continuous groups originated and elaborated by an outstanding mathematician of the nineteenth century, Sophus Lie. Lie group analysis provides general methods for integration of linear and nonlinear ordinary differential equations analytically using their symmetries. Lie group methods are also efficient in finding exact solutions to nonlinear partial differential equations.

Professor of Stanford University Brian Cantwell states in the preface to his recent book [3]: *"It is my firm belief that any graduate program in science or engineering needs to include a broad-based course on dimensional analysis and Lie groups. Symmetry analysis should be as familiar to the student as Fourier analysis, especially when so many unsolved problems are strongly nonlinear."*

Few would disagree with this statement. However, Lie group analysis has not enjoyed widespread acceptance in the past and the subject is still neglected in university programs. Moreover, there is a growing tendency in the modern literature to augment the old tradition to neglect Lie group methods and to write university texts on differential equations in a *cookbook style* containing numerous *ad hoc* recipes for integrating various special types of equations by means of artificial substitutions instead of using symmetries and dealing with Lie's few standard equations. Indeed, *"often the less there is to justify a traditional custom, the harder it is to get rid of it"* (M. Twain).

We present in this chapter a simple introduction to the basic concepts of the Lie group approach to ordinary differential equations. For a detailed discussion of the material outlined here, the reader is referred to [17].

6.2   Transformation groups

6.2.1   One-parameter groups on the plane

Let us consider a change of the variables x, y involving a parameter a :

$$T_a : \quad \bar{x} = \varphi(x, y, a), \quad \bar{y} = \psi(x, y, a), \tag{6.2.1}$$

with functions φ and ψ such that

$$T_0 : \quad \varphi(x, y, 0) = x, \quad \psi(x, y, 0) = y. \tag{6.2.2}$$

It is assumed that $\varphi(x, y, a)$ and $\psi(x, y, a)$ are functionally independent, i.e., their Jacobian does not vanish (see Section 1.2.8, Theorem 1.2.4):

$$\begin{vmatrix} \varphi_x & \varphi_y \\ \psi_x & \psi_y \end{vmatrix} \neq 0.$$

One can treat the equations T_a (6.2.1) also as a transformation that carries any point $P = (x, y)$ of the (x, y)-plane into a new position $\bar{P} = (\bar{x}, \bar{y})$ and write $\bar{P} = T_a(P)$. Accordingly, the inverse transformation T_a^{-1} given by

$$T_a^{-1} : \quad x = \varphi^{-1}(\bar{x}, \bar{y}, a), \quad y = \psi^{-1}(\bar{x}, \bar{y}, a) \tag{6.2.3}$$

returns $\bar{P}$ into the original position P, i.e.,

$$T_a^{-1}(\bar{P}) = P.$$

Furthermore, Eqs. (6.2.2) mean that T_0 is the identical transformation:

$$T_0(P) = P.$$

Let T_a and T_b be two transformations (6.2.1) with different values a and b of the parameter. Their *composition* (or *product*) $T_b T_a$ is defined as the consecutive application of these transformations and is given by

$$\bar{\bar{x}} = \varphi(\bar{x}, \bar{y}, b) = \varphi\Big(\varphi(x, y, a), \psi(x, y, a), \; b\Big),$$

$$\bar{\bar{y}} = \psi(\bar{x}, \bar{y}, b) = \psi\Big(\varphi(x, y, a), \psi(x, y, a), \; b\Big). \tag{6.2.4}$$

The geometric interpretation of the product is as follows. Since T_a carries the point P to the point $\bar{P} = T_a(P)$, which T_b carries to the new position $\bar{\bar{P}} = T_b(\bar{P})$, the product $T_b T_a$ is destined to carry P directly to its final location $\bar{\bar{P}}$, without a stopover at $\bar{P}$. Thus, (6.2.4) means that

$$\bar{\bar{P}} \overset{\text{def}}{=} T_b(\bar{P}) = T_b T_a(P).$$

Definition 6.2.1. The one-parameter family G of transformations (6.2.1) obeying the initial condition (6.2.2) is called a *one-parameter group* if G contains inverse (6.2.3) and the composition $T_b T_a$ of all its elements:

$$T_b T_a = T_{a+b}.$$

The latter condition, invoking (6.2.4), is written:

$$\varphi\Big(\varphi(x, y, a), \psi(x, y, a), \; b\Big) = \varphi(x, y, a + b),$$

$$\psi\Big(\varphi(x, y, a), \psi(x, y, a), \; b\Big) = \psi(x, y, a + b). \tag{6.2.5}$$

6.2.2 Group generator and the Lie equations

The expansion of the functions $\varphi(x, y, a)$ and $\psi(x, y, a)$ into Taylor's series in a near $a = 0$, taking into account the initial condition (6.2.2), yields the *infinitesimal transformation*

$$\bar{x} \approx x + \xi(x, y)a, \quad \bar{y} \approx y + \eta(x, y)a, \tag{6.2.6}$$

where

$$\xi(x, y) = \frac{\partial \varphi(x, y, a)}{\partial a}\Big|_{a=0}, \quad \eta(x, y) = \frac{\partial \psi(x, y, a)}{\partial a}\Big|_{a=0}. \tag{6.2.7}$$

The vector (ξ, η) with components (6.2.7) is the tangent vector (at the point (x, y)) to the curve described by the transformed points $(\bar{x}, \bar{y})$, and is therefore called the *tangent vector field* of the group G.

The tangent vector field (6.2.7) is associated with the first-order differential operator

$$X = \xi(x, y)\frac{\partial}{\partial x} + \eta(x, y)\frac{\partial}{\partial y} \tag{6.2.8}$$

called the *generator* of the group G.

Given an infinitesimal transformation (6.2.6), or generator (6.2.8), transformations (6.2.1) are defined by integrating the following system of ordinary differential equations called the *Lie equations*:

$$\frac{d\varphi}{da} = \xi(\varphi, \psi), \quad \varphi|_{a=0} = x,$$
$$\frac{d\psi}{da} = \eta(\varphi, \psi), \quad \psi|_{a=0} = y. \qquad (6.2.9)$$

These equations are written also as follows:

$$\frac{d\bar{x}}{da} = \xi(\bar{x}, \bar{y}), \quad \bar{x}|_{a=0} = x,$$
$$\frac{d\bar{y}}{da} = \eta(\bar{x}, \bar{y}), \quad \bar{y}|_{a=0} = y. \qquad (6.2.10)$$

For example, the group of rotations defined by

$$\bar{x} = x \cos a + y \sin a, \quad \bar{y} = y \cos a - x \sin a, \qquad (6.2.11)$$

has the infinitesimal transformation

$$\bar{x} \approx x + ya, \quad \bar{y} \approx y - xa$$

and the generator

$$X = y \frac{\partial}{\partial x} - x \frac{\partial}{\partial y}, \qquad (6.2.12)$$

respectively. You can easily verify the Lie equations:

$$\frac{d\bar{x}}{da} = \bar{y}, \quad \bar{x}|_{a=0} = x,$$
$$\frac{d\bar{y}}{da} = -\bar{x}, \quad \bar{y}|_{a=0} = y.$$

Example 6.2.1. Let us find the one-parameter group given by its infinitesimal transformation

$$\bar{x} \approx x + ax^2, \quad \bar{y} \approx y + axy,$$

or, equivalently, by the following generator:

$$X = x^2 \frac{\partial}{\partial x} + xy \frac{\partial}{\partial y}. \qquad (6.2.13)$$

The Lie equations (6.2.10) have the form

$$\frac{d\bar{x}}{da} = \bar{x}^2, \quad \bar{x}|_{a=0} = x,$$
$$\frac{d\bar{y}}{da} = \bar{x}\bar{y}, \quad \bar{y}|_{a=0} = y.$$

The differential equations of this system are easily solved and yield

$$\bar{x} = -\frac{1}{a + C_1}, \quad \bar{y} = \frac{C_2}{a + C_1}.$$

The initial conditions imply that $C_1 = -1/x$, $C_2 = -y/x$. Consequently, we arrive at the following one-parameter group known as the projective transformation group:

$$\bar{x} = \frac{x}{1 - ax}, \quad \bar{y} = \frac{y}{1 - ax}. \tag{6.2.14}$$

6.2.3 Exponential map

One can also solve the Lie equations (6.2.10) by looking for the solution in the form of infinite power series. Then the group transformation (6.2.1) for a generator X (6.2.8) is given by the following *exponential map*:

$$\bar{x} = e^{aX}(x), \quad \bar{y} = e^{aX}(y), \tag{6.2.15}$$

where

$$e^{aX} = 1 + \frac{a}{1!}X + \frac{a^2}{2!}X^2 + \cdots + \frac{a^s}{s!}X^s + \cdots. \tag{6.2.16}$$

Example 6.2.2. Consider again generator (6.2.13):

$$X = x^2 \frac{\partial}{\partial x} + xy \frac{\partial}{\partial y}.$$

According to (6.2.15)—(6.2.16), we should find $X^s(x)$ and $X^s(y)$ for all $s = 1, 2, \ldots$. We calculate several terms, e.g.

$$X(x) = x^2, \quad X^2(x) = X(X(x)) = X(x^2) = 2!x^3, \quad X^3(x) = X(2!x^3) = 3!x^4,$$

make a guess:

$$X^s(x) = s!x^{s+1}$$

and proof the latter equation by induction:

$$X^{s+1}(x) = X(s!x^{s+1}) = (s+1)!x^2 x^s = (s+1)!x^{s+2}.$$

Likewise, we calculate

$$X(y) = xy, \quad X^2(y) = X(xy) = yX(x) + xX(y) = yx^2 + xxy = 2!yx^2,$$

$$X^3(y) = 2![yX(x^2) + x^2 X(y)] = 2![y(2x^3) + x^2 xy] = 3!yx^3,$$

make a guess

$$X^s(y) = s!yx^s$$

and prove it by induction:

$$X^{s+1}(y) = s!X(yx^s) = s![syx^{s+1} + x^s(xy)] = (s+1)!yx^{s+1}.$$

Substitution of the above expressions in the exponential map yields:

$$e^{aX}(x) = x + ax^2 + \cdots + a^s x^{s+1} + \cdots.$$

We rewrite the right-hand side in the form $x(1+ax+\cdots+a^s x^s+\cdots)$ and observe that the series in brackets is the Taylor expansion of the function $1/(1 - ax)$ provided that $|ax| < 1$. Consequently,

$$\bar{x} = e^{aX}(x) = \frac{x}{1 - ax}.$$

Likewise, one obtains

$$e^{aX}(y) = y + ayx + a^2 yx^2 + \cdots + a^s yx^s + \cdots$$

$$= y(1 + ax + \cdots + a^s x^s + \cdots).$$

Hence,

$$\bar{y} = e^{aX}(y) = \frac{y}{1 - ax}.$$

Thus, we have arrived at the projective transformation (6.2.14):

$$\bar{x} = \frac{x}{1 - ax}, \quad \bar{y} = \frac{y}{1 - ax}.$$

6.2.4   Invariants and invariant equations

Definition 6.2.2. A function $F(x, y)$ is called an invariant of the group G of transformations (6.2.1) if $F(\bar{x}, \bar{y}) = F(x, y)$, i.e.,

$$F(\varphi(x, y, a), \psi(x, y, a)) = F(x, y) \qquad (6.2.17)$$

identically in the variables x, y and the group parameter a.

Theorem 6.2.1. A function $F(x, y)$ is an invariant of the group G if and only if it solves the following first-order linear partial differential equation:

$$XF \equiv \xi(x, y)\frac{\partial F}{\partial x} + \eta(x, y)\frac{\partial F}{\partial y} = 0. \qquad (6.2.18)$$

Proof. Let $F(x, y)$ be an invariant. Let us take the Taylor expansion of $F(\varphi(x, y, a), \psi(x, y, a))$ with respect to a :

$$F(\varphi(x, y, a), \psi(x, y, a)) \approx F(x + a\xi, y + a\eta) \approx F(x, y) + a\left(\xi\frac{\partial F}{\partial x} + \eta\frac{\partial F}{\partial y}\right),$$

or

$$F(\bar{x}, \bar{y}) = F(x, y) + aXF + o(a),$$

and substitute it into Eq. (6.2.17):

$$F(x, y) + aXF + o(a) = F(x, y).$$

It follows that $aXF + o(a) = 0$, whence $XF = 0$, i.e., Eq. (6.2.18).

Conversely, let $F(x, y)$ be a solution of Eq. (6.2.18). Assuming that the function $F(x, y)$ is analytic and using its Taylor expansion, one can extend the exponential map (6.2.15) to the function $F(x, y)$ as follows:

$$F(\overline{x}, \overline{y}) = e^{aX} F(x, y) \overset{\text{def}}{=} \left(1 + \frac{a}{1!} X + \frac{a^2}{2!} X^2 + \cdots + \frac{a^s}{s!} X^s + \cdots \right) F(x, y).$$

Since $XF(x, y) = 0$, one has $X^2 F = X(XF) = 0, \ldots, X^s F = 0$. We conclude that $F(\overline{x}, \overline{y}) = F(x, y)$, i.e., Eq. (6.2.17) thus proving the theorem.

It follows from Theorem 6.2.1 that every one-parameter group of transformations in the plane has one independent invariant, which can be taken to be the left-hand side of any first integral $h(x, y) = C$ of the characteristic equation for (6.2.18):

$$\frac{dx}{\xi(x, y)} = \frac{dy}{\eta(x, y)}. \tag{6.2.19}$$

Any other invariant F is then a function of h, i.e., $F(x, y) = \Phi(h(x, y))$.

Example 6.2.3. Consider the group with generator (6.2.13),

$$X = x^2 \frac{\partial}{\partial x} + xy \frac{\partial}{\partial y}.$$

The characteristic equation (6.2.19) is written

$$\frac{dx}{x} = \frac{dy}{y}$$

and yields the first integral $h = x/y$. Hence, the general invariant is given by $F(x, y) = \Phi(x/y)$ with an arbitrary function Φ of one variable.

Groups of transformations (6.2.1) and the related concepts discussed above can be generalized in an obvious way to the multi-dimensional case by considering groups of transformations

$$\overline{x}^i = f^i(x, a), \quad i = 1, \ldots, n, \tag{6.2.20}$$

in the n-dimensional space of points $x = (x^1, \ldots, x^n)$. The generator of transformations (6.2.20) is written

$$X = \xi^i(x) \frac{\partial}{\partial x^i}, \tag{6.2.21}$$

where

$$\xi^i(x) = \frac{\partial f^i(x, a)}{\partial a} \Big|_{a=0}.$$

The Lie equations (6.2.10) become

$$\frac{d\overline{x}^i}{da} = \xi^i(\overline{x}), \quad \overline{x}^i \big|_{a=0} = x^i. \tag{6.2.22}$$

The exponential map is written:

$$\bar{x}^i = e^{aX}(x^i), \quad i = 1, \ldots, n, \tag{6.2.23}$$

where

$$e^{aX} = 1 + \frac{a}{1!}X + \frac{a^2}{2!}X^2 + \cdots + \frac{a^s}{s!}X^s + \cdots. \tag{6.2.24}$$

Definition 6.2.2 of invariant functions has the same formulation in the case of several variables. Namely, an invariant is defined by the equation $F(\bar{x}) = F(x)$. The invariant test is again given by Theorem 6.2.1 with the evident replacement of Eq. (6.2.18) by its n-dimensional version:

$$\sum_{i=1}^{n} \xi^i(x) \frac{\partial F}{\partial x^i} = 0. \tag{6.2.25}$$

Then $n - 1$ functionally independent first integrals $\psi_1(x), \ldots, \psi_{n-1}(x)$ of the characteristic system for Eq. (6.2.25):

$$\frac{dx^1}{\xi^1(x)} = \frac{dx^2}{\xi^2(x)} = \cdots = \frac{dx^n}{\xi^n(x)} \tag{6.2.26}$$

provides a basis of invariants. Namely, any invariant $F(x)$ is given by

$$F(x) = \Phi\big(\psi_1(x), \ldots, \psi_{n-1}(x)\big). \tag{6.2.27}$$

Let us dwell on this higher-dimensional case and consider a system of equations

$$F(x) = 0, \ldots, F_s = 0, \quad s < n. \tag{6.2.28}$$

We shall assume that the rank of the matrix $(\partial F_k/\partial x^i)$ is equal to s at all points x satisfying the system of equations (6.2.28). The system of equations (6.2.28) then defines an $(n - s)$-dimensional surface M.

Definition 6.2.3. The system of equations (6.2.28) is said to be invariant with respect to the group G of transformations (6.2.20) if each point x on the surface M is moved by G along M, i.e., $x \in M$ implies $\bar{x} \in M$. The surface M is called an invariant surface.

Theorem 6.2.2. The surface M given by the system of equations (6.2.28) is invariant with respect to the group G of transformations (6.2.20) with generator X (6.2.21) if and only if

$$X(F_k)\Big|_M = 0, \quad k = 1, \ldots, s. \tag{6.2.29}$$

6.2.5 Canonical variables

The following simple statement resulting from Lemma 4.2.1 has many applications, e.g. in integration of differential equations.

Theorem 6.2.3. Every one-parameter group of transformations (6.2.1) with generator (6.2.8),

$$X = \xi(x, y)\frac{\partial}{\partial x} + \eta(x, y)\frac{\partial}{\partial y}$$

can be reduced, by a suitable change of variables

$$t = t(x, y), \quad u = u(x, y), \tag{6.2.30}$$

to the group of translations $\bar{t} = t + a$, $\bar{u} = u$ with the generator

$$X = \frac{\partial}{\partial t}. \tag{6.2.31}$$

The variables t and u are called *canonical variables*.

Proof. The change of variables (6.2.30) transforms the differential operator (6.2.8) as follows (see Eq. (4.2.3)):

$$X = X(t(x, y))\frac{\partial}{\partial t} + X(u(x, y))\frac{\partial}{\partial u}. \tag{6.2.32}$$

Hence, one arrives at operator (6.2.31) if one defines t and u by solving the following first-order linear partial differential equations:

$$X(t) \equiv \xi(x, y)\frac{\partial t}{\partial x} + \eta(x, y)\frac{\partial t}{\partial y} = 1,$$

$$X(u) \equiv \xi(x, y)\frac{\partial u}{\partial x} + \eta(x, y)\frac{\partial u}{\partial y} = 0. \tag{6.2.33}$$

Example 6.2.4. Let us find canonical variables for the dilation group with the generator

$$X = x\frac{\partial}{\partial x} + y\frac{\partial}{\partial y}. \tag{6.2.34}$$

The first equation of (6.2.33) has the form

$$X(t) = x\frac{\partial t}{\partial x} + y\frac{\partial t}{\partial y} = 1.$$

Since it is sufficient to find any particular solution to this equation, we can look, e.g. for a solution $t = t(x)$ depending only on x. Then the above equation reduces to the ordinary differential equation $x\, t'(x) = 1$, whence $t = \ln|x|$. The second equation of (6.2.33) has the form

$$X(u) = x\frac{\partial u}{\partial x} + y\frac{\partial u}{\partial y} = 0.$$

The characteristic equation

$$\frac{\mathrm{d}x}{x} = \frac{\mathrm{d}y}{y}$$

has the first integral $y/x = C$, and hence $u = y/x$ solves the equation $X(u) = 0$. Thus, we have the following canonical variables:

$$t = \ln|x|, \quad u = \frac{y}{x}.$$

In these variables, the dilation group

$$\bar{x} = x\,\mathrm{e}^a, \quad \bar{y} = y\,\mathrm{e}^a$$

reduces to the translation group

$$\bar{t} = t + a, \quad \bar{u} = u.$$

Indeed,

$$\bar{t} = \ln|\bar{x}| = \ln(|x|\,\mathrm{e}^a) = \ln|x| + a = t + a,$$

$$\bar{u} = \frac{\bar{y}}{\bar{x}} = \frac{y\,\mathrm{e}^a}{x\,\mathrm{e}^a} = \frac{y}{x} = u.$$

6.3   Symmetries of first-order equations

6.3.1   First prolongation of group generators

The transformation of derivatives under the group transformations (6.2.1), regarded as a change of variables, is given in Section 1.4.5. In particular, the transformation of the first derivative is given by the formula

$$\bar{y}' \equiv \frac{\mathrm{d}\bar{y}}{\mathrm{d}\bar{x}} = \frac{D_x(\psi)}{D_x(\varphi)}. \tag{6.3.1}$$

Thus, starting from the group G of transformations (6.2.1) we have obtained the group $G_{(1)}$ consisting of transformations (6.2.1) and (6.3.1) in the space of the variables (x, y, y'). The group $G_{(1)}$ is called the *first prolongation* of G. Substituting into (6.3.1) the infinitesimal transformation (6.2.6), $\varphi = x + a\xi(x, y)$, $\psi = y + a\eta(x, y)$, and neglecting the higher order terms in a one obtains the following infinitesimal transformations of y':

$$\bar{y}' = \frac{y' + a\,D_x(\eta)}{1 + a\,D_x(\xi)} \approx [y' + aD(\eta)][1 - aD(\xi)] \approx y' + [D(\eta) - y'D(\xi)]a,$$

or

$$\bar{y}' \approx y' + a\zeta_1,$$

where

$$\zeta_1 = D_x(\eta) - y' D_x(\xi) = \eta_x + (\eta_y - \xi_x)y' - y'^2 \xi_y. \tag{6.3.2}$$

Therefore, the generator X of the group G after the prolongation becomes:

$$X = \xi \frac{\partial}{\partial x} + \eta \frac{\partial}{\partial y} + \zeta_1 \frac{\partial}{\partial y'}. \tag{6.3.3}$$

Equations (6.3.2) and (6.3.3) are referred to as the *first prolongation formula* and the *first prolongation of the infinitesimal generator* (6.2.13), respectively.

6.3.2 Symmetry group: definition and main property

Definition 6.3.1. The group G of transformations (6.2.1) is called a symmetry group of a first-order ordinary differential equation

$$\frac{dy}{dx} = f(x, y), \tag{6.3.4}$$

or that Eq. (6.3.4) admits the group G if the form of the differential equation (6.3.4) remains the same after the change of variables (6.2.1), i.e.,

$$\frac{d\bar{y}}{d\bar{x}} = f(\bar{x}, \bar{y}),$$

where the function f is the same as in the original equation (6.3.4). In other words, G is a symmetry group for Eq. (6.3.4) if the frame (see Section 1.4.4) of Eq. (6.3.4) is invariant, in the sense of Definition 6.2.3, with respect to the first prolongation $G_{(1)}$ of the group G.

The main property of a symmetry group first proved by S. Lie (see, e.g. [26], Chapter 16, Section 1, Theorem 1) is that G is a symmetry group if and only if it converts any classical solution of Eq. (6.3.4) into a classical solution of the same equation.

The generator X of a group admitted by a differential equation is termed an *admitted operator* or an *infinitesimal symmetry* of Eq. (6.3.4).

Example 6.3.1. It is evident that the equation

$$y' = f(y)$$

does not alter after the transformation $\bar{x} = x + a$ since the equation does not explicitly contain the independent variable x. Therefore, the symmetry of this differential equation is given by the *group of translations* along the x-axis, $\bar{x} = x + a$, with the generator

$$X = \frac{\partial}{\partial x}.$$

Likewise, the equation

$$y' = f(x)$$

admits the group of translations along the y-axis, $\bar{y} = y + a$, with the generator

$$X = \frac{\partial}{\partial y} \, .$$

Example 6.3.2. The equation

$$y' = f\left(\frac{y}{x}\right) \tag{6.3.5}$$

admits the *group of homogeneous dilations* (scaling transformations)

$$\bar{x} = x e^a, \quad \bar{y} = y e^a$$

with generator (6.2.34),

$$X = x \frac{\partial}{\partial x} + y \frac{\partial}{\partial y} \, .$$

Example 6.3.3. Consider the following Riccati equation:

$$y' + y^2 - \frac{2}{x^2} = 0. \tag{6.3.6}$$

Its left-hand side is a rational function in the variables x, y, y'. Therefore, one can try to find an admitted group in the form of a dilation:

$$\bar{x} = kx, \quad \bar{y} = ly.$$

Since

$$\bar{y}' + \bar{y}^2 - \frac{2}{\bar{x}^2} = \frac{l}{k} y' + l^2 y^2 - \frac{1}{k^2} \frac{2}{x^2} \, ,$$

the invariance condition requires that

$$\bar{y}' + \bar{y}^2 - \frac{2}{\bar{x}^2} = \lambda \cdot \left(y' + y^2 - \frac{2}{x^2} \right),$$

where

$$\lambda = \frac{l}{k} = l^2 = \frac{1}{k^2} \, .$$

The latter equations yield $l = 1/k$, where $k > 0$ is an arbitrary parameter. Setting $k = e^a$, we get the non-homogeneous dilation group:

$$\bar{x} = x e^a, \quad \bar{y} = y e^{-a}.$$

Hence, the Riccati equation (6.3.6) has the following infinitesimal symmetry:

$$X = x \frac{\partial}{\partial x} - y \frac{\partial}{\partial y} \, . \tag{6.3.7}$$

6.3.3 Equations with a given symmetry

Solution of this problem is based on Definition 6.3.1 of a symmetry group, prolongation formulae (6.3.2)—(6.3.3), and the invariant surface test (6.2.29). Namely, in order to find the general first-order differential equation (6.3.4),

$$y' = f(x, y), \tag{6.3.8}$$

admitting a given operator

$$X = \xi(x, y)\frac{\partial}{\partial x} + \eta(x, y)\frac{\partial}{\partial y}, \tag{6.3.9}$$

we prolong the operator X by means of formula (6.3.2):

$$X = \xi\frac{\partial}{\partial x} + \eta\frac{\partial}{\partial y} + \zeta_1\frac{\partial}{\partial y'}$$

and write the invariant surface test (6.2.29):

$$\zeta_1\big|_{y'=f(x,y)} = \xi\frac{\partial f}{\partial x} + \eta\frac{\partial f}{\partial y}.$$

Substituting here the expression for ζ_1 given by the prolongation formula (6.3.3), we obtain the following quasi-linear first-order partial differential equation for determining $f(x, y)$:

$$\xi\frac{\partial f}{\partial x} + \eta\frac{\partial f}{\partial y} = \eta_x + (\eta_y - \xi_x)f - \xi_y f^2. \tag{6.3.10}$$

Upon solving Eq. (6.3.10), we obtain all equations (6.3.8) admitting the group generated by operator (6.3.9).

Example 6.3.4. Let us find the equations admitting the operator

$$X = x\frac{\partial}{\partial x} - y\frac{\partial}{\partial y}. \tag{6.3.11}$$

Equation (6.3.10) is written

$$x\frac{\partial f}{\partial x} - y\frac{\partial f}{\partial y} = -2f. \tag{6.3.12}$$

The characteristic system (4.4.7) for Eq. (6.3.12) has the form

$$\frac{dx}{x} = -\frac{dy}{y} = -\frac{df}{2f}.$$

Taking its two first integrals $\psi_1 = xy$ and $\psi_2 = x^2 f$, we obtain the general solution to Eq. (6.3.12):

$$f = x^{-2} F(xy).$$

Hence, the equations admitting operator (6.3.11) have the form

$$x^2 y' = F(xy). \tag{6.3.13}$$

Example 6.3.5. Let us find the equations admitting the group of rotations (6.2.11) with generator (6.2.12),

$$X = y\frac{\partial}{\partial x} - x\frac{\partial}{\partial y}. \tag{6.3.14}$$

Equation (6.3.10) is written

$$y\frac{\partial f}{\partial x} - x\frac{\partial f}{\partial y} = -(1 + f^2). \tag{6.3.15}$$

The characteristic system (4.4.7) for Eq. (6.3.15) has the form

$$\frac{dx}{y} = -\frac{dy}{x} = -\frac{df}{1 + f^2}.$$

The first equation, $x\,dx + y\,dy = 0$, yields $x^2 + y^2 =$ const. Hence, we take

$$\psi_1 = \sqrt{x^2 + y^2}.$$

Thus, on the characteristic equations we have $x^2 + y^2 = a^2 =$ const. By virtue of this relation, the second equation is rewritten (cf. Example 4.4.1)

$$\frac{df}{1 + f^2} = \frac{dy}{\sqrt{a^2 - y^2}},$$

whence, upon integration,

$$\arctan f = \arctan(y/x) + \arctan C. \tag{6.3.16}$$

It remains to solve this equation with respect to the constant of integration C and identify C with a first integral ψ_2. Then the solution to Eq. (6.3.15) is obtained explicitly by letting $\psi_2 = F(\psi_1)$. Thus, we write Eq. (6.3.16) in the form

$$\arctan f = \arctan(y/x) + \arctan F(\psi_1)$$

and solve it with respect to f by using the elementary formula

$$\tan(\alpha + \beta) = \frac{\tan\alpha + \tan\beta}{1 - \tan\alpha\,\tan\beta}$$

to obtain:

$$f = \frac{(y/x) + F(\psi_1)}{1 - (y/x)\,F(\psi_1)} = \frac{y + xF(\sqrt{x^2 + y^2})}{x - yF(\sqrt{x^2 + y^2})}.$$

Hence, the equations admitting operator (6.3.14) have the form

$$y' = \frac{y + xF(\sqrt{x^2 + y^2})}{x - yF(\sqrt{x^2 + y^2})}. \tag{6.3.17}$$

Table 6.3.1 shows us some nonlinear first-order equationg with a known symmetry.

Table 6.3.1 Some nonlinear first-order equations
with a known symmetry

No.	Equation	Symmetry
1	$y' = F(y)$	$X = \frac{\partial}{\partial x}$
	$y' = F(kx + ly)$	$X = l\frac{\partial}{\partial x} - k\frac{\partial}{\partial y}$
2	$y' = F(y/x)$	$X = x\frac{\partial}{\partial x} + y\frac{\partial}{\partial y}$
3	$y' = x^{k-1}F(y/x^k)$	$X = x\frac{\partial}{\partial x} + ky\frac{\partial}{\partial y}$
4	$xy' = F(xe^{-y})$	$X = x\frac{\partial}{\partial x} + \frac{\partial}{\partial y}$
5	$y' = yF(ye^{-x})$	$X = \frac{\partial}{\partial x} + y\frac{\partial}{\partial y}$
6	$y' = \frac{y}{x} + xF(y/x)$	$X = \frac{\partial}{\partial x} + \frac{y}{x}\frac{\partial}{\partial y}$
7	$xy' = y + F(y/x)$	$X = x^2\frac{\partial}{\partial x} + xy\frac{\partial}{\partial y}$
8	$y' = \dfrac{y}{x + F(y/x)}$	$X = xy\frac{\partial}{\partial x} + y^2\frac{\partial}{\partial y}$
9	$y' = \dfrac{y}{x + F(y)}$	$X = y\frac{\partial}{\partial x}$
10	$xy' = y + F(x)$	$X = x\frac{\partial}{\partial y}$
11	$xy' = \dfrac{y}{\ln x + F(y)}$	$X = xy\frac{\partial}{\partial x}$
12	$xy' = y[\ln y + F(x)]$	$X = xy\frac{\partial}{\partial y}$

6.4 Integration of first-order equations using symmetries

6.4.1 Lie's integrating factor

Consider a first-order equation written in the symmetric form (3.2.3):

$$M(x, y)\mathrm{d}x + N(x, y)\mathrm{d}y = 0. \tag{6.4.1}$$

Lie showed that if

$$X = \xi(x, y)\frac{\partial}{\partial x} + \eta(x, y)\frac{\partial}{\partial y}$$

is a symmetry for Eq. (6.4.1) and if $\xi M + \eta N \neq 0$, then the function

$$\mu = \frac{1}{\xi M + \eta N} \tag{6.4.2}$$

is an integrating factor for Eq. (6.4.1). It is called *Lie's integrating factor*.

Example 6.4.1. The equation $2xydx + (y^2 - 3x^2)dy = 0$ from Example 3.2.2 is homogeneous (Example 3.1.1), i.e., has form (6.3.5) and admits

$$X = x\frac{\partial}{\partial x} + y\frac{\partial}{\partial y}.$$

Lie's formula (6.4.2) provides the integrating factor

$$\mu = \frac{1}{2x^2y + y^3 - 3x^2y} = \frac{1}{y^3 - x^2y}.$$

Example 6.4.2. Let us solve the Riccati equation (6.3.6) by using Lie's integrating factor. We rewrite Eq. (6.3.6) in the differential form (6.4.1):

$$dy + \left(y^2 - \frac{2}{x^2}\right)dx = 0, \tag{6.4.3}$$

and use symmetry (6.3.7), $X = x\dfrac{\partial}{\partial x} - y\dfrac{\partial}{\partial y}$. Substituting

$$\xi = x, \quad \eta = -y, \quad M = y^2 - \frac{2}{x^2}, \quad N = 1$$

into (6.4.2), we obtain the integrating factor

$$\mu = \frac{x}{x^2y^2 - xy - 2}.$$

After multiplication by this factor, Eq. (6.4.3) becomes

$$\frac{xdy}{x^2y^2 - xy - 2} + \frac{1}{x^2y^2 - xy - 2}\frac{x^2y^2 - 2}{x}dx = 0. \tag{6.4.4}$$

This equation is exact, i.e., its left-hand side can be written in the form $d\Phi$. The function $\Phi(x, y)$ can be determined by the general procedure described in Section 3.2.2. In this particular case, we can use the following simple calculations. Noting that

$$\frac{x^2y^2 - 2}{x} = y + \frac{x^2y^2 - xy - 2}{x},$$

we rewrite the left-hand side of Eq. (6.4.4) in the form

$$\frac{xdy + ydx}{x^2y^2 - xy - 2} + \frac{dx}{x} = \frac{d(xy)}{x^2y^2 - xy - 2} + \frac{dx}{x}.$$

Denoting $z = xy$ and using the decomposition

$$\frac{1}{z^2 - z - 2} = \frac{1}{3}\left(\frac{1}{z - 2} - \frac{1}{z + 1}\right),$$

we obtain

$$\int \frac{dz}{z^2 - z - 2} = \frac{1}{3} \ln \frac{z-2}{z+1}$$

and hence Eq. (6.4.4) is written

$$\frac{x dy + y dx}{x^2 y^2 - xy - 2} + \frac{dx}{x} = d\left(\ln x + \frac{1}{3} \ln \frac{xy-2}{xy+1} \right) = 0.$$

The integration yields

$$\frac{xy - 2}{xy + 1} = \frac{C}{x^3}, \quad C \neq 0.$$

Solving for y, we arrive at the solution of the Riccati equation (6.3.6):

$$y = \frac{2x^3 + C}{x(x^3 - C)}.$$

Example 6.4.3. Consider the following equation (see also Eq. (6.4.13)):

$$y' = \frac{y}{x} + \frac{y^2}{x^3}.$$

It is shown in Example 6.4.6 that this equation has the symmetry

$$X_1 = x^2 \frac{\partial}{\partial x} + xy \frac{\partial}{\partial y}.$$

Proceeding as in Example 6.3.3 from Section 6.3.2, one can see that our equation admits also the following dilation generator:

$$X_2 = x \frac{\partial}{\partial x} + 2y \frac{\partial}{\partial y}.$$

Writing the equation in this Example in the differential form

$$(x^2 y + y^2) dx - x^3 dy = 0$$

and substituting the coordinates of the operators X_1 and X_2 into (6.4.2), we obtain two linearly independent integrating factors:

$$\mu_1 = \frac{1}{x^2(x^2 y + y^2) - x^4 y} = \frac{1}{x^2 y^2}, \quad \mu_2 = \frac{1}{x(x^2 y + y^2) - 2x^3 y} = \frac{1}{xy(y - x^2)}.$$

Equation (3.2.13), $\mu_1/\mu_2 = C$, is written $y - x^2 = Cxy$ and yields the solution $y = x^2/(1 - Cx)$. Compare with (6.4.15) in Example 6.4.6.

6.4.2   Integration using canonical variables

To integrate a linear or nonlinear first-order ordinary differential equation

$$y' = f(x, y) \tag{6.4.5}$$

with a known infinitesimal symmetry

$$X = \xi(x, y)\frac{\partial}{\partial x} + \eta(x, y)\frac{\partial}{\partial y} \tag{6.4.6}$$

by the *method of canonical variables*, you need to make the following steps.

First step: Find canonical variables t, u by solving Eqs. (6.2.33) for the given generator (6.4.6).

Second step: Rewrite Eq. (6.4.5) in the canonical variables t and u by letting u be the new dependent variable of the independent variable t, i.e., letting $u = u(t)$ and expressing the old derivative $y' = dy/dx$ via new variables t, u and the derivative $u' = du/dt$. Then Eq. (6.4.5) will have the integrable form

$$\frac{du}{dt} = g(u). \tag{6.4.7}$$

Third step: Integrate Eq. (6.4.7), substitute into its solution $u = \phi(t, C)$ the expressions $t = t(x, y)$ and $u = u(x, y)$ thus obtaining the solution of Eq. (6.4.5).

Example 6.4.4. One can integrate by quadrature any equation of form (6.3.5) using its infinitesimal symmetry (6.2.34). To be specific, let us integrate the following particular equation of form (6.3.5):

$$y' = \frac{y}{x} + \frac{y^3}{x^3}. \tag{6.4.8}$$

First step: The canonical variables for the infinitesimal symmetry (6.2.34) are found in Example 6.2.4. They are

$$t = \ln|x|, \quad u = \frac{y}{x}.$$

Second step: Let us rewrite Eq. (6.4.8) in the canonical variables t and u. Since $y = xu$ and $dt/dx = 1/x$, we have denoting $u' = du/dt$:

$$y' \equiv \frac{dy}{dx} = \frac{d(xu)}{dx} = u + x\frac{du}{dx} = u + x\frac{du}{dt}\frac{dt}{dx} = u + xu'\frac{1}{x} = u + u'.$$

Thus, Eq. (6.4.8) becomes a particular equation of form (6.4.7):

$$\frac{du}{dt} = u^3.$$

Third step: Integration of the above equation yields

$$u = \pm \frac{1}{\sqrt{C - 2t}},$$

whence, upon substituting $t = \ln|x|$ and $y = xu$:

$$y = \pm \frac{x}{\sqrt{C - \ln x^2}}.$$

Example 6.4.5. Let us integrate the Riccati equation (6.3.6),

$$\frac{dy}{dx} + y^2 - \frac{2}{x^2} = 0,$$

by applying the method of canonical variables to symmetry (6.3.7),

$$X = x \frac{\partial}{\partial x} - y \frac{\partial}{\partial y}.$$

First step: Solution of Eqs. (6.2.33) with the above operator X provides the canonical variables, $t = \ln|x|$ and $u = xy$.

Second step: We have

$$\frac{dy}{dx} = \frac{d}{dx}\left(\frac{u}{x}\right) = -\frac{u}{x^2} + \frac{1}{x}\frac{du}{dx} = -\frac{u}{x^2} + \frac{1}{x}\frac{du}{dt}\frac{dt}{dx} = -\frac{u}{x^2} + \frac{u'}{x^2}.$$

Therefore,

$$\frac{dy}{dx} + y^2 - \frac{2}{x^2} = \frac{u'}{x^2} - \frac{u}{x^2} + \frac{u^2}{x^2} - \frac{2}{x^2} = \frac{1}{x^2}\left(u' + u^2 - u - 2\right) = 0.$$

Thus, the Riccati equation is rewritten in the canonical variables in the following integrable form (6.4.7):

$$\frac{du}{dt} = -(u^2 - u - 2).$$

Third step: Let us integrate the above equation. Separating the variables:

$$\frac{du}{u^2 - u - 2} = -dt,$$

decomposing here the rational fraction into elementary fractions:

$$\frac{1}{u^2 - u - 2} = \frac{1}{3}\left(\frac{1}{u - 2} - \frac{1}{u + 1}\right),$$

and integrating we have

$$\ln \frac{u - 2}{u + 1} = -3t + \ln C.$$

Now solve this equation with respect to u,

$$u = \frac{C + 2e^{3t}}{e^{3t} - C},$$

substitute $t = \ln|x|$, $u = xy$ and obtain the solution of the Riccati equation (6.3.6) given in Example 6.4.2:

$$y = \frac{2x^3 + C}{x(x^3 - C)}, \quad C = \text{const.} \tag{6.4.9}$$

Note that the above calculations require that both $xy - 2$ and $xy + 1$ do not vanish. Therefore, we should add to (6.4.9) the singular solutions of Eq. (6.3.6):

$$y = \frac{2}{x} \quad \text{and} \quad y = -\frac{1}{x}. \tag{6.4.10}$$

Example 6.4.6. The equation

$$y' = \frac{y}{x} + \frac{1}{x} F\left(\frac{y}{x}\right), \tag{6.4.11}$$

with an arbitrary function F, admits generator (6.2.13)

$$X = x^2 \frac{\partial}{\partial x} + xy \frac{\partial}{\partial y} \tag{6.4.12}$$

of the projective transformation group (6.2.14),

$$\bar{x} = \frac{x}{1 - ax}, \quad \bar{y} = \frac{y}{1 - ax}.$$

Indeed, the equations

$$D_x(\bar{x}) = \frac{1}{(1 - ax)^2}, \quad D_x(\bar{y}) = \frac{(1 - ax)y' + ay}{(1 - ax)^2}$$

yield

$$\bar{y}' = \frac{d\bar{y}}{d\bar{x}} = \frac{D_x(\bar{y})}{D_x(\bar{x})} = (1 - ax)y' + ay.$$

Therefore,

$$\bar{y}' - \frac{\bar{y}}{\bar{x}} - \frac{1}{\bar{x}} F\left(\frac{\bar{y}}{\bar{x}}\right) = (1 - ax)y' + ay - \frac{y}{x} - \frac{1 - ax}{x} F\left(\frac{y}{x}\right)$$

or

$$\bar{y}' - \frac{\bar{y}}{\bar{x}} - \frac{1}{\bar{x}} F\left(\frac{\bar{y}}{\bar{x}}\right) = (1 - ax)\left[y' - \frac{y}{x} - \frac{1}{x} F\left(\frac{y}{x}\right)\right].$$

Hence, Eq. (6.4.11) yields

$$\bar{y}' - \frac{\bar{y}}{\bar{x}} - \frac{1}{\bar{x}} F\left(\frac{\bar{y}}{\bar{x}}\right) = 0.$$

Let us integrate a particular equation of form (6.4.11) by taking $F(\sigma) = \sigma^2$, i.e., the following equation (see Example 6.4.3):

$$y' = \frac{y}{x} + \frac{y^2}{x^3}.$$ (6.4.13)

First step: Eqs. (6.2.33) with operator (6.4.12) are written:

$$X(t) = x^2\frac{\partial t}{\partial x} + xy\frac{\partial t}{\partial y} = 1, \quad X(u) = x^2\frac{\partial u}{\partial x} + xy\frac{\partial u}{\partial y} = 0.$$

Taking a particular solution of these equations with $t = t(x)$, one obtains the canonical variables

$$t = -\frac{1}{x}, \quad u = \frac{y}{x}.$$ (6.4.14)

Second step: We have

$$y' = \frac{d(xu)}{dx} = u + x\frac{du}{dx} = u + x\frac{du}{dt}\frac{dt}{dx} = u + xu'\frac{1}{x^2} = u + \frac{1}{x}u',$$

and Eq. (6.4.13) takes the following integrable form (6.4.7):

$$\frac{du}{dt} = u^2.$$

Third step: Excluding the solution $u = 0$ and integrating we have

$$u = -\frac{1}{C + t}.$$

Substituting here $t = -1/x, u = y/x$ and adding the solution $u = 0$ we arrive at the general solution to Eq. (6.4.13) given by

$$y = \frac{x^2}{1 - Cx} \quad \text{and} \quad y = 0.$$ (6.4.15)

6.4.3 Invariant solutions

An essential feature of a symmetry group G of an ordinary differential equation is that it converts any solution (integral curve) of the equation in question into a solution. In other words, the symmetry transformations merely permute the integral curves among themselves. It may happen that some of the integral curves are individually unaltered under G. Such integral curves are termed *invariant solutions*.

Example 6.4.7. Consider again the Riccati equation (6.3.6)

$$\frac{dy}{dx} + y^2 - \frac{2}{x^2} = 0$$

and find its invariant solutions with respect to the infinitesimal symmetry

$$X = x\frac{\partial}{\partial x} - y\frac{\partial}{\partial y}.$$

The invariant test (6.2.18), $X(J) = 0$, provides a single independent invariant xy. Therefore, $xy = $ const. is the only relation written in terms of invariants. Hence, the general form of invariant solutions is $xy = \lambda$, or $y = \lambda/x$, $\lambda = $ const. The substitution into the Riccati equation reduces the latter to an algebraic, namely the quadratic equation, $\lambda^2 - \lambda - 2 = 0$, whence $\lambda_1 = 2$ and $\lambda_2 = -1$. Thus, the invariant solutions are identical with two singular solutions (6.4.10):

$$y_1 = \frac{2}{x} \quad \text{and} \quad y_2 = -\frac{1}{x}.$$

6.4.4 General solution provided by invariant solutions

The technique of invariant solutions furnishes a simple way for obtaining the general integral of first-order ordinary differential equations with two known infinitesimal symmetries. Consider the following example.

Example 6.4.8. Consider Eq. (6.4.11) with $F(\sigma) = \sigma^n$, i.e., the equation

$$y' = \frac{y}{x} + \frac{y^n}{x^{n+1}}.$$

It admits, along with the projective group with generator (6.4.12), the dilation group with the generator

$$X = (n-1)x\frac{\partial}{\partial x} + ny\frac{\partial}{\partial y}.$$

Let us take, e.g. $n = 2$. Thus, we consider Eq. (6.4.13),

$$y' = \frac{y}{x} + \frac{y^2}{x^3},$$

with two known infinitesimal symmetries:

$$X_1 = x^2\frac{\partial}{\partial x} + xy\frac{\partial}{\partial y}, \quad X_2 = x\frac{\partial}{\partial x} + 2y\frac{\partial}{\partial y}.$$

The equation $X_2(J) = 0$ provides one independent invariant y/x^2. Consequently, the invariant solution is obtained by letting $y/x^2 = \lambda$, or $y = \lambda x^2$ with an arbitrary constant $\lambda \neq 0$. Then our differential equation reduces to

$$y' - \frac{y}{x} - \frac{y^2}{x^3} = 2\lambda x - \lambda x - \lambda^2 x = \lambda(1 - \lambda)x = 0.$$

Hence $\lambda = 1$, and the invariant solution simply is

$$y = x^2. \tag{6.4.16}$$

Let us take now the projective transformation (6.2.14)

$$\bar{x} = \frac{x}{1 - ax}, \quad \bar{y} = \frac{y}{1 - ax}$$

generated by X_1, rewrite the invariant solution (6.4.16) in the new variables in the form

$$\bar{y} = \bar{x}^2$$

and substitute here the expressions for $\bar{x}$ and $\bar{y}$:

$$\frac{y}{1 - ax} = \frac{x^2}{(1 - ax)^2}.$$

Denoting the parameter a by C, one obtains the general solution (6.4.15):

$$y = \frac{x^2}{1 - Cx}.$$

6.5 Second-order equations

6.5.1 Second prolongation of group generators. Calculation of symmetries

The infinitesimal symmetries of ordinary differential equations of the second and higher orders can be found by solving the so-called *determining equations*. The student who wants to further develop his analytical skills in applying the techniques of Lie symmetry analysis can find enough material in the literature given in the bibliography. Furthermore, examples presented here will prepare the reader, up to certain extend, to use computer algebra packages for calculating symmetries.

In this section, We will illustrate the method of determining equations by calculating symmetries of second-order equations

$$y'' = f(x, y, y'). \tag{6.5.1}$$

We look for an admissible infinitesimal generator

$$X = \xi(x, y)\frac{\partial}{\partial x} + \eta(x, y)\frac{\partial}{\partial y}, \tag{6.5.2}$$

with coefficients ξ and η to be found from the following equation known as the *determining equation*:

$$X\left(y'' - f(x, y, y')\right)\Big|_{y''=f} \equiv \left(\zeta_2 - \zeta_1 f_{y'} - \xi f_x - \eta f_y\right)\Big|_{y''=f} = 0, \tag{6.5.3}$$

where the symbol $|_{y''=f}$ means that y'' in the corresponding expression is replaced by the right-hand side of Eq. (6.5.1). Here, ζ_1 and ζ_2 are given by the following *prolongation formulae*:

$$\zeta_1 = D_x(\eta) - y'D_x(\xi) = \eta_x + (\eta_y - \xi_x)y' - \xi_y y'^2,$$

$$\zeta_2 = D_x(\zeta_1) - y''D_x(\xi) = \eta_{xx} + (2\eta_{xy} - \xi_{xx})y' \qquad (6.5.4)$$

$$+(\eta_{yy} - 2\xi_{xy})y'^2 - \xi_{yy} y'^3 + (\eta_y - 2\xi_x - 3\xi_y y')y''.$$

Upon substituting expressions (6.5.4) into Eq. (6.5.3), one obtains

$$\eta_{xx} + (2\eta_{xy} - \xi_{xx})y' + (\eta_{yy} - 2\xi_{xy})y'^2 - y'^3\xi_{yy} - \xi f_x - \eta f_y$$

$$+(\eta_y - 2\xi_x - 3y'\xi_y)f - [\eta_x + (\eta_y - \xi_x)y' - y'^2\xi_y]f_{y'} = 0. \qquad (6.5.5)$$

Equation (6.5.5) involves all three variables x, y and y', but y' does not occur in ξ and η. Consequently, the determining equation (6.5.5) decomposes into several equations thus becoming an over-determined system of differential equations for two unknown functions ξ and η. After solving this system, one finds all infinitesimal symmetries for Eq. (6.5.1).

Example 6.5.1. Let us find the infinitesimal symmetries of the equation

$$y'' = \frac{y'}{y^2} - \frac{1}{xy}. \qquad (6.5.6)$$

Substituting $f = y'y^{-2} - (xy)^{-1}$ into the determining equation (6.5.5) we have

$$\eta_{xx} + (2\eta_{xy} - \xi_{xx})y' + (\eta_{yy} - 2\xi_{xy})y'^2 - y'^3\xi_{yy} - \frac{\xi}{x^2y} + \left(2\frac{y'}{y^3} - \frac{1}{xy^2}\right)\eta$$

$$+(\eta_y - 2\xi_x - 3y'\xi_y)\left(\frac{y'}{y^2} - \frac{1}{xy}\right) - \frac{1}{y^2}[\eta_x + (\eta_y - \xi_x)y' - y'^2\xi_y] = 0.$$

This equation should be satisfied identically in the variables x, y, and y'. Since its left-hand side is a cubic polynomial in y', we equate to zero the coefficients of $y'^3, y'^2, \ldots$ and obtain the following four equations:

$$(y')^3 : \xi_{yy} = 0,$$

$$(y')^2 : y^2 (\eta_{yy} - 2\xi_{xy}) - 2\xi_y = 0,$$

$$(y')^1 : y^3 (2\eta_{xy} - \xi_{xx}) - y\xi_x + 2\eta + 3(y^2/x)\xi_y = 0,$$

$$(y')^0 : x^2y^2\eta_{xx} - x^2\eta_x + xy(2\xi_x - \eta_y) - x\eta - y\xi = 0.$$

The first two equations yield, upon integration with respect to y :

$$\xi = p(x)y + a(x), \quad \eta = -p(x)\ln(y^2) + p'(x)y^2 + q(x)y + b(x).$$

We substitute these expressions for ξ and η into the third and fourth equations. The left-hand sides of these equations will contain, along with polynomials in y, also the terms with $\ln(y^2)$. Equating the latter to zero, we get $p(x) = 0$. Hence, $\xi = a(x)$, $\eta = q(x)y + b(x)$. Now the third and fourth equations readily yield

$$\xi = C_1 x^2 + C_2 x, \quad \eta = \left(C_1 x + \frac{1}{2}C_2\right)y.$$

We conclude that the general solution of the determining equation provides the following infinitesimal symmetry of Eq. (6.5.6):

$$X = \left(C_1 x^2 + C_2 x\right)\frac{\partial}{\partial x} + \left(C_1 x + \frac{1}{2}C_2\right)y\frac{\partial}{\partial y}$$

or

$$X = C_1\left(x^2\frac{\partial}{\partial x} + xy\frac{\partial}{\partial y}\right) + C_2\left(x\frac{\partial}{\partial x} + \frac{y}{2}\frac{\partial}{\partial y}\right) = C_1 X_1 + C_2 X_2,$$

where X_1 and X_2 are the following two linearly independent (basic) infinitesimal symmetries of Eq. (6.5.6):

$$X_1 = x^2\frac{\partial}{\partial x} + xy\frac{\partial}{\partial y}, \quad X_2 = x\frac{\partial}{\partial x} + \frac{y}{2}\frac{\partial}{\partial y}. \tag{6.5.7}$$

Example 6.5.2. Let us find the infinitesimal symmetries of the equation

$$y'' + e^{3y}y'^4 + y'^2 = 0. \tag{6.5.8}$$

Substituting $f = -(e^{3y}y'^4 + y'^2)$ into the determining equation (6.5.5) we have

$$\eta_{xx} + (2\eta_{xy} - \xi_{xx})y' + (\eta_{yy} - 2\xi_{xy})y'^2 - y'^3\xi_{yy}$$
$$+3e^{3y}y'^4\,\eta - (\eta_y - 2\xi_x - 3y'\xi_y)(e^{3y}y'^4 + y'^2)$$
$$+[\eta_x + (\eta_y - \xi_x)y' - y'^2\xi_y](4e^{3y}y'^3 + 2y') = 0.$$

The left-hand side of this equation is a polynomial of fifth degree in y'. We proceed as in the previous example, i.e., equate to zero the coefficients of $y'^5, y'^4, \ldots$ and obtain the following four independent equations:

$$(y')^5 : \xi_y = 0,$$
$$(y')^4 : 3(\eta_y + \eta) - 2\xi_x = 0,$$
$$(y')^3 : \eta_x = 0,$$
$$(y')^1 : \xi_{xx} = 0.$$

The coefficients for $(y')^2$ and $(y')^0$ vanish together with the coefficients of $(y')^4$ and $(y')^1$, respectively. The above four differential equations for $\xi(x,y)$ and $\eta(x,y)$ are readily solved and yield (see Problem 7.16):

$$\xi = C_2 + 3C_3\,x, \quad \eta = 2C_3 + C_1 e^{-y},$$

where C_1, C_2, C_3 are arbitrary constants. Hence, the operator

$$X = \xi(x, y)\frac{\partial}{\partial x} + \eta(x, y)\frac{\partial}{\partial y}$$

admitted by Eq. (6.5.8) is given by

$$X = C_1 X_1 + C_2 X_2 + C_3 X_3,$$

where

$$X_1 = e^{-y}\frac{\partial}{\partial y}, \quad X_2 = \frac{\partial}{\partial x}, \quad X_3 = 3x\frac{\partial}{\partial x} + 2\frac{\partial}{\partial y}. \tag{6.5.9}$$

In other words, Eq. (6.5.8) admits a three-dimensional vector space L_3 spanned by operators (6.5.9).

6.5.2   Lie algebras

The above examples can serve to illustrate the general property of determining equations. Namely, the set of all solutions of these equations constitute what is called a *Lie algebra* defined as follows.

Consider any first-order linear partial differential operators

$$X_1 = \xi_1(x, y)\frac{\partial}{\partial x} + \eta_1(x, y)\frac{\partial}{\partial y}, \quad X_2 = \xi_2(x, y)\frac{\partial}{\partial x} + \eta_2(x, y)\frac{\partial}{\partial y}. \tag{6.5.10}$$

Definition 6.5.1. The *commutator* $[X_1, X_2]$ of operators (6.5.10) is a linear partial differential operator defined by the formula

$$[X_1, X_2] = X_1 X_2 - X_2 X_1,$$

or equivalently

$$[X_1, X_2] = \left(X_1(\xi_2) - X_2(\xi_1)\right)\frac{\partial}{\partial x} + \left(X_1(\eta_2) - X_2(\eta_1)\right)\frac{\partial}{\partial y}. \tag{6.5.11}$$

Definition 6.5.2. Let L_r be an r-dimensional linear space spanned by any r linearly independent operators of form (6.5.10), i.e., the set of the operators

$$X = C_1 X_1 + \cdots + C_r X_r, \quad C_1, \ldots, C_r = \text{const.}$$

The space L_r is called a *Lie algebra* if it is closed under the commutator, i.e., $[X, Y] \in L_r$ whenever $X, Y \in L_r$. This is equivalent to the condition that $[X_i, X_j] \in L_r$ $(i, j = 1, \ldots, r)$, i.e., if

$$[X_i, X_j] = c_{ij}^k X_k, \quad c_{ij}^k = \text{const.} \tag{6.5.12}$$

The operators $X_1, \ldots, X_r$ provide a basis of the Lie algebra L_r. We also say that L_r is a Lie algebra spanned by X_i $(i = 1, \ldots, r)$.

Example 6.5.3. Consider operators (6.5.7). Using (6.5.11), we obtain the commutator $[X_1, X_2] = -X_1$. Hence, operators (6.5.7) span a two-dimensional Lie algebra L_2. Accordingly, we say that Eq. (6.5.6) admits a two-dimensional Lie algebra.

Definition 6.5.3. Let L_r be a Lie algebra spanned by X_i, $i = 1, \ldots, r$. A subspace L_s of the vector space L_r spanned by a subset of the basic operators, e.g. by $X_1, \ldots, X_s$, $s < r$, is called a *subalgebra* of L_r if

$$[X, Y] \in L_s \quad \text{for any} \quad X, Y \in L_s,$$

i.e., if

$$[X_i, X_j] \in L_s, \ i, j = 1, \ldots, s.$$

Furthermore, L_s is called an *ideal* of L_r if

$$[X, Y] \in L_s \quad \text{whenever} \quad X \in L_s, \quad Y \in L_r,$$

i.e., if

$$[X_i, X_j] \in L_s, \ i = 1, \ldots, s; j = 1, \ldots, r.$$

A convenient way to expose a Lie algebra, subalgebra and other properties is to dispose the commutators in a *commutator table* whose entry at the intersection of the X_i row with the X_j column is $[X_i, X_j]$. Since commutator (6.5.11) is antisymmetric, the commutator table will be antisymmetric as well, with zeros on the main diagonal.

Example 6.5.4. Consider operators (6.5.9). Using Definition 6.5.11 of the commutator, one can readily set up the following commutator table.

	X_1	X_2	X_3
X_1	0	0	$2X_1$
X_2	0	0	$3X_2$
X_3	$-2X_1$	$-3X_2$	0

It follows from the above table that operators (6.5.9) span a three-dimensional Lie algebra L_3 admitted by Eq. (6.5.8). The table also shows that any two operators, namely, (X_1, X_2), (X_1, X_3) or (X_2, X_3), span a two-dimensional subalgebra. Furthermore, the commutator table also shows that the two-dimensional subalgebra spanned by X_1 and X_2 is an ideal of the Lie algebra L_3, whereas, e.g. the operators X_1 and X_3 do not span an ideal of the Lie algebra L_3.

6.5.3 Standard forms of two-dimensional Lie algebras

Lie's method of integration of second-order ordinary differential equations discussed in the next section is based on so-called *canonical coordinates* in two-dimensional Lie algebras. These variables provide, for every L_2, the simplest form of its basis and therefore reduce a differential equation admitting the L_2, to an integrable form. The basic statements are as follows.

Theorem 6.5.1. Any two dimensional Lie algebra can be transformed, by a proper choice of its basis and suitable variables t, u, called *canonical variables*, to one of the four non-similar standard forms presented in Table 6.5.1.

Table 6.5.1 Structure and standard forms of L_2

Type	Structure of L_2	Standard form of L_2
I	$[X_1, X_2] = 0, \ \xi_1\eta_2 - \eta_1\xi_2 \neq 0$	$X_1 = \dfrac{\partial}{\partial t}, \ X_2 = \dfrac{\partial}{\partial u}$
II	$[X_1, X_2] = 0, \ \xi_1\eta_2 - \eta_1\xi_2 = 0$	$X_1 = \dfrac{\partial}{\partial u}, \ X_2 = t\dfrac{\partial}{\partial u}$
III	$[X_1, X_2] = X_1, \ \xi_1\eta_2 - \eta_1\xi_2 \neq 0$	$X_1 = \dfrac{\partial}{\partial u}, \ X_2 = t\dfrac{\partial}{\partial t} + u\dfrac{\partial}{\partial u}$
IV	$[X_1, X_2] = X_1, \ \xi_1\eta_2 - \eta_1\xi_2 = 0$	$X_1 = \dfrac{\partial}{\partial u}, \ X_2 = u\dfrac{\partial}{\partial u}$

Remark 6.5.1. In types III and IV, the condition $[X_1, X_2] = X_1$ can be satisfied by a proper change of the basis in L_2 provided that $[X_1, X_2] \neq 0$.

6.5.4   Lie's integration method

Lie proved Theorem 6.5.1 in order to integrate all second-order equations

$$y'' = f(x, y, y') \tag{6.5.13}$$

admitting a two-dimensional Lie algebra. Lie's method consists in classifying these equations into four types in accordance with Table 6.5.1. Namely, introducing canonical variables, t, u, one reduces the admitted Lie algebra L_2 to one of the *standard forms* given in Table 6.5.1. Then one rewrites Equation (6.5.13) in the canonical variables. The resulting equation

$$u'' = g(t, u, u') \tag{6.5.14}$$

will have one of the four integrable *canonical forms* given in Table 6.5.2.

Thus, the method is as follows. Provided that we know an admitted algebra L_2 with a basis (6.5.10), the integration requires the following steps.

First step: Determine the type of L_2 according to the *Structure* column of Table 6.5.1. A change of the basis of L_2 may be required to accord the expression of the commutators for types III and IV (see Remark 6.5.1).

Second step: Find canonical variables by solving the following equations in accordance with the type:

$$\textbf{Type I}: X_1(t) = 1, \ X_2(t) = 0; \quad X_1(u) = 0, \ X_2(u) = 1.$$

$$\textbf{Type II}: X_1(t) = 0, \ X_2(t) = 0; \quad X_1(u) = 1, \ X_2(u) = t.$$

$$\textbf{Type III}: X_1(t) = 0, \ X_2(t) = t; \quad X_1(u) = 1, \ X_2(u) = u.$$

$$\textbf{Type IV}: X_1(t) = 0, \ X_2(t) = 0; \quad X_1(u) = 1, \ X_2(u) = u.$$

$$(6.5.15)$$

Then rewrite the differential equation in the canonical variables choosing t as a new independent variable and u as a dependent one. It will have one of the integrable forms given in Table 6.5.2. Integrate the equation.

Table 6.5.2 Four types of second-order equations admitting L_2

Type	Standard form of L_2	Canonical form of the equation
I	$X_1 = \dfrac{\partial}{\partial t}, \ X_2 = \dfrac{\partial}{\partial u}$	$u'' = f(u')$
II	$X_1 = \dfrac{\partial}{\partial u}, \ X_2 = t\dfrac{\partial}{\partial u}$	$u'' = f(t)$
III	$X_1 = \dfrac{\partial}{\partial u}, \ X_2 = t\dfrac{\partial}{\partial t} + u\dfrac{\partial}{\partial u}$	$u'' = \dfrac{1}{t}f(u')$
IV	$X_1 = \dfrac{\partial}{\partial u}, \ X_2 = u\dfrac{\partial}{\partial u}$	$u'' = f(t)u'$

Third step: Rewrite the resulting solution in the original variables x, y, thus completing the integration procedure.

Example 6.5.5. The equation

$$y'' = yy'^2 - xy'^3 \tag{6.5.16}$$

admits the two-dimensional Lie algebra with the basis

$$X_1 = y\frac{\partial}{\partial x}, \quad X_2 = x\frac{\partial}{\partial x}. \tag{6.5.17}$$

First step: Operators (6.5.17) satisfy the equations

$$[X_1, X_2] = X_1, \quad \xi_1\eta_2 - \eta_1\xi_2 = 0,$$

and hence, L_2 has type IV in Table 6.5.1.

Second step: Equations (6.5.15) for type IV:

$$X_1(t) = 0, \ X_2(t) = 0; \ X_1(u) = 1, \ X_2(u) = u,$$

yield the canonical variables

$$t = y, \quad u = \frac{x}{y}.$$

From the definition of t, we have the change of the total differentiation:

$$D_x = y' D_t$$

and, using it in differentiating the equation $u = x/y$, we obtain

$$y - xy' = y^2 y' u'.$$

Solving the latter equation for y' we have

$$y' = \frac{y}{x + y^2 u'},$$

or rewriting the right-hand side in the new variables,

$$y' = \frac{1}{u + tu'}.$$

Differentiating this equation again, we obtain

$$y'' = -y' \frac{2u' + tu''}{(u + tu')^2} = -\frac{2u' + tu''}{(u + tu')^3}.$$

Consequently, Eq. (6.5.16) assumes the following linear form in accordance with Table 6.5.2:

$$u'' = -\left(t + \frac{2}{t}\right) u'. \qquad (6.5.18)$$

Denoting $u' = v$, we rewrite Eq. (6.5.18) as a first-order equation:

$$\frac{dv}{dt} = -\left(t + \frac{2}{t}\right) v,$$

whence

$$\ln v = \ln C_1 + \ln(t^{-2}) - \frac{t^2}{2},$$

or

$$v = \frac{C_1}{t^2} e^{-t^2/2}.$$

Hence, we have the equation

$$u' = \frac{C_1}{t^2} e^{-t^2/2}.$$

Its integration yields at the following solution of Eq. (6.5.18):

$$u = C_2 + C_1 \int \frac{1}{t^2} e^{-t^2/2} dt.$$

Third step: Returning to the original variables, we arrive at the following implicit representation of the general solution to Eq. (6.5.16):

$$x = y \left(C_2 + C_1 \int \frac{1}{y^2} e^{-y^2/2} dy \right).$$

Example 6.5.6. We know from Example 6.5.1 that the nonlinear equation (6.5.6):

$$y'' = \frac{y'}{y^2} - \frac{1}{xy}$$

admits the two-dimensional Lie algebra L_2 with basis (6.5.7). Therefore, we can apply Lie's integration method.

First step: We will use the basic infinitesimal symmetries in form (6.5.7'):

$$X_1 = x^2 \frac{\partial}{\partial x} + xy \frac{\partial}{\partial y}, \quad X_2 = -x \frac{\partial}{\partial x} - \frac{y}{2} \frac{\partial}{\partial y}.$$

Then $[X_1, X_2] = X_1$ (cf. Example 6.5.3) and $\xi_1 \eta_2 - \eta_1 \xi_2 = x^2 y/2 \neq 0$. Hence, the Lie algebra L_2 has the structure of type III in Table 6.5.1.

Second step: Let us find the canonical variables and integrate our equation. The system of equations $X_1(t) = 0$, $X_2(t) = t$ for the new variable t yields

$$t = \left(\frac{y}{x}\right)^2, \tag{6.5.19}$$

and the system $X_1(u) = 1$, $X_2(u) = u$ for u yields

$$u = -\frac{1}{x}. \tag{6.5.20}$$

In the canonical variables t, u, the operators X_1, X_2 in (6.5.7') become

$$X_1 = \frac{\partial}{\partial u}, \quad X_2 = t \frac{\partial}{\partial t} + u \frac{\partial}{\partial u}.$$

Under the change of the independent variable (6.5.19), the total differentiation D_x in x defined by (1.2.66) transforms into the total differentiation D_t in t defined by the following equation:

$$D_x = D_x \left(\frac{y^2}{x^2}\right) D_t = 2 \frac{y(xy' - y)}{x^3} D_t,$$

or

$$D_x = 2u(t - \sqrt{t}\, y') D_t. \tag{6.5.21}$$

Now we differentiate both sides of Eq. (6.5.20) by using Eq. (6.5.21) and invoking that the left-hand side of (6.5.20) depends on t and its right-hand side depends on x, to obtain

$$2u(t - \sqrt{t}\, y') D_t(u) = D_x \left(-\frac{1}{x}\right) = \frac{1}{x^2} = u^2,$$

or

$$t - \sqrt{t}\, y' = \frac{u}{2u'}.$$

For calculation of the transformation of the second derivative, it is convenient to write the transformation of the total differentiation and of the first derivative as

$$D_x = \frac{u^2}{u'} D_t$$

and

$$y' = \sqrt{t} - \frac{u}{2\sqrt{t}\,u'},$$

respectively. Then one readily obtains the following transformation of the second derivative:

$$y'' = \frac{u^3}{4t\sqrt{t}\,u'^2} + \frac{u^3 u''}{2\sqrt{t}\,u'^3}.$$

Equations (6.5.19)—(6.5.20) yield

$$\frac{1}{xy} = \frac{u^2}{\sqrt{t}}.$$

Furthermore, the expression for y' and Eqs. (6.5.19)—(6.5.20) yield

$$\frac{y'}{y^2} = \frac{u^2}{\sqrt{t}} - \frac{u^3}{2t\sqrt{t}\,u'}.$$

After substituting the above expressions, Eq. (6.5.6) assumes the integrable form:

$$u'' = -\frac{1}{t}u'\left(u' + \frac{1}{2}\right). \tag{6.5.22}$$

Is Eq. (6.5.22) equivalent to the original equation (6.5.6)? More specifically, the question is whether or not all solutions of Eq. (6.5.6) are obtained from the solutions of (6.5.22) by the change of variables (6.5.19)~(6.5.20), and vice versa. The answer is not self-evident since the variable t defined by (6.5.19) involves the dependent variable y of the original equation (6.5.6) and therefore t can be regarded as a new independent variable only if (6.5.6) does not have solutions along which t is not identically constant. The direct inspection shows, however, that (6.5.6) has indeed such *singular* solutions where $t = $ const., namely the solutions given by the straight lines:

$$y = Kx, \quad K = \text{const.}$$

All the other solutions of Eq. (6.5.6) are obtained from solutions of (6.5.22) by the change of variables (6.5.19)—(6.5.20).

Furthermore, one should also inspect whether or not all solutions of Eq. (6.5.22) are related with solutions of (6.5.6). We notice that Eq. (6.5.22) is obviously satisfied by $u' = 0$ as well as by $u' = -1/2$, the corresponding solutions being $u = A$, $u = C - \frac{t}{2}$, where A and C are arbitrary constants. According to (6.5.20), the first of the above solutions, $u = A$, means $x = $ const.

Hence, it is not related to any solution of Eq. (6.5.6) and should be ignored. But the second solution, $u = C - \dfrac{t}{2}$, provides a solution of Eq. (6.5.6). Namely, substituting there the expressions (6.5.19) and (6.5.20) for t and u, respectively, one obtains the following solution of Eq. (6.5.6):

$$y = \pm\sqrt{2x + Cx^2}.$$

Now we integrate Eq. (6.5.22) excluding the above singular solutions, i.e., assuming $u' \neq 0$ and $u' + 1/2 \neq 0$. Then (6.5.22) yields

$$\ln|K_1\sqrt{t}| = -\int \frac{du'}{u'(2u'+1)} = \int \frac{2du'}{2u'+1} - \int \frac{du'}{u'} = \ln\left|\frac{2u'+1}{u'}\right|,$$

or

$$K_1\sqrt{t} = \frac{2u'+1}{u'},$$

whence

$$u' = \frac{1}{2(C_1\sqrt{t}-1)},$$

where $C_1 = K_1/2 \neq 0$. Finally, the elementary integration yields

$$u = \frac{1}{C_1^2}\left(C_1\sqrt{t} + \ln|C_1\sqrt{t}-1| + C_2\right).$$

Third step: Let us rewrite the solution in the original variables. Replacing in the last equation t and u by expressions (6.5.19) and (6.5.20), respectively, one arrives at the following implicit representation of the solution $y(x)$ of Eq. (6.5.6) involving two constants, $C_1 \neq 0$ and C_2:

$$C_1 y + C_2 x + x\ln\left|C_1\frac{y}{x} - 1\right| + C_1^2 = 0.$$

Adding to the latter two singular solutions discussed above one ultimately arrives at the *general solution* of Eq. (6.5.6) represented by the following three formulae with arbitrary constants K, C, C_1, C_2 with $C_1 \neq 0$:

$$y = Kx, \tag{6.5.23}$$

$$y = \pm\sqrt{2x + Cx^2}, \tag{6.5.24}$$

$$C_1 y + C_2 x + x\ln\left|C_1\frac{y}{x} - 1\right| + C_1^2 = 0. \tag{6.5.25}$$

The fact that the general solution of Eq. (6.5.6) is given by three distinctly different formulae does not contradict Theorem 3.1.2 on the uniqueness of the solution to any initial value problem. Indeed, the solution of the following exercise shows that the initial conditions themselves single out from (6.5.23)— (6.5.25) precisely one solution formula.

Exercise 6.5.1. Solve the following Cauchy problems for Eq. (6.5.6):

$$\text{(i)} \quad y'' = \frac{y'}{y^2} - \frac{1}{xy}, \quad y\big|_{x=1} = 1, \; y'\big|_{x=1} = 1;$$

$$\text{(ii)} \quad y'' = \frac{y'}{y^2} - \frac{1}{xy}, \quad y\big|_{x=1} = 1, \; y'\big|_{x=1} = 0;$$

$$\text{(iii)} \quad y'' = \frac{y'}{y^2} - \frac{1}{xy}, \quad y\big|_{x=1} = 1, \; y'\big|_{x=1} = 2.$$

Solution. Problem (i): Substituting $x = 1, y = 1, y' = 1$ into all three solution formulae (6.5.23)—(6.5.25) one can verify that the initial conditions (i) can be satisfied only by (6.5.23) with $K = 1$.

Problem (ii): Similar reasoning for $x = 1, y = 1, y' = 0$ singles out the second solution formula (6.5.24) with the plus sign and with $C = -1$.

Problem (iii): Likewise, the substitution $x = 1, y = 1, y' = 2$ singles out the solution formula (6.5.25) with $C_1 = 2, C_2 = -6$.

Thus, the solutions to the above Cauchy problems are given by

$$\text{(i)} \; y = x, \quad \text{(ii)} \; y = \sqrt{2x - x^2}, \quad \text{(iii)} \; 2y - 6x + x \ln \left| 2\frac{y}{x} - 1 \right| + 4 = 0.$$

6.5.5 Integration of linear equations with a known particular solution

Let us assume that a particular solution $y = z(x)$ to a linear equation

$$y'' + a(x)y' + b(x)y = 0 \tag{6.5.26}$$

is known. Thus, $z''(x) + a(x)z'(x) + b(x)z(x) = 0$ identically in x. We will discuss here two different methods for obtaining the general solution of Eq. (6.5.26) using its particular solution $z(x)$.

First method is transformation to the simple form provided by Theorem 3.3.1. According to this theorem, the transformation

$$t = \int \frac{e^{-\int a(x)dx}}{z^2(x)} \, dx, \quad u = \frac{y}{z(x)} \tag{6.5.27}$$

reduces Eq. (6.5.26) to the simplest second-order linear equation

$$u'' = 0. \tag{6.5.28}$$

It follows $u = C_1 t + C_2$, whence the general solution to Eq. (6.5.26):

$$y = z(x)\left[C_1 \int \frac{e^{-\int a(x)dx}}{z^2(x)} \, dx + C_2 \right]. \tag{6.5.29}$$

Second method is based on the fact that Eq. (6.5.26) is invariant under the transformation $\bar{y} = y + az(x)$, and hence admits the generator

$$X_1 = z(x)\frac{\partial}{\partial y}. \tag{6.5.30}$$

Since Eq. (6.5.26) is homogeneous, it admits also the generator

$$X_2 = y\frac{\partial}{\partial y}. \tag{6.5.31}$$

Operators (6.5.30) and (6.5.31) span a two-dimensional Lie algebra of type IV, and hence Eq. (6.5.26) can be solved by Lie's integration method discussed in Section 6.5.4. Namely, the canonical variables for operators (6.5.30)~(6.5.31) are

$$t = x, \quad u = \frac{y}{z(x)}. \tag{6.5.32}$$

In these variables, Eq. (6.5.26) is written in the integrable form:

$$u'' + \left[a(x) + 2\frac{z'(x)}{z(x)}\right]u' = 0, \tag{6.5.33}$$

whence

$$u = C_1 \int \frac{e^{-\int a(x)dx}}{z^2(x)}\, dx + C_2.$$

Thus, we arrive again to solution (6.5.29).

In practice, it is better to use one of the methods described above rather than the final formula (6.5.29) for the solution.

Example 6.5.7. Consider the equation

$$y'' = xy' - y. \tag{6.5.34}$$

One can readily find its particular solution $z(x) = x$ by looking for a polynomial solution $y = A_0 + A_1 x + A_2 x^2 + \cdots$. Let us apply the first method of integration. Transformation (6.5.27) yields

$$t = \int \frac{e^{x^2/2}}{x^2}\, dx, \quad u = \frac{y}{x}.$$

In these variables, Eq. (6.5.34) is written $u'' = 0$, and hence $u = C_1 t + C_2$. Substituting here the expressions for t and u, we obtain the following general solution to Eq. (6.5.34):

$$y = \left[C_1 \int \frac{e^{x^2/2}}{x^2}\, dx + C_2\right]x. \tag{6.5.35}$$

Let us apply the second method. Using the Lie algebra L_2 of type IV spanned by operators (6.5.30) and (6.5.31),

$$X_1 = x\frac{\partial}{\partial y}, \quad X_2 = y\frac{\partial}{\partial y},$$

one obtains the canonical variables $t = x, u = y/x$. In these variables, Eq. (6.5.34) is written

$$u'' = \left(x - \frac{2}{x}\right)u'.$$

After the standard substitution $u' = v$ and separation of variables, it becomes

$$\frac{dv}{v} = \left(x - \frac{2}{x}\right)dx$$

and yields

$$v = C_1 \frac{e^{x^2/2}}{x^2}.$$

It follows, upon integration:

$$u = C_1 \int \frac{e^{x^2/2}}{x^2}\,dx + C_2.$$

Finally, $y = xu$ yields the general solution (6.5.35).

6.5.6   Lie's linearization test

Example 6.5.8. To illustrate the problem, consider the nonlinear equation

$$y'' = 2\left(\frac{y'^2}{y} - \frac{xy'}{1+x^2}\right). \tag{6.5.36}$$

We obtained it from the simplest linear equation (6.5.28), $u'' = 0$, by the following change of variables:

$$t = \frac{1}{y}, \quad u = \arctan x. \tag{6.5.37}$$

Indeed, applying the transformation formulae (1.4.13)~(1.4.14) to Eqs. (6.5.37) we have

$$D_x = -\frac{y'}{y^2}D_t, \quad -\frac{y'}{y^2}D_t(u) = D_x(\arctan x), \tag{6.5.38}$$

whence

$$u' = -\frac{y^2}{(1+x^2)\,y'}. \tag{6.5.39}$$

Differentiating both sides of Eq. (6.5.39) by using (6.5.38), we obtain

$$\frac{y'^2}{y^4}u'' - \frac{y''}{y^2}u' + 2\frac{y'^2}{y^3}u' = -\frac{2x}{(1+x^2)^2}.$$

Whence, invoking Eqs. (6.5.28) and (6.5.39), we ultimately arrive at Eq. (6.5.36). Subjecting the general solution $u = C_1 t - C_2$ of the linear equation (6.5.28) to the change of variables (6.5.37), we have

$$\arctan x = \frac{C_1}{y} - C_2,$$

and hence, we obtain the general solution to the nonlinear equation (6.5.36):

$$y = \frac{C_1}{C_2 + \arctan x}. \tag{6.5.40}$$

Linearizable equations such as (6.5.36) occur in applications quite often. Therefore, it is important to have a general test for identifying linearizable equations. S. Lie [25] solved this problem for second-order ordinary differential equations. Namely, he found the general form of all second-order equations (6.5.13) that can be reduced to the linear equation

$$\frac{\mathrm{d}^2 u}{\mathrm{d}t^2} = A(t)\frac{\mathrm{d}u}{\mathrm{d}t} + B(t)u + C(t) \tag{6.5.41}$$

by a change of variables

$$t = \varphi(x, y), \quad u = \psi(x, y). \tag{6.5.42}$$

He showed first of all that the linearizable second-order equations should be at most cubic in the first-order derivative. This statement can be obtained by using the transformation of derivatives under a change of variables. Recall that any linear equation can be reduced to the simplest equation (6.5.28) by a change of both independent and dependent variables variables. We also know from Lemma 3.3.1 the linear equations are *equivalent by function* to Eq. (3.3.12). Here, we will write Eq. (3.3.12) in the variables t and u :

$$u'' + \alpha(t)\, u = 0, \tag{6.5.43}$$

where u'' is the second derivative of u with respect to t. Then, using the transformation of derivatives given by Theorem 1.4.1 in Section 1.4.5, we arrive at the following result.

Lemma 6.5.1. All second-order equations obtained from Eq. (6.5.43) by a change of variables (6.5.42) are at most cubic in the first-order derivative, i.e., belong to the family of equations of the form

$$y'' + F_3(x, y)y'^3 + F_2(x, y)y'^2 + F_1(x, y)y' + F(x, y) = 0, \tag{6.5.44}$$

where

$$F_3(x, y) = \frac{\varphi_y \psi_{yy} - \psi_y \varphi_{yy} + \alpha \psi \varphi_y^3}{\varphi_x \psi_y - \varphi_y \psi_x},$$

$$F_2(x, y) = \frac{\varphi_x \psi_{yy} - \psi_x \varphi_{yy} + 2(\varphi_y \psi_{xy} - \psi_y \varphi_{xy}) + 3\alpha \psi \varphi_x \varphi_y^2}{\varphi_x \psi_y - \varphi_y \psi_x},$$

$$F_1(x, y) = \frac{\varphi_y \psi_{xx} - \psi_y \varphi_{xx} + 2(\varphi_x \psi_{xy} - \psi_x \varphi_{xy}) + 3\alpha \psi \varphi_x^2 \varphi_y}{\varphi_x \psi_y - \varphi_y \psi_x}, \qquad (6.5.45)$$

$$F(x, y) = \frac{\varphi_x \psi_{xx} - \psi_x \varphi_{xx} + \alpha \psi \varphi_x^3}{\varphi_x \psi_y - \varphi_y \psi_x}.$$

However, not every equation of form (6.5.44) with arbitrary coefficients $F_3(x, y), \ldots, F(x, y)$ is linearizable. The linearization is possible if and only if the over-determined system of nonlinear partial differential equations (6.5.45) for two function $\varphi(x, y)$ and $\psi(x, y)$ with given $F_3(x, y), \ldots, F(x, y)$ is integrable. Lie [25] provided the compatibility (i.e., integrability) conditions for system (6.5.45). Lie's linearization test can be formulated as follows (see also [21]).

Theorem 6.5.2. Equation (6.5.44) is linearizable if and only if its coefficients satisfy the following equations:

$$3(F_3)_{xx} - 2(F_2)_{xy} + (F_1)_{yy} = (3F_1 F_3 - F_2^2)_x - 3(FF_3)_y - 3F_3 F_y + F_2(F_1)_y,$$

$$3F_{yy} - 2(F_1)_{xy} + (F_2)_{xx} = 3(FF_3)_x + (F_1^2 - 3FF_2)_y + 3F(F_3)_x - F_1(F_2)_x.$$

Lie's linearization test is simple and convenient in practice. Consider examples.

Example 6.5.9. The equation

$$y'' + F(x, y) = 0$$

has form (6.5.44) with $F_3 = F_2 = F_1 = 0$. The linearization test yields $F_{yy} = 0$. Hence, the equation $y'' + F(x, y) = 0$ cannot be linearized unless it is already linear.

Example 6.5.10. The equations

$$y'' - \frac{1}{x}(y' + y'^3) = 0$$

and

$$y'' + \frac{1}{x}(y' + y'^3) = 0$$

also have form (6.5.44). Their coefficients are $F_3 = F_1 = -1/x, F_2 = F = 0$ and $F_3 = F_1 = 1/x, F_2 = F = 0$, respectively. The linearization test shows that the first equation is linearizable, whereas the second one is not.

If the coefficients $F_3(x,y), \ldots, F(x,y)$ of Eq. (6.5.44) satisfy the linearization test given in Theorem 6.5.2, then a change of variables (6.5.42) reducing
Eq. (6.5.44) to a linear equation of form (6.5.43) can be obtained by solving
the over-determined system of differential equations (6.5.45) for $\varphi(x,y)$ and
$\psi(x,y)$ with the known functions $F_3(x,y), \ldots, F(x,y)$.

Example 6.5.11. Consider the equation

$$y'' - \frac{1}{x}(y' + y'^{\,3}) = 0 \qquad (6.5.46)$$

from the previous example. We know that its coefficients

$$F_3 = F_1 = -\frac{1}{x}, \qquad F_2 = F = 0$$

satisfies the conditions of Theorem 6.5.2. Let us show that Eq. (6.5.46) can be
transformed into the simplest linear equation $u'' = 0$ and find the linearizing
map. Thus, we assume that $\alpha(t) = 0$ in (6.5.43). Then Eqs. (6.5.45) for
determining $\varphi(x,y)$ and $\psi(x,y)$ are written

$$\varphi_y \psi_{yy} - \psi_y \varphi_{yy} = -\frac{1}{x}(\varphi_x \psi_y - \varphi_y \psi_x),$$

$$\varphi_x \psi_{yy} - \psi_x \varphi_{yy} + 2(\varphi_y \psi_{xy} - \psi_y \varphi_{xy}) = 0, \qquad (6.5.47)$$

$$\varphi_y \psi_{xx} - \psi_y \varphi_{xx} + 2(\varphi_x \psi_{xy} - \psi_x \varphi_{xy}) = -\frac{1}{x}(\varphi_x \psi_y - \varphi_y \psi_x),$$

$$\varphi_x \psi_{xx} - \psi_x \varphi_{xx} = 0.$$

In order to linearize Eq. (6.5.46) it suffices to find any particular functions
$\varphi(x,y)$ and $\psi(x,y)$ which solve system (6.5.47) and are functionally independence, i.e., the Jacobian does not vanish:

$$\varphi_x \psi_y - \varphi_y \psi_x \neq 0. \qquad (6.5.48)$$

Therefore, we will satisfy the last equation of system (6.5.47) by letting $\varphi_x = 0$,
i.e., $\varphi = \varphi(y)$. Then condition (6.5.48) requires that $\psi_x \neq 0$ and $\varphi_y \neq 0$. To
further simplify calculations, set $\varphi = y$. Then the second equation of system
(6.5.47) yields $\psi_{xy} = 0$, whence

$$\psi = a(x) + b(y).$$

Now the first equation of (6.5.47) yields the equation with separated variables:

$$b''(y) = \frac{1}{x}a'(x).$$

It follows that

$$b''(y) = \frac{1}{x}a'(x) = \lambda, \qquad \lambda = \text{const}.$$

These equations yield

$$a(x) = \frac{\lambda}{2} x^2 + C_1 x + C_2, \quad b(y) = \frac{\lambda}{2} y^2 + K_1 y + K_2.$$

Noting that the third equation of (6.5.47) is satisfied identically and setting $\lambda = 2$, $C_1 = C_2 = K_1 = K_2 = 0$, we arrive at the following change of variables (6.5.42):

$$t = y, \quad u = x^2 + y^2. \tag{6.5.49}$$

This change of variables reduces Eq. (6.5.46) to the linear equation $u'' = 0$. Writing the general solution of the latter equation in the form $u + At + B = 0$ and using (6.5.49), we obtain the following implicit solution of the non-linear equation (6.5.46):

$$x^2 + y^2 + Ay + B = 0, \quad A, B = \text{const.} \tag{6.5.50}$$

Note that the change of variables (6.5.49) is illegal for the solution $y = \text{const.}$ of (6.5.46). Therefore, the general solution of Eq. (6.5.46) is obtained by adding to (6.5.50) the singular solution $y = \text{const.}$

Example 6.5.12. The equation $y'' + y'^2 = f(x)$ has form (6.5.44) with $F_3 = F_1 = 0$, $F_2 = 1$, $F = -f(x)$ and satisfies Theorem 6.5.2. Let us check if our equation is linearizable by a change of the dependent variable only. In other words, let us seek the linearizing transformation (6.5.42) in the form $t = x$, $u = \psi(x, y)$. Then the first equation of (6.5.45) is valid identically while the remaining threes equations yield:

$$\psi_{yy} = \psi_y, \quad \psi_{xy} = 0, \quad \psi_{xx} + f(x)\psi_y + \alpha(x)\psi = 0.$$

The first two equations yield $\psi = g(x) + Ce^y$. Setting, e.g. $C = 1$, $g(x) = 0$, we obtain $\psi = e^y$. Then the third equation yields $\alpha = -f(x)$. Thus, the change of variables $t = x$, $u = e^y$ linearizes the equation in question and maps it to $u'' = f(x)u$.

6.6   Higher-order equations

6.6.1   Invariant solutions. Derivation of Euler's ansatz

The concept of group invariant solutions introduced in Section 6.4.3 for first-order equations is applicable to higher-order equations as well.

Example 6.6.1. The general homogeneous linear ordinary differential equation with constant coefficients (3.4.6),

$$y^{(n)} + a_1 y^{(n-1)} + \cdots + a_{n-1} y' + a_n y = 0, \quad a_1, \ldots, a_n = \text{const.},$$

admits the translation group with the generator

$$X_1 = \frac{\partial}{\partial x}$$

due to the constant coefficients, and the multiplication of y by any parameter, i.e., the dilation group with the generator

$$X_2 = y\frac{\partial}{\partial y}$$

due to the homogeneity of the equation under consideration. Therefore, the equation admits the linear combination of these generators, $X = X_1 + \lambda X_2$:

$$X = \frac{\partial}{\partial x} + \lambda y\frac{\partial}{\partial y}, \quad \lambda = \text{const.}$$

The equation $dy/y = \lambda \, dx$ yields the invariant $u = ye^{-\lambda x}$. The invariant solutions are obtained by setting $u = C$. Hence, *Euler's ansatz* (3.3.25):

$$y = Ce^{\lambda x}.$$

Since $y' = C\lambda e^{\lambda x}, \ldots, y^{(n)} = C\lambda^n e^{\lambda x}$, substitution into Eq. (3.4.6) yields an algebraic equation, namely, the characteristic equation (3.4.7):

$$\lambda^n + a_1\lambda^{n-1} + \cdots + a_{n-1}\lambda + a_n = 0.$$

Example 6.6.2. Consider Euler's equation (3.4.11) from Section 3.4.4:

$$x^n \frac{d^n y}{dx^n} + a_1 x^{n-1} \frac{d^{n-1} y}{dx^{n-1}} + \cdots + a_{n-1} x \frac{dy}{dx} + a_n y = 0,$$

where $a_1, \ldots, a_n = \text{const.}$ It is double homogeneous (see Definition 3.1.3), i.e., admits the dilation groups with the generators

$$X_1 = x\frac{\partial}{\partial x}, \quad X_2 = y\frac{\partial}{\partial y}.$$

We proceed as in Example 6.6.1 and find invariant solutions by taking the linear combination $X = X_1 + \lambda X_2$:

$$X = x\frac{\partial}{\partial x} + \lambda y\frac{\partial}{\partial y}, \quad \lambda = \text{const.}$$

The characteristic equation $dy/y = \lambda \, dx/x$ yields the invariant $u = y x^{-\lambda}$. The invariant solutions are obtained by setting $u = C$, whence

$$y = Cx^\lambda. \tag{6.6.1}$$

Differentiating and multiplying by x, we have

$$xy' = C\lambda x^\lambda, \quad x^2y'' = C\lambda^2 x^\lambda, \ldots, \quad x^n y^{(n)} = C\lambda^n x^\lambda.$$

Substituting into Eq. (3.4.11) and dividing by the common factor Cx^λ we obtain the following *characteristic equation* for Euler's equation (3.4.11):

$$\lambda^n + a_1 \lambda^{n-1} + \cdots + a_{n-1}\lambda + a_n = 0.$$

It is identical with the characteristic equation (3.4.7) for the equation with constant coefficients (cf. Section 3.4.4).

6.6.2   Integrating factor (N.H. Ibragimov, 2006)

It is customary to consider integrating factors exclusively for first-order ordinary differential equations. Furthermore, in the classical approach to integrating factors, differential equations are written as differential forms (see Section 3.2.3). We will discuss here an alternative approach to integrating factors developed in [22]. The new approach allows us to determine integrating factors for higher-order equations and systems.

Let $u = (u^1, \ldots, u^m)$ denote $m \geq 1$ dependent variables with successive derivatives $u_{(1)} = \{du^\alpha/dx\}, u_{(2)} = \{d^2u^\alpha/dx^2\}, \ldots$ with respect to the single independent variable x. The total differentiation (1.4.9) has the form

$$D_x = \frac{\partial}{\partial x} + u^\alpha_{(1)}\frac{\partial}{\partial u^\alpha} + u^\alpha_{(2)}\frac{\partial}{\partial u^\alpha_{(1)}} + \cdots . \tag{6.6.2}$$

The higher-order variational derivatives (cf. Eqs. (1.5.4) and (2.6.24)) with one independent variable and several dependent variables are written:

$$\frac{\delta}{\delta u^\alpha} = \frac{\partial}{\partial u^\alpha} - D_x\frac{\partial}{\partial u^\alpha_x} + D^2_x\frac{\partial}{\partial u^\alpha_{xx}} - D^3_x\frac{\partial}{\partial u^\alpha_{xxx}} + \cdots . \tag{6.6.3}$$

The new approach to integrating factors for ordinary differential equations of any order with one or several dependent variables is based on the following statements (for more details and the proofs, see [21], Section 8.4).

Lemma 6.6.1. Let $F(x, u, u_{(1)}, \ldots, u_{(s)}) \in \mathcal{A}$. The equation $D_x(F) = 0$ holds identically in all variables $x, u, u_{(1)}, \ldots, u_{(s)}$ and $u_{(s+1)}$ if and only if $F = C = $ const.

Lemma 6.6.2. A differential function $F(x, u, u_{(1)}, \ldots, u_{(s)}) \in \mathcal{A}$ with one independent variable x is a total derivative:

$$F = D_x(\Phi), \quad \Phi(x, u, u_{(1)}, \ldots, u_{(s-1)}) \in \mathcal{A}, \tag{6.6.4}$$

if and only if the following equations hold identically in $x, u, u_{(1)}, \ldots$:

$$\frac{\delta F}{\delta u^\alpha} = 0, \quad \alpha = 1, \ldots, m. \tag{6.6.5}$$

In the case of equations with one dependent variable y, we will use the usual notation y', y'', etc. for the successive derivatives. The total differentiation (6.6.2) and the variational derivative (6.6.3) are written

$$D_x = \frac{\partial}{\partial x} + y'\frac{\partial}{\partial y} + y''\frac{\partial}{\partial y'} + \cdots + y^{(s+1)}\frac{\partial}{\partial y^{(s)}} + \cdots$$

and

$$\frac{\delta}{\delta y} = \frac{\partial}{\partial y} - D_x\frac{\partial}{\partial y'} + D_x^2\frac{\partial}{\partial y''} - D_x^3\frac{\partial}{\partial y'''} + \cdots, \qquad (6.6.6)$$

respectively. In this case, Lemmas 6.6.1 and 6.6.2 are formulated as follows.

Lemma 6.6.3. Let $f(x, y, y', \ldots, y^{(s)}) \in \mathcal{A}$. If the equation $D_x(f) = 0$ holds identically in all variables $x, y, y', \ldots, y^{(s)}$, and $y^{(s+1)}$, then $f = $ const.

Lemma 6.6.4. A differential function $f(x, y, y', \ldots, y^{(s)}) \in \mathcal{A}$ with one independent variable x is a total derivative, i.e.,

$$f = D_x(\phi), \quad \phi(x, y, y', \ldots, y^{(s-1)}) \in \mathcal{A}, \qquad (6.6.7)$$

if and only if the following equation holds identically in $x, y, y', \ldots$:

$$\frac{\delta f}{\delta y} \equiv \frac{\partial f}{\partial y} - D_x\frac{\partial f}{\partial y'} + D_x^2\frac{\partial f}{\partial y''} - D_x^3\frac{\partial f}{\partial y'''} + \cdots + (-1)^s D_x^s\frac{\partial f}{\partial y^{(s)}} = 0. \quad (6.6.8)$$

Definition 6.6.1. Consider sth-order ordinary differential equations

$$a(x, y, y', \ldots, y^{(s-1)})\, y^{(s)} + b(x, y, y', \ldots, y^{(s-1)}) = 0. \qquad (6.6.9)$$

A function $\mu(x, y, y', \ldots, y^{(s-1)})$ is called an integrating factor for Eq. (6.6.9) if the multiplication by μ converts the left-hand side of Eq. (6.6.9) into a total derivative of some function $\phi(x, y, y', \ldots, y^{(s-1)})$:

$$\mu a\, y^{(s)} + \mu b = D_x(\phi). \qquad (6.6.10)$$

Knowledge of an integrating factor allows one to reduce the order of Eq. (6.6.9). Indeed, it follows from Eqs. (6.6.9)$\sim$(6.6.10) that $D_x(\phi) = 0$, and Lemma 6.6.1 yields a first integral for Eq. (6.6.9):

$$\phi(x, y, y', \ldots, y^{(s-1)}) = C. \qquad (6.6.11)$$

Definition 6.6.1 can be readily extended to systems of ordinary differential equations of any order.

Theorem 6.6.1. The integrating factors for Eq. (6.6.9) are determined by the equation

$$\frac{\delta}{\delta y}(\mu a\, y^{(s)} + \mu b) = 0, \qquad (6.6.12)$$

where $\delta/\delta y$ is the variational derivative (6.6.6). Equation (6.6.12) involves the variables $x, y, y', \ldots, y^{(2s-2)}$ and is satisfied identically in these variables.

Proof. Equation (6.6.12) is obtained from Lemma 6.6.2. The highest derivative that may appear after the variational differentiation (6.6.6) has the order $2s - 1$. It occurs in the terms

$$(-1)^s D_x^s(\mu a) \quad \text{and} \quad (-1)^{s-1} D_x^{s-1}\left[y^{(s)} \frac{\partial(\mu a)}{\partial y^{(s-1)}}\right].$$

Dropping the terms that certainly do not involve $y^{(2s-1)}$, we have

$$(-1)^s D_x^s(\mu a) = -(-1)^{s-1} D_x^{s-1}\left[y^{(s)} \frac{\partial(\mu a)}{\partial y^{(s-1)}}\right] + \cdots.$$

Thus, the terms containing $y^{(2s-1)}$ annihilate each other, and Eq. (6.6.12) involves only the variables $x, y, y', \ldots, y^{(2s-2)}$. This completes the proof.

For the first-order equation

$$a(x, y)y' + b(x, y) = 0, \tag{6.6.13}$$

Eq. (6.6.12) is written:

$$\frac{\delta}{\delta y}(\mu a\, y' + \mu b) = y'\, (\mu a)_y + (\mu b)_y - D_x(\mu a) = 0.$$

Since $D_x(\mu a) = (\mu a)_x + y'(\mu a)_y$, we obtain the following equation for determining the integrating factor for first-order equation (6.6.13):

$$(\mu b)_y - (\mu a)_x = 0. \tag{6.6.14}$$

Equation (6.6.14) is identical with Eq. (3.2.12) where $N = a, M = b$.

Example 6.6.3. Let us consider, from the new point of view, Eq. (3.2.14) from Example 3.2.2. We rewrite Eq. (3.2.14) in the form

$$(y^2 - 3x^2)y' + 2xy = 0,$$

and have

$$\frac{\delta}{\delta y}\left[(y^2 - 3x^2)y' + 2xy\right] = 2yy' + 2x - D_x(y^2 - 3x^2) = 8x.$$

Thus, condition (6.6.12) is not satisfied, and hence $(y^2 - 3x^2)y' + 2xy$ is not a total derivative. On the other hand, upon multiplying by $\mu = 1/y^4$, we obtain the equation

$$\left(\frac{1}{y^2} - \frac{3x^2}{y^4}\right)y' + \frac{2x}{y^3} = 0 \tag{6.6.15}$$

satisfying condition (6.6.12). Indeed,

$$\frac{\delta}{\delta y}\left[\left(\frac{1}{y^2} - \frac{3x^2}{y^4}\right)y' + \frac{2x}{y^3}\right] = -\frac{2y'}{y^3} + 12\frac{x^2y'}{y^5} - \frac{6x}{y^4} - D_x\left(\frac{1}{y^2} - \frac{3x^2}{y^4}\right) = 0.$$

Now, we write

$$\frac{y'}{y^2} = D_x\left(-\frac{1}{y}\right), \quad -3x^2\frac{y'}{y^4} = x^2 D_x\left(\frac{1}{y^3}\right) = D_x\left(\frac{x^2}{y^3}\right) - \frac{2x}{y^3}$$

and obtain

$$\left(\frac{1}{y^2} - \frac{3x^2}{y^4}\right)y' + \frac{2x}{y^3} = D_x\left(\frac{x^2}{y^3} - \frac{1}{y}\right).$$

Hence, the solution of our differential equation is given implicitly by

$$\frac{x^2}{y^3} - \frac{1}{y} = C \quad \text{or} \quad x^2 - y^2 = Cy^3.$$

Consider now the second-order equations

$$a(x, y, y')y'' + b(x, y, y') = 0. \tag{6.6.16}$$

The integrating factors μ depend on x, y, y', and Eq. (6.6.12) for determining $\mu(x, y, y')$ is written:

$$\frac{\delta}{\delta y}(\mu a\, y'' + \mu b) = y''\,(\mu a)_y + (\mu b)_y - D_x\left[y''\,(\mu a)_{y'} + (\mu b)_{y'}\right] + D_x^2(\mu a) = 0.$$

We have

$$D_x(\mu a) = y''(\mu a)_{y'} + y'(\mu a)_y + (\mu a)_x,$$

$$D_x^2(\mu a) = y'''(\mu a)_{y'} + y'^2\,(\mu a)_{y'y'} + 2y'y''(\mu a)_{yy'} + 2y''(\mu a)_{xy'}$$
$$+ y''(\mu a)_y + y'^2(\mu a)_{yy} + 2y'(\mu a)_{xy} + (\mu a)_{xx},$$

$$D_x\left(y''(\mu a)_{y'}\right) = y'''(\mu a)_{y'} + y'^2\,(\mu a)_{y'y'} + y'y''(\mu a)_{yy'} + y''(\mu a)_{xy'},$$

$$D_x\left((\mu b)_{y'}\right) = y''\,(\mu b)_{y'y'} + y'\,(\mu b)_{yy'} + (\mu b)_{xy'},$$

and hence,

$$\frac{\delta}{\delta y}(\mu a\, y'' + \mu b) = y''[y'\,(\mu a)_{yy'} + (\mu a)_{xy'} + 2(\mu a)_y - (\mu b)_{y'y'}] + y'^2\,(\mu a)_{yy}$$

$$+ 2y'(\mu a)_{xy} + (\mu a)_{xx} - y'(\mu b)_{yy'} - (\mu b)_{xy'} + (\mu b)_y.$$

Since this expression should vanish identically in x, y, y' and y'', we arrive at the following statement.

Theorem 6.6.2. The integrating factors $\mu(x, y, y')$ for the second-order equation (6.6.16) are determined by the following system of two equations:

$$y'\,(\mu a)_{yy'} + (\mu a)_{xy'} + 2(\mu a)_y - (\mu b)_{y'y'} = 0, \tag{6.6.17}$$

$$y'^2\,(\mu a)_{yy} + 2y'(\mu a)_{xy} + (\mu a)_{xx} - y'(\mu b)_{yy'} - (\mu b)_{xy'} + (\mu b)_y = 0. \tag{6.6.18}$$

Theorem 6.6.2 shows that the second-order equations, unlike the first-order ones, may have no integrating factors. Indeed, the integrating factor $\mu(x,y)$ for any first-order equation is determined by the single first-order linear partial differential equation (6.6.14) which always has infinite number of solutions. In the case of second-order equations (6.6.16), one unknown function $\mu(x,y,y')$ should satisfy two second-order linear partial differential equations (6.6.17)~(6.6.18). An integrating factor exists provided that the overdetermined system (6.6.17)~(6.6.18) is compatible.

Remark 6.6.1. If a second-order equation (6.6.16) has two integrating factors that lead to two distinctly different first integrals (6.6.11), then the general solution to Eq. (6.6.16) can be found without integration.

Example 6.6.4. Consider the following second-order equation:

$$y'' + \frac{y'^2}{y} + 3\frac{y'}{x} = 0. \qquad (6.6.19)$$

One can verify that its left-hand side does not satisfy condition (6.6.5), and hence it is not a total derivative. Let us calculate an integrating factor. Equation (6.6.19) has form (6.6.16) with

$$a = 1, \quad b = \frac{y'^2}{y} + 3\frac{y'}{x}.$$

For the sake of simplicity, we will look for the integrating factors of the form $\mu = \mu(x,y)$. Equation (6.6.17) is written $2\mu_y - (\mu b)_{y'y'} = 0$. Since $(\mu b)_{y'y'} = 2\mu/y$, we obtain $\mu_y = \mu/y$, whence $\mu = \phi(x)y$. Thus, we have

$$\mu = \phi(x)y, \quad \mu_{yy} = 0, \quad \mu_{xy} = \phi', \quad \mu_{xx} = \phi''y, \quad \mu b = \phi y'^2 + 3\frac{\phi}{x}yy',$$

$$(\mu b)_y = 3\frac{\phi}{x}y', \quad (\mu b)_{yy'} = 3\frac{\phi}{x}, \quad (\mu b)_{xy'} = 2\phi'y' + 3\left(\frac{\phi'}{x} - \frac{\phi}{x^2}\right)y.$$

Substitution into Eq. (6.6.18) leads to the following Euler's equation:

$$x^2\phi'' - 3x\phi' + 3\phi = 0.$$

Integrating it by the standard change of the independent variable, $t = \ln|x|$, we obtain two independent solutions, $\phi = x$ and $\phi = x^3$. Thus, Eq. (6.6.19) has two integrating factors:

$$\mu_1 = xy, \quad \mu_2 = x^3y, \qquad (6.6.20)$$

and can be solved without an additional integration (see Remark 6.6.1).

Indeed, multiplying Eq. (6.6.19) by the first integrating factor, we have

$$xy\left(y'' + \frac{y'^2}{y} + 3\frac{y'}{x}\right) = xyy'' + xy'^2 + 3yy' = 0.$$

Substituting $xyy'' = D_x(xyy') - yy' - xy'^2$, we reduce it to $D_x(xyy') + 2yy' = D_x(xyy' + y^2) = 0$, whence

$$xyy' + y^2 = C_1. \qquad (6.6.21)$$

Likewise, the second integrating factor (6.6.20) yields

$$x^3 yy' = C_2. \qquad (6.6.22)$$

Eliminating y' from Eqs. (6.6.21)—(6.6.22), we obtain the following general solution to Eq. (6.6.19):

$$y = \pm\sqrt{C_1 - \frac{C_2}{x^2}}. \qquad (6.6.23)$$

Application of the linearization test (Section 6.5.6) shows that Eq. (6.6.19) is linearizable. Therefore, it admits an 8-dimensional Lie algebra, and hence can be integrated by the group method of Section 6.5.4. Consider now an equation without symmetries.

Example 6.6.5. Consider the following non-linear second-order equation:

$$y'' - \frac{1}{y} y'^2 - \frac{x+x^2}{y} y' + 2x + 1 = 0. \qquad (6.6.24)$$

One can verify, by solving the determining equations, that Eq. (6.6.24) has no point symmetries, and hence cannot be integrated by Lie's method. Therefore, let us apply the method of integrating factors. Eq. (6.6.24) has form (6.6.16) with

$$a = 1, \quad b = -\frac{y'^2}{y} - \frac{x+x^2}{y} y' + 2x + 1.$$

For the sake of simplicity, we will look for the integrating factors depending on two variables. Let us take, e.g. $\mu = \mu(x, y)$. Then Eq. (6.6.17) yields $\mu_y = -\mu/y$, whence $\mu = p(x)/y$. Now Eq. (6.6.18) is written:

$$\frac{2p}{y^3} y'^2 - \frac{2p'}{y^2} y' + \frac{p''}{y} + y' H_{yy'} + H_{xy'} - H_y = 0, \qquad (6.6.25)$$

where

$$H = \frac{p}{y^2} y'^2 + (x+x^2) \frac{p}{y^2} y' - (2x + 1) \frac{p}{y}.$$

The reckoning shows that Eq. (6.6.25) reduces to the simple equation

$$\frac{p''(x)}{y} + (x+x^2) \frac{p'(x)}{y^2} = 0,$$

whence, separating the variables, we have $p'(x) = 0$, i.e., $p = \text{const.}$ Letting $p = 1$, we arrive at the following integrating factor for Eq. (6.6.24):

$$\mu = \frac{1}{y}.$$

Upon multiplying by μ, Eq. (6.6.24) becomes

$$\frac{y''}{y} - \frac{1}{y^2} y'^2 - \frac{x + x^2}{y^2} y' + \frac{2x + 1}{y} = 0. \tag{6.6.26}$$

Its left-hand side can easily be written as a total derivative. Indeed,

$$\frac{y''}{y} = D_x \left(\frac{y'}{y} \right) + \frac{y'^2}{y^2}, \qquad -\frac{x + x^2}{y^2} y' = D_x \left(\frac{x + x^2}{y} \right) - \frac{1 + 2x}{y}$$

and Eq. (6.6.26) is written

$$D_x \left(\frac{y' + x + x^2}{y} \right) = 0. \tag{6.6.27}$$

Hence,

$$\frac{y' + x + x^2}{y} = C_1,$$

or

$$y' + x + x^2 = C_1 y, \quad C_1 = \text{const.} \tag{6.6.28}$$

Integrating the non-homogeneous linear first-order equation (6.6.28) we obtain the following general solution to Eq. (6.6.24):

$$y = C_2 \, e^{C_1 x} - e^{C_1 x} \int (x + x^2) \, e^{-C_1 x} \, dx, \quad C_1, C_2 = \text{const.} \tag{6.6.29}$$

Working out the integral in (6.6.29), one can express the solution in elementary functions. Namely,

$$y = C_2 - \frac{1}{2} x^2 - \frac{1}{3} x^3 \tag{6.6.30}$$

if $C_1 = 0$ and

$$y = C_2 \, e^{C_1 x} + \frac{1}{C_1^3} \left[C_1^2 x^2 + (2 + C_1) C_1 x + 2 + C_1 \right] \tag{6.6.31}$$

if $C_1 \neq 0$.

Example 6.6.6. Consider the following system of the first-order equations:

$$\begin{aligned} F_1(x, y, z, y', z') &\equiv xzy' - 2xyz' + yz = 0, \\ F_2(x, y, z, y', z') &\equiv xy' + 2x^2 zz' + 2xz^2 + y = 0. \end{aligned} \tag{6.6.32}$$

Let us check conditions (6.6.5). Setting $u^1 = y$, $u^2 = z$, we have

$$\frac{\delta F_1}{\delta y} = -3xz', \qquad \frac{\delta F_1}{\delta z} = 3(y + xy').$$

Hence, $F_1(x, y, z, y', z')$ is not a total derivative. On the other hand,

$$\frac{\delta F_2}{\delta y} = 0, \quad \frac{\delta F_2}{\delta z} = 0,$$

and hence F_2 is a total derivative. Proceeding as in Example 6.6.3, we obtain

$$F_2 = D_x(xy + x^2 z^2). \tag{6.6.33}$$

Let us find an integrating factor of the form $\mu = \mu(x, y, z)$ for F_1. The reckoning shows that the first equation in (6.6.5) is written

$$\frac{\delta(\mu F_1)}{\delta y} = \frac{\partial(\mu F_1)}{\partial y} - D_x\left(\mu \frac{\partial F_1}{\partial y'}\right)$$

$$= -3xz'\mu - xz\mu_x + (yz - 2xyz')\mu_y - xzz'\mu_z = 0.$$

Since μ does not depend on z', the above equation splits into two equations:

$$2y\mu_y + z\mu_z + 3\mu = 0, \quad x\mu_x - y\mu_y = 0.$$

Solving this system of the first-order linear partial differential equations, we obtain

$$\mu = \frac{1}{z^3} \phi\left(\frac{xy}{z^2}\right), \tag{6.6.34}$$

where ϕ is an arbitrary function. One can verify that μ given by (6.6.34) solves also the second equation in (6.6.5),

$$\frac{\delta(\mu F_1)}{\delta z} = \frac{\partial(\mu F_1)}{\partial z} - D_x\left(\mu \frac{\partial F_1}{\partial z'}\right) = 0,$$

and hence provides an integrating factor for F_1. Since the function ϕ in (6.6.34) is arbitrary, we set $\phi = 1$, multiply F_1 by $\mu = z^{-3}$ and obtain

$$\mu F_1 = \frac{xy'}{z^2} - 2\frac{xyz'}{z^3} + \frac{y}{z^2} = D_x\left(\frac{xy}{z^2}\right). \tag{6.6.35}$$

Substituting (6.6.33) and (6.6.35) into Eqs. (6.6.32), we obtain the following first integrals:

$$xy + x^2 z^2 = C_1, \quad \frac{xy}{z^2} = C_2.$$

Solving for y and z, we obtain the general solution to system (6.6.32):

$$y = \frac{C_1 C_2}{x(C_2 + x^2)}, \quad z = \pm\sqrt{\frac{C_1}{C_2 + x^2}}.$$

6.6.3   Linearization of third-order equations

Lie's method of integration is applicable to higher-order equations as well (see, e.g. [26], [31], [17]). However, our concern here is on linearization rather than on integration of higher-order equations.

It is advantageous, e.g. for calculation of invariants, to write the general linear homogeneous equation (3.4.3) in the following standard form involving the binomial coefficients:

$$y^{(n)} + nc_1(x)y^{(n-1)} + \frac{n!c_2(x)}{(n-2)!2!}y^{(n-2)} + \cdots + nc_{n-1}(x)y' + c_n(x)y = 0. \quad (6.6.36)$$

Equivalence transformations for higher-order equations (3.4.3) are provided by the same transformations (3.3.8)—(3.3.9) as for the second-order equations:

$$\bar{x} = \phi(x), \quad \phi'(x) \neq 0,$$

$$y = \sigma(x)\bar{y}, \quad \sigma \neq 0.$$

An analog of Theorem 3.3.1 for higher-order linear equations was discovered in the nineteenth century. Namely, J. Cockle in 1876 and E. Laguerre in 1879 proved for $n = 3$ and for arbitrary n, respectively, that the two terms of orders next below the highest can be simultaneously removed in any equation (6.6.36). Their result can be formulated as follows (for the proof, see [21], Section 10.2.1, and the references therein).

Theorem 6.6.3. Any equation (6.6.36) can be reduced to the form

$$y^{(n)} + \frac{n!\,c_3(x)}{3!\,(n-3)!}\,y^{(n-3)} + \cdots + nc_{n-1}(x)\,y' + c_n(x)\,y = 0 \quad (6.6.37)$$

by appropriate equivalence transformations (3.3.8)—(3.3.9). Determination of the equivalence transformations requires integration of a second-order ordinary differential equation independently on the order n of Eq. (6.6.36).

Form (6.6.37) is called *Laguerre's canonical form* of the linear homogeneous nth-order equation. In particular, the canonical form (6.6.37) of the third-order equation is

$$y''' + \alpha(x)y = 0. \quad (6.6.38)$$

Recently we[1] found all linearizable third-order equations

$$y''' = f(x, y, y', y''). \quad (6.6.39)$$

We formulate here the basic theorems on linearization of third-order equations and illustrate them by examples. We will use the canonical form (6.6.38) of the linear third-order equations guaranteed by Theorem 6.6.3.

---

[1]N.H. Ibragimov and S.V. Meleshko, *J. Math. Anal. Appl.*, 308(1), 2005, 266-289.

Lemma 6.6.5. The third-order equations (6.6.39) obtained from a linear equation (6.6.38) by a change of variables (6.5.42) with $\varphi_y = 0$,

$$t = \varphi(x), \quad u = \psi(x, y), \tag{6.6.40}$$

belong to the family of equations of the form

$$y''' + (A_1 y' + A_0)y'' + B_3 y'^3 + B_2 y'^2 + B_1 y' + B_0 = 0, \tag{6.6.41}$$

where $A_0 = A_0(x, y), \ldots, B_3 = B_3(x, y)$. The third-order equations (6.6.39) obtained from a linear equation (6.6.38) by a change of variables

$$t = \varphi(x, y), \quad u = \psi(x, y), \quad \text{where } \varphi_y \neq 0, \tag{6.6.42}$$

belong to the family of equations of the form

$$y''' + \frac{1}{y' + r}\left[-3(y'')^2 + (C_2 y'^2 + C_1 y' + C_0)y''\right.$$
$$\left. + D_5 y'^5 + D_4 y'^4 + D_3 y'^3 + D_2 y'^2 + D_1 y' + D_0\right] = 0, \tag{6.6.43}$$

where $C_0 = C_0(x, y), \ldots, D_5 = D_5(x, y)$, and $r = r(x, y)$.

Eqs. (6.6.41) and (6.6.43) with arbitrary coefficients $A_0, \ldots, B_3$ and $C_0, \ldots, D_5$, respectively, provide two *candidates for linearization*.

Theorem 6.6.4. Equation (6.6.41) is linearizable if and only if its coefficients satisfy the following five equations:

$$A_{0y} - A_{1x} = 0, \quad (3B_1 - A_0^2 - 3A_{0x})_y = 0, \tag{6.6.44}$$

$$3B_2 = 3A_{1x} + A_0 A_1, \quad 9B_3 = 3A_{1y} + A_1^2, \tag{6.6.45}$$

$$27B_{0yy} = (9B_1 - 6A_{0x} - 2A_0^2)A_{1x} + 9(B_{1x} - A_1 B_0)_y + 3A_0 B_{1y}. \tag{6.6.46}$$

Provided that conditions (6.6.44)—(6.6.46) are satisfied, the linearizing transformation (6.6.40) is defined by a third-order ordinary differential equation for the function $\varphi(x)$, namely, by the Riccati equation

$$6\frac{d\chi}{dx} - 3\chi^2 = 3B_1 - A_0^2 - 3A_{0x} \tag{6.6.47}$$

for

$$\chi = \frac{\varphi_{xx}}{\varphi_x}, \tag{6.6.48}$$

and by the following integrable system of partial differential equations for $\psi(x, y)$:

$$3\psi_{yy} = A_1 \psi_y, \quad 3\psi_{xy} = (3\chi + A_0)\psi_y, \tag{6.6.49}$$

$$\psi_{xxx} = 3\chi \psi_{xx} + B_0 \psi_y - \frac{1}{6}(3A_{0x} + A_0^2 - 3B_1 + 9\chi^2)\psi_x - \Omega, \tag{6.6.50}$$

where χ is given by (6.6.48) and Ω is the following expression:

$$\Omega = \frac{1}{6}\left(A_{0xx} + 2A_0 A_{0x} + 6B_{0y} - 3B_{1x} + \frac{4}{9}A_0^3 - 2A_0 B_1 + 2A_1 B_0\right). \quad (6.6.51)$$

Finally, the coefficient α of the resulting linear equation (6.6.38) is given by

$$\alpha = \Omega\,\varphi_x^{-3}. \quad (6.6.52)$$

Example 6.6.7. The equation

$$y''' - \left(\frac{6}{y}y' + \frac{3}{x}\right)y'' + 6\left(\frac{y'^3}{y^2} + \frac{y'^2}{xy} + \frac{y'}{x^2} + \frac{y}{x^3}\right) = 0 \quad (6.6.53)$$

is an equation of form (6.6.41) with the coefficients

$$A_1 = -\frac{6}{y}, \quad A_0 = -\frac{3}{x}, \quad B_3 = \frac{6}{y^2}, \quad B_2 = \frac{6}{xy}, \quad B_1 = \frac{6}{x^2}, \quad B_0 = \frac{6y}{x^3}. \quad (6.6.54)$$

One can readily verify that coefficients (6.6.54) obey conditions (6.6.44)—(6.6.46). We have

$$3B_1 - A_0^2 - 3A_{0x} = 0, \quad (6.6.55)$$

and Eq. (6.6.47) is written

$$2\frac{d\chi}{dx} - \chi^2 = 0.$$

Let us take its simplest solution $\chi = 0$. Then, invoking (6.6.48), we let $\varphi = x$. Now Eqs. (6.6.49) are written

$$\frac{\partial \ln |\psi_y|}{\partial y} = -\frac{2}{y}, \quad \frac{\partial \ln |\psi_y|}{\partial x} = -\frac{1}{x}$$

and yield

$$\psi_y = \frac{K}{xy^2}, \quad K = \text{const.}$$

Hence,

$$\psi = -\frac{K}{xy} + f(x).$$

One can take any particular solution. We set $K = -1, f(x) = 0$ and take

$$\psi = \frac{1}{xy}.$$

Invoking (6.6.55) and noting that (6.6.51) yields $\Omega = 0$, one can readily verify that the function $\psi = 1/(xy)$ solves Eq. (6.6.50) as well. Since $\Omega = 0$, Eq. (6.6.52) gives $\alpha = 0$. Hence, the transformation

$$t = x, \quad u = \frac{1}{xy} \quad (6.6.56)$$

maps Eq. (6.6.53) to the linear equation

$$u''' = 0.$$

Example 6.6.8. Consider the following equation of form (6.6.41):

$$y''' + \frac{3}{y} y' y'' - 3y'' - \frac{3}{y} y'^2 + 2y' - y = 0. \tag{6.6.57}$$

One can readily verify that its coefficients

$$A_1 = \frac{3}{y}, \ A_0 = -3, \ B_3 = 0, \ B_2 = -\frac{3}{y}, \ B_1 = 2, \ B_0 = -y$$

obey the linearization conditions (6.6.44)—(6.6.46). Furthermore,

$$3B_1 - A_0^2 - 3A_{0x} = -3$$

and Eq. (6.6.47) is written

$$6\frac{d\chi}{dx} - 3\chi^2 = -3.$$

We take its evident solution $\chi = 1$ and obtain from (6.6.48) the equation $\varphi'' = \varphi'$, whence

$$\varphi = e^x.$$

Eqs. (6.6.49) have the form

$$\frac{\partial \ln |\psi_y|}{\partial y} = \frac{1}{y}, \quad \psi_{xy} = 0$$

and can be readily solved. We take the simplest solution $\psi = y^2$ and obtain the following change of variables (6.6.40):

$$t = e^x, \quad u = y^2. \tag{6.6.58}$$

Substituting $\Omega = -2$ and $\varphi_x = e^x = t$ into Eq. (6.6.52), we obtain $\alpha(t) = -2t^{-3}$.

Thus, Eq. (6.6.57) is mapped by transformation (6.6.58) to the linear equation

$$u''' - \frac{2}{t^3} u = 0. \tag{6.6.59}$$

Theorem 6.6.5. Equation (6.6.43),

$$y''' + \frac{1}{y' + r} \Big[- 3(y'')^2 + (C_2 y'^2 + C_1 y' + C_0)y''$$

$$+ D_5 y'^5 + D_4 y'^4 + D_3 y'^3 + D_2 y'^2 + D_1 y' + D_0 \Big] = 0,$$

is linearizable if and only if its coefficients obey the following equations:

$$C_0 = 6r \frac{\partial r}{\partial y} - 6 \frac{\partial r}{\partial x} + r C_1 - r^2 C_2, \tag{6.6.60}$$

$$6\frac{\partial^2 r}{\partial y^2} = \frac{\partial C_2}{\partial x} - \frac{\partial C_1}{\partial y} + r\frac{\partial C_2}{\partial y} + C_2\frac{\partial r}{\partial y}, \tag{6.6.61}$$

$$18D_0 = 3r^2\left[r\frac{\partial C_1}{\partial y} - 2\frac{\partial C_1}{\partial x} - r\frac{\partial C_2}{\partial x} + 3r^2\frac{\partial C_2}{\partial y} - 12\frac{\partial^2 r}{\partial x\partial y}\right] - 54\left(\frac{\partial r}{\partial x}\right)^2$$
$$+6r\left[3\frac{\partial^2 r}{\partial x^2} + 15\frac{\partial r}{\partial x}\frac{\partial r}{\partial y} - 6r\left(\frac{\partial r}{\partial y}\right)^2 + (3C_1 - rC_2)\frac{\partial r}{\partial x}\right]$$
$$+r^2\left[9(rC_2 - 2C_1)\frac{\partial r}{\partial y} - 2C_1^2 + 2rC_1C_2\right.$$
$$\left. +4r^2C_2^2 + 18r^2D_4 - 72r^3D_5\right], \tag{6.6.62}$$

$$18D_1 = 9r^2\frac{\partial C_1}{\partial y} - 12r\frac{\partial C_1}{\partial x} - 27r^2\frac{\partial C_2}{\partial x} + 33r^3\frac{\partial C_2}{\partial y} - 36r\frac{\partial^2 r}{\partial x\partial y} + 18\frac{\partial^2 r}{\partial x^2}$$
$$+ 6(3C_1 + 4rC_2)\frac{\partial r}{\partial x} - 3r(6C_1 + 7rC_2)\frac{\partial r}{\partial y} + 18r\left(\frac{\partial r}{\partial y}\right)^2 - 18\frac{\partial r}{\partial x}\frac{\partial r}{\partial y}$$
$$- 4rC_1^2 - 2r^2C_1C_2 + 20r^3C_2^2 + 72r^3D_4 - 270r^4D_5, \tag{6.6.63}$$

$$9D_2 = 3r\frac{\partial C_1}{\partial y} - 3\frac{\partial C_1}{\partial x} - 21r\frac{\partial C_2}{\partial x} + 21r^2\frac{\partial C_2}{\partial y} + 15C_2\frac{\partial r}{\partial x}$$
$$-15\,r\,C_2\frac{\partial r}{\partial y} - C_1^2 - 5rC_1C_2 + 14r^2C_2^2 + 54r^2D_4 - 180r^3D_5, \tag{6.6.64}$$

$$3D_3 = 3r\frac{\partial C_2}{\partial y} - 3\frac{\partial C_2}{\partial x} - C_1C_2 + 2rC_2^2 + 12rD_4 - 30r^2D_5, \tag{6.6.65}$$

$$54\frac{\partial D_4}{\partial x} = 18\frac{\partial^2 C_1}{\partial y^2} + 3C_2\frac{\partial C_1}{\partial y} - 72\frac{\partial^2 C_2}{\partial x\partial y} - 39C_2\frac{\partial C_2}{\partial x}$$
$$+18r\frac{\partial^2 C_2}{\partial y^2} - 3rC_2\frac{\partial C_2}{\partial y} + \left(72\frac{\partial C_2}{\partial y} + 33C_2^2\right)\frac{\partial r}{\partial y} + 108D_4\frac{\partial r}{\partial y}$$
$$+270D_5\frac{\partial r}{\partial x} + 378\,r\frac{\partial D_5}{\partial x} - 108r^2\frac{\partial D_5}{\partial y} - 540rD_5\frac{\partial r}{\partial y}$$
$$+36\,r\,C_1D_5 - 8\,r\,C_2^3 - 36\,r\,C_2D_4 + 108\,r^2\,C_2D_5 + 54\,r\,H, \tag{6.6.66}$$

and

$$\frac{\partial H}{\partial x} = 3H\frac{\partial r}{\partial y} + r\frac{\partial H}{\partial y}, \tag{6.6.67}$$

where

$$H = \frac{\partial D_4}{\partial y} - 2\frac{\partial D_5}{\partial x} - 3r\frac{\partial D_5}{\partial y} - 5D_5\frac{\partial r}{\partial y} - 2rC_2D_5$$
$$+\frac{1}{3}\left[\frac{\partial^2 C_2}{\partial y^2} + 2C_2\frac{\partial C_2}{\partial y} - 2C_1D_5 + 2C_2D_4\right] + \frac{4}{27}C_2^3. \tag{6.6.68}$$

If conditions (6.6.60)—(6.6.67) are satisfied, transformation (6.6.42),

$$t = \varphi(x, y), \quad u = \psi(x, y), \quad \varphi_y \neq 0,$$

mapping equation (6.6.43) to a linear equation (6.6.38) is obtained by solving the following compatible system of equations for the functions $\varphi(x, y)$ and $\psi(x, y)$:

$$\frac{\partial \varphi}{\partial x} = r\frac{\partial \varphi}{\partial y}, \qquad \frac{\partial \psi}{\partial x} = -\frac{\partial \varphi}{\partial y}W + r\frac{\partial \psi}{\partial y}, \tag{6.6.69}$$

$$6\frac{\partial \varphi}{\partial y}\frac{\partial^3 \varphi}{\partial y^3} = 9\left(\frac{\partial^2 \varphi}{\partial y^2}\right)^2 + \left[15rD_5 - 3D_4 - C_2^2 - 3\frac{\partial C_2}{\partial y}\right]\left(\frac{\partial \varphi}{\partial y}\right)^2, \tag{6.6.70}$$

$$\frac{\partial^3 \psi}{\partial y^3} = W D_5 \frac{\partial \varphi}{\partial y} + \frac{1}{6}\left[15\,r\,D_5 - C_2^2 - 3D_4 - 3\frac{\partial C_2}{\partial y}\right]\frac{\partial \psi}{\partial y}$$
$$-\frac{1}{2}H\psi + 3\frac{\partial^2 \varphi}{\partial y^2}\frac{\partial^2 \psi}{\partial y^2}\left(\frac{\partial \varphi}{\partial y}\right)^{-1} - \frac{3}{2}\left(\frac{\partial^2 \varphi}{\partial y^2}\right)^2\frac{\partial \psi}{\partial y}\left(\frac{\partial \varphi}{\partial y}\right)^{-2}, \tag{6.6.71}$$

where the function W is defined by the equations

$$3\frac{\partial W}{\partial x} = \left[C_1 - rC_2 + 6\frac{\partial r}{\partial y}\right]W, \qquad 3\frac{\partial W}{\partial y} = C_2 W. \tag{6.6.72}$$

The coefficient α of the resulting linear equation (6.6.38) is given by (cf. (6.6.52))

$$\alpha = \frac{H}{2(\varphi_y)^3}, \tag{6.6.73}$$

where H is the function defined in (6.6.68).

Example 6.6.9. Consider the nonlinear equation

$$y''' + \frac{1}{y'}\left[-3y''^2 - xy'^5\right] = 0. \tag{6.6.74}$$

It has form (6.6.43) with the following coefficients:

$$r = 0, \quad C_0 = C_1 = C_2 = 0,$$
$$D_0 = D_1 = D_2 = D_3 = D_4 = 0, \quad D_5 = -x. \tag{6.6.75}$$

Let us test Eq. (6.6.74) for linearization by using Theorem 6.6.5. It is manifest that coefficients (6.6.75) satisfy Eqs. (6.6.60)—(6.6.66). Furthermore, Eq. (6.6.67) also holds since (6.6.68) yields

$$H = 2. \tag{6.6.76}$$

Thus, Eq. (6.6.74) is linearizable, and we can proceed further. Eqs. (6.6.72) are written

$$\frac{\partial W}{\partial x} = 0, \quad \frac{\partial W}{\partial y} = 0$$

and yield $W = $ const. Therefore, Eqs. (6.6.69) have the form

$$\frac{\partial \varphi}{\partial x} = 0, \quad \frac{\partial \psi}{\partial x} = -W \frac{\partial \varphi}{\partial y}$$

and hence,

$$\varphi = \varphi(y), \quad \psi = -W x \, \varphi'(y) + \omega(y). \tag{6.6.77}$$

Now, the third-order equations (6.6.70) and (6.6.71) yield the ordinary differential equation

$$\varphi''' = \frac{3}{2} \frac{\varphi''^2}{\varphi'} \tag{6.6.78}$$

for $\varphi(y)$ and the partial differential equation

$$\frac{\partial^3 \psi}{\partial y^3} = 3 \frac{\varphi''}{\varphi'} \frac{\partial^2 \psi}{\partial y^2} - \frac{3}{2} \frac{\varphi''^2}{\varphi'^2} \frac{\partial \psi}{\partial y} - \psi - W x \varphi' \tag{6.6.79}$$

for $\psi(x, y)$, respectively. Using the expression for ψ given in (6.6.77) and Eq. (6.6.78) for φ, we reduce Eq. (6.6.79) to

$$3 \frac{\varphi''}{\varphi'} \omega'' - \frac{3}{2} \frac{\varphi''^2}{\varphi'^2} \omega' - \omega - \omega''' = 0.$$

Hence, one can satisfy Eq. (6.6.79) by letting $\omega(y) = 0$. Then the construction of the linearizing transformation requires integration of Eq. (6.6.78) known in the literature as the Schwarzian equation. Its general solution is provided by the straight lines

$$\varphi = ky + l, \quad k, l = \text{const.}, \tag{6.6.80}$$

and the hyperbolas

$$\varphi = a + \frac{1}{b - cy}, \quad a, b, c = \text{const.} \tag{6.6.81}$$

Let us take the simplest solution $\varphi = y$ of form (6.6.80). Then (6.6.73) yields $\alpha = 1$. Furthermore, we set $W = -1, \omega = 0$ in (6.6.77) and arrive at the change of variables

$$t = y, \quad u = x, \tag{6.6.82}$$

reducing (6.6.74) to the following linear equation:

$$u''' + u = 0. \tag{6.6.83}$$

6.7 Nonlinear superposition

6.7.1 Introduction

Sophus Lie had an extraordinary geometric imagination that simplified analytical calculations and often led him to new theoretical concepts. Lie's generalization of linear equations and the related theory of nonlinear superpositions considered in this section provide a good example.

Recall that solution of a homogeneous linear partial differential equation and integration of its characteristic system (a system of ordinary differential equations) are equivalent problems. Furthermore, any system of ordinary differential equations with n dependent variables x^i,

$$\frac{dx^i}{dt} = f^i(t, x), \quad i = 1, \ldots, n, \tag{6.7.1}$$

can be regarded as the characteristic system for the partial differential equation with $n + 1$ independent variables, t and $x = (x^1, \ldots, x^n)$:

$$\frac{\partial u}{\partial t} + f^1(t, x)\frac{\partial u}{\partial x^1} + \cdots + f^n(t, x)\frac{\partial u}{\partial x^n} = 0. \tag{6.7.2}$$

Lie noticed that the classical theories of linear ordinary differential equations

$$\frac{dx^i}{dt} = a_{i1}(t)x^1 + \cdots + a_{in}(t)x^n, \quad i = 1, \ldots, n, \tag{6.7.3}$$

as well as of the associated partial differential equations (6.7.2),

$$\frac{\partial u}{\partial t} + \sum_{i,k=1}^{n} a_{ki}(t)x^i\frac{\partial u}{\partial x^k} = 0, \tag{6.7.4}$$

are due to the fact that the n^2 operators $X_{ik} = x^i\partial/\partial x^k$ generate a finite continuous group with n variables x^i, namely the linear homogeneous group. This observation led him to believe that the main features of linear equations (6.7.3) can be extended to the vast class of nonlinear equations having the form of a *generalized separation of variables*:

$$\frac{dx^i}{dt} = T_1(t)\xi_1^i(x) + \cdots + T_r(t)\xi_r^i(x), \quad i = 1, \ldots, n, \tag{6.7.5}$$

provided that the operators

$$X_\alpha = \xi_\alpha^i(x)\frac{\partial}{\partial x^i}, \quad \alpha = 1, \ldots, r, \tag{6.7.6}$$

span a finite-dimensional Lie algebra. The coefficients $T_\alpha(t)$ are arbitrary functions of the variable t. Systems (6.7.5) are considered together with the equivalent linear partial differential equations

$$A[u] \equiv \frac{\partial u}{\partial t} + [T_1(t)X_1 + \cdots + T_r(t)X_r] u = 0. \tag{6.7.7}$$

Sophus Lie wrote in 1893 the following (see [15] and the references therein):

"*I have already outlined a general integration theory for the equation $A[u] = 0$. The theory is based, as I explained in 'General studies on differential equations admitting a finite continuous group' (Mathematische Annalen, 25(1), 1885, p. 128), on the fact that it is possible to find the general integrals of systems (6.7.5) if one knows a certain finite number of their particular solutions*

$$x_1 = (x_1^1, \ldots, x_1^n), \quad \ldots, \quad x_m = (x_m^1, \ldots, x_m^n). \tag{6.7.8}$$

I add that the expressions for the general solutions $x = (x^1, \ldots, x^n)$, as functions of quantities (6.7.8), are obtained by solving certain equations,

$$J_i(x^1, \ldots, x^n; x_1^1, \ldots, x_1^n; \ldots, x_m^1, \ldots, x_m^n) = C_i,$$

with respect to $x^1, \ldots, x^n$, where the J_i designate what I call invariants of $m+1$ points x^i, $x_1^i, \ldots, x_m^i$ with respect to the group generated by $X_1, \ldots, X_r$.

E. Vessiot, whose recent thesis constitutes such an important progress in the theory of linear differential equations, has come to a lucky idea to look for all ordinary differential equations possessing fundamental systems of integrals. Alf Guldberg is also occupied by the same question.

It is a very interesting problem to seek, together with Vessiot and Guldberg, all systems (6.7.1) whose general solutions $x = (x^1, \ldots, x^n)$ can be expressed via m particular solutions only, viz.

$$x^i = \varphi^i(x_1^1, \ldots, x_1^n; \ldots, x_m^1, \ldots, x_m^n; C_1, \ldots, C_n), \quad i = 1, \ldots, n. \tag{6.7.9}$$

Since these authors did not find, however, all my systems (6.7.5) admitting in effect the required property, it seems to me that their investigations must have a gap. I presume that these authors introduced implicitly an essential restriction that the most general formulae (6.7.9) are deducible from a given system of these formulae by merely changing arbitrary constants.

If I am correct, I have the fortune to be the first to prove rigorously and simply that my systems (6.7.5) are the only ones possessing the required property."

6.7.2   Main theorem on nonlinear superposition

The above discussion leads to the following definition.

Definition 6.7.1. A system of ordinary differential equations (6.7.1),

$$\frac{dx^i}{dt} = f^i(t, x), \quad i = 1, \ldots, n,$$

is said to possess a *fundamental system of solutions* if its general solution can be represented in form (6.7.9) involving a finite number m of particular solutions (6.7.8) and n arbitrary constants $C_1, \ldots, C_n$. Expression (6.7.9) of the general solution is termed a *nonlinear superposition*, and the particular solutions (6.7.8) are referred to as a *fundamental system of solutions* for Eqs. (6.7.1).

The following result due to Lie (1893) identifies those equations (6.7.1) possessing fundamental systems of solutions.

Theorem 6.7.1. Equations (6.7.1) possess a fundamental system of solutions if and only if they have form (6.7.5),

$$\frac{dx^i}{dt} = T_1(t)\xi_1^i(x) + \cdots + T_r(t)\xi_r^i(x), \quad i = 1, \ldots, n,$$

where the coefficients $\xi_\alpha^i(x)$ satisfy the condition that the operators

$$X_\alpha = \xi_\alpha^i(x)\frac{\partial}{\partial x^i}, \quad \alpha = 1, \ldots, r,$$

span a Lie algebra L_r of a finite dimension r. The number m of necessary particular solutions (6.7.8) is estimated by

$$nm \geqslant r. \tag{6.7.10}$$

Finally, superposition formulae (6.7.9)

$$x^i = \varphi^i(x_1^1, \ldots, x_1^n; \ldots, x_m^1, \ldots, x_m^n; C_1, \ldots, C_n), \quad i = 1, \ldots, n,$$

are defined implicitly by n equations

$$J_i(x^1, \ldots, x^n; x_1^1, \ldots, x_1^n; \ldots, x_m^1, \ldots, x_m^n) = C_i, \quad i = 1, \ldots, n, \tag{6.7.11}$$

where J_i are functionally independent (with respect to x^i) invariants of the $(m+1)$-point representation $V_\alpha = X_\alpha + X_\alpha^{(1)} + \cdots + X_\alpha^{(m)}$ of operators (6.7.6). In other words, $J_i(x_1, \ldots, x_m)$ solve the equations

$$\xi_\alpha^i(x)\frac{\partial J}{\partial x^i} + \xi_\alpha^i(x_1)\frac{\partial J}{\partial x_1^i} + \cdots + \xi_\alpha^i(x_m)\frac{\partial J}{\partial x_m^i} = 0, \quad \alpha = 1, \ldots, r, \tag{6.7.12}$$

and satisfy the condition $\det(\partial J_i/\partial x^k) \neq 0$.

Proof (see the English edition of [15] and the related references therein). Let Eqs. (6.7.1),

$$\frac{dx^i}{dt} = f^i(t, x), \quad i = 1, \ldots, n,$$

possess a fundamental system of solutions. The superposition formulae (6.7.9) can be written, upon solving them with respect to C_i, in the form

$$J_i(x^1, \ldots, x^n; x_1^1, \ldots, x_1^n; \ldots, x_m^1, \ldots, x_m^n) = C_i, \quad i = 1, \ldots, n.$$

Differentiating these identities with respect to t, one obtains the equations

$$\frac{\partial J_i}{\partial x^k}\frac{dx^k}{dt} + \frac{\partial J_i}{\partial x_1^k}\frac{dx_1^k}{dt} + \cdots + \frac{\partial J_i}{\partial x_m^k}\frac{dx_m^k}{dt} = 0, \quad i = 1, \ldots, n,$$

whence, invoking Eqs. (6.7.1):

$$f^k(t,x)\frac{\partial J_i}{\partial x^k} + f^k(t,x_1)\frac{\partial J_i}{\partial x_1^k} + \cdots + f^k(t,x_m)\frac{\partial J_i}{\partial x_m^k} = 0. \qquad (6.7.13)$$

Noting that Eqs. (6.7.13) hold for any system of $m+1$ solutions

$$x = (x^1,\ldots,x^n),\ x_1 = (x_1^1,\ldots,x_1^n),\ \ldots,\ x_m = (x_m^1,\ldots,x_m^n), \qquad (6.7.14)$$

and taking into account that the initial values of the latter are arbitrary, we conclude that (6.7.13) are satisfied identically in $nm+n+1$ variables $x_1^i,\ldots,x_m^i, x^i$ and t.

Now we introduce the operator

$$Y = f^k(t,x)\frac{\partial}{\partial x^k} \qquad (6.7.15)$$

and denote by $Y^{(1)},\ldots,Y^{(m)}$ the operators obtained from Y by replacing x by $x_1,\ldots,x_m$, respectively. Then Eqs. (6.7.13) are written:

$$Y(J_i) + Y^{(1)}(J_i) + \cdots + Y^{(m)}(J_i) = 0, \quad i = 1,\ldots,n.$$

In other words, the linear partial differential equation of the first order

$$U(J) \equiv [Y + Y^{(1)} + \cdots + Y^{(m)}](J) = 0 \qquad (6.7.16)$$

must have n independent solutions $J_1,\ldots,J_n$ (6.7.11) that are free from t. Here

$$U \equiv Y + Y^{(1)} + \cdots + Y^{(m)} = f^k(t,x)\frac{\partial}{\partial x^k} + f^k(t,x_1)\frac{\partial}{\partial x_1^k} + \cdots + f^k(t,x_m)\frac{\partial}{\partial x_m^k}$$

is a differential operator with $nm+n$ variables x^i, x_μ^i ($\mu = 1,\ldots,m$) given by (6.7.14). The variable t, involved in the coefficients of U, is regarded as a parameter.

By letting t assume a series of fixed values t_σ, one singles out from (6.7.16) a certain number of linear partial differential equations that are free from the parameter t. The equations of this series have $J_1,\ldots,J_n$ as their common solutions since the latter do not explicitly depend upon the parameter t, and hence solve Eq. (6.7.16) for any t, e.g. for all assigned values $t = t_\sigma$. According to the classical theory of systems of linear partial differential equations of the first order, the series of equations in question can be replaced by a system of a finite number s of independent equations

$$U_\sigma(J) \equiv [Y_\sigma + Y_\sigma^{(1)} + \cdots + Y_\sigma^{(m)}](J) = 0, \quad \sigma = 1,\ldots,s, \qquad (6.7.17)$$

where Y_σ and $Y_\sigma^{(\mu)}$ are obtained by letting $t = t_\sigma$ in Y and $Y^{(\mu)}$, respectively. System (6.7.17) has at most $nm+n-s$ functionally independent solutions. On the other hand, n independent solutions already exist, namely $J_1,\ldots,J_n$.

Hence, $nm+n-s \geq n$. In other words, the number of Eqs. (6.7.17) is estimated by $s \leq nm$. Note that the operators U_σ are completely free from the variable t, and that the general equation (6.7.16) must be a consequence of Eqs. (6.7.17).

Furthermore, one can replace (6.7.17) by a complete system adding to the operators U_σ all independent commutators $[U_\sigma, U_\tau]$. Since the operators $Y, Y^{(1)}, \ldots, Y^{(m)}$ involve distinctly different types of variables x, we have

$$[U_\sigma, U_\tau] = [Y_\sigma, Y_\tau] + [Y_\sigma^{(1)}, Y_\tau^{(1)}] + \cdots + [Y_\sigma^{(m)}, Y_\tau^{(m)}]. \qquad (6.7.18)$$

Let the resulting complete system comprise r independent equations:

$$V_\alpha(J) \equiv [X_\alpha + X_\alpha^{(1)} + \cdots + X_\alpha^{(m)}](J) = 0, \quad \alpha = 1, \ldots, r, \qquad (6.7.19)$$

where X_α and $X_\alpha^{(\mu)}$ are operators of the type Y_σ and $Y_\sigma^{(\mu)}$, respectively. Accordingly, V_α obey the commutator relations similar to (6.7.18):

$$[V_\alpha, V_\beta] = [X_\alpha, X_\beta] + [X_\alpha^{(1)}, X_\beta^{(1)}] + \cdots + [X_\alpha^{(m)}, X_\beta^{(m)}].$$

Note that system (6.7.19) has at least n solutions, namely $J_1, \ldots, J_n$. Hence, as discussed above, $r \leq nm$. Since system (6.7.19) is complete, it obeys the relations $[V_\alpha, V_\beta] = h_{\alpha\beta}^\gamma V_\gamma$ (summation in $\gamma = 1, \ldots, r$) :

$$[X_\alpha, X_\beta] + [X_\alpha^{(1)}, X_\beta^{(1)}] + \cdots + [X_\alpha^{(m)}, X_\beta^{(m)}] = h_{\alpha\beta}^\gamma (X_\gamma + X_\gamma^{(1)} + \cdots + X_\gamma^{(m)}).$$

Whence, separating the variables:

$$[X_\alpha, X_\beta] = \sum_{\gamma=1}^r h_{\alpha\beta}^\gamma X_\gamma; \qquad [X_\alpha^{(\mu)}, X_\beta^{(\mu)}] = h_{\alpha\beta}^\gamma X_\gamma^{(\mu)}, \quad \mu = 1, \ldots, m.$$

At the beginning, the coefficients $h_{\alpha\beta}^\gamma$ might depend upon all $nm + n$ variables (6.7.14). However, in each of the final $m + 1$ equations, the $h_{\alpha\beta}^\gamma$ can involve only one kind of n variables. This is possible only if $h_{\alpha\beta}^\gamma$ are free from all $nm+n$ variables, i.e., they are constants $c_{\alpha\beta}^\gamma$. Thus, X_α satisfy the equations

$$[X_\alpha, X_\beta] = c_{\alpha\beta}^\gamma X_\gamma, \quad \alpha, \beta = 1, \ldots, r, \qquad (6.7.20)$$

and hence span a Lie algebra L_r of the dimension $r \leq nm$.

Since s equations (6.7.17) are contained in r equations (6.7.19), their operators are linearly connected as follows:

$$U_\sigma = h_\sigma^\beta V_\beta, \quad \sigma = 1, \ldots, s.$$

Here the coefficients h_σ^β are, for the time being, functions of the variables (6.7.14). But one can use expansions (6.7.17) and (6.7.19) of U_σ and V_α, respectively, to obtain $m + 1$ separated equations:

$$Y_\sigma = h_\sigma^\beta X_\beta; \qquad Y_\sigma^{(\mu)} = h_\sigma^\beta X_\beta^{(\mu)}, \quad \mu = 1, \ldots, m.$$

It follows, as above, that h_σ^β can be only constants C_σ^β. Thus, Y_σ are linear combinations of X_β with constant coefficients:

$$Y_\sigma = C_\sigma^\beta X_\beta, \qquad \sigma = 1, \ldots, s, \qquad (6.7.21)$$

and hence belong to the Lie algebra spanned by $X_1, \ldots, X_r$.

Since Eq. (6.7.16) with an arbitrary value of t is a consequence of Eqs. (6.7.17), we have $U = \omega_1 U_1 + \cdots + \omega_s U_s$, or in the separated form:

$$Y = \omega_1 Y_1 + \cdots + \omega_s Y_s; \quad Y^{(\mu)} = \omega_1 Y_1^{(\mu)} + \cdots + \omega_s Y_s^{(\mu)}, \ \mu = 1, \ldots, m.$$

The latter equations yield that all ω_σ are free from variables (6.7.14), $x, x_1, \ldots, x_m$, but may depend on t. Hence, $Y = \omega_1(t) Y_1 + \cdots + \omega_s(t) Y_s$ or, invoking (6.7.21):

$$Y = T_1(t) X_1 + \cdots + T_r(t) X_r. \qquad (6.7.22)$$

Here X_α have form (6.7.6) and by (6.7.20) span a Lie algebra L_r. Invoking definition (6.7.15) of Y, we see that if equations $dx^i/dt = f^i(t, x)$ possess a fundamental system of solutions they have Lie's form (6.7.5):

$$\frac{dx^i}{dt} = T_1(t) \xi_1^i(x) + \cdots + T_r(t) \xi_r^i(x).$$

We also obtained estimation (6.7.10), $r \leq nm$.

Conversely, all equations of form (6.7.5) have the required property. Indeed, one can take the number m to be sufficiently large so that Eq. (6.7.16), or in our case the equation

$$\sum_{\alpha=1}^r T_\alpha(t) [X_\alpha + X_\alpha^{(1)} + \cdots + X_\alpha^{(m)}](J) = 0,$$

has at least n solutions $J_1, \ldots, J_n$ that are functionally independent (with respect to $x^1, \ldots, x^n$) and do not involve t. Then the equations

$$[X_\alpha + X_\alpha^{(1)} + \cdots + X_\alpha^{(m)}](J) = 0, \quad \alpha = 1, \ldots, r, \qquad (6.7.23)$$

with an appropriately chosen m, are independent. Furthermore, system (6.7.23) of r equations with $nm + n$ independent variables is complete thanks to the commutator relations (6.7.20). It has $mn + n - r$ solutions. By taking a sufficiently large m, one can find at least n solutions

$$J_i(x^1, \ldots, x^n; x_1^1, \ldots, x_1^n; \ldots, x_m^1, \ldots, x_m^n), \quad i = 1, \ldots, n,$$

that are functionally independent with respect to $x^1, \ldots, x^n$. Now, one can convert the derivation of Eqs. (6.7.13) and show that J_i hold constant values whenever the $mn + n$ variables (6.7.14) solve Eqs. (6.7.5). This completes the proof.

Remark 6.7.1. Theorem 6.7.1 does not exclude the possibility to have, for a given system of Eqs. (6.7.5), several distinctly different representations (6.7.11) of the general solution as well as different numbers m of the particular solutions (6.7.8) involved. See further Examples 6.7.6 and 6.7.7.

Remark 6.7.2. The Lie algebra L_r spanned by operators (6.7.6) will be referred to as the *Vessiot-Guldberg-Lie algebra* for Eqs. (6.7.5).

6.7.3   Examples of nonlinear superposition

Example 6.7.1. Let us consider the single homogeneous linear equation $dx/dt = A(t)x$. Here, $r = 1$ and $X = xd/dx$ (cf. Eq. (6.7.3) with $n = 1$). We take the two-point representation V of X (see (6.7.19) with $m = 1$):

$$V = x\frac{\partial}{\partial x} + x_1\frac{\partial}{\partial x_1}$$

and its invariant $J(x, x_1) = x/x_1$. Eq. (6.7.11) has the form $x/x_1 = C$. Hence, $m = 1$ and formula (6.7.9) is the linear superposition $x = Cx_1$. Condition (6.7.10) is satisfied as an equality.

Lie's generalization (6.7.5) of this simplest example is the equation with separated variables:

$$\frac{dx}{dt} = T(t)h(x).$$

Here, $r = 1$ and $X = h(x)d/dx$. Taking the two-point representation

$$V = h(x)\frac{\partial}{\partial x} + h(x_1)\frac{\partial}{\partial x_1},$$

and integrating the characteristic Equation $dx/h(x) = dx_1/h(x_1)$, one obtains the invariant $J(x, x_1) = H(x) - H(x_1)$, where $H(x) = \int(1/h(x))dx$. Eq. (6.7.11) has the form $H(x) - H(x_1) = C$. Hence, $m = 1$ and formula (6.7.9) provides the nonlinear superposition $x = H^{-1}(H(x_1) + C)$.

Example 6.7.2. The non-homogeneous linear equation

$$\frac{dx}{dt} = A(t)x + B(t)$$

has form (6.7.5) with $T_1 = B(t)$ and $T_2 = A(t)$. The Vessiot-Guldberg-Lie algebra (6.7.6) is an L_2 spanned by the operators

$$X_1 = \frac{d}{dx}, \quad X_2 = x\frac{d}{dx}.$$

Substituting $n = 1$ and $r = 2$ in (6.7.10), $m \geq 2$, we see that expression (6.7.9) for the general solution requires at least two particular solutions. In fact, this number is sufficient. Indeed, let us take the three-point representation (6.7.12) of the basic operators X_1 and X_2 :

$$V_1 = \frac{\partial}{\partial x} + \frac{\partial}{\partial x_1} + \frac{\partial}{\partial x_2}, \qquad V_2 = x\frac{\partial}{\partial x} + x_1\frac{\partial}{\partial x_1} + x_2\frac{\partial}{\partial x_2},$$

and show that they admit one invariant. To find it, we first solve the characteristic system for the equation $V_1(J) = 0$, namely, $dx = dx_1 = dx_2$. Integration yields two independent invariants, e.g., $u = x - x_1$ and $v = x_2 - x_1$. Hence, the common invariant $J(x, x_1, x_2)$ for two operators, V_1 and V_2, can be obtained by taking it in the form $J = J(u, v)$ and solving the equation $\tilde{V}_2(J(u, v)) = 0$, where the action of V_2 is restricted to the space of the variables u, v by using the formula $\tilde{V}_2 = V_2(u)\partial/\partial u + V_2(v)\partial/\partial v$. Noting that $V_2(u) = x - x_1 \equiv u$ and $V_2(v) = x_2 - x_1 \equiv v$, we have

$$\tilde{V}_2 = u\frac{\partial}{\partial u} + v\frac{\partial}{\partial v}.$$

Hence, the invariant is $J(u, v) = u/v$, or returning to the original variables, $J(x, x_1, x_2) = (x - x_1)/(x_2 - x_1)$. Thus, we have $m = 2$, and Eq. (6.7.11) is written $(x - x_1)/(x_2 - x_1) = C$, or $(x - x_1) = C(x_2 - x_1)$. Formula (6.7.9) is then the linear superposition $x = x_1 + C(x_2 - x_1) \equiv (1 - C)x_1 + Cx_2$.

Example 6.7.3. An example of a nonlinear equation with a fundamental system of solutions is the Riccati equation

$$\frac{dx}{dt} = P(t) + Q(t)x + R(t)x^2. \tag{6.7.24}$$

It has Lie's form (6.7.5) with $r = 3$ and with the following operators (6.7.6):

$$X_1 = \frac{d}{dx}, \qquad X_2 = x\frac{d}{dx}, \qquad X_3 = x^2\frac{d}{dx}. \tag{6.7.25}$$

The latter span the Lie algebra L_3 of the projective group. Estimation (6.7.10), $m \geq 3$, shows that expression (6.7.9) requires at least three particular solutions. Let us check that this number ($m = 3$) is sufficient for producing the general solution of the Riccati equation. Namely, let us take the four-point representation (6.7.12) of operators (6.7.25):

$$V_1 = \frac{\partial}{\partial x} + \frac{\partial}{\partial x_1} + \frac{\partial}{\partial x_2} + \frac{\partial}{\partial x_3}, \qquad V_2 = x\frac{\partial}{\partial x} + x_1\frac{\partial}{\partial x_1} + x_2\frac{\partial}{\partial x_2} + x_3\frac{\partial}{\partial x_3},$$

$$V_3 = x^2\frac{\partial}{\partial x} + x_1^2\frac{\partial}{\partial x_1} + x_2^2\frac{\partial}{\partial x_2} + x_3^2\frac{\partial}{\partial x_3}, \tag{6.7.26}$$

and show that they admit one invariant involving x. To find it, we proceed as in Example 6.7.2. The characteristic system associated with $V_1(J) = 0$ yields three invariants, $u_1 = x_1 - x$, $u_2 = x_1 - x_2$, and $u_3 = x_3 - x_2$. Now we restrict V_2 to these invariants to obtain

$$\widetilde{V}_2 = u_1 \frac{\partial}{\partial u_1} + u_2 \frac{\partial}{\partial u_2} + u_3 \frac{\partial}{\partial u_3}.$$

It provides two independent invariants, e.g.:

$$v = \frac{u_2}{u_1} \equiv \frac{x_1 - x_2}{x_1 - x}, \qquad w = \frac{u_3}{u_2} \equiv \frac{x_3 - x_2}{x_1 - x_2}.$$

Finally, noting that the common invariant should be of the form $J(v, w)$, we calculate the action of V_3 on the variables v and w to obtain

$$\widetilde{V}_3 = (x_1 - x_2) \left[(1 - v) \frac{\partial}{\partial v} + (w - 1) w \frac{\partial}{\partial w} \right].$$

Hence, the equation $\widetilde{V}_3 \big(J(v, w) \big) = 0$ is equivalent to

$$(1 - v) \frac{\partial J}{\partial v} + (w - 1) w \frac{\partial J}{\partial w} = 0.$$

The characteristic equation

$$\frac{dv}{1 - v} = \frac{dw}{w(w - 1)} \equiv \frac{dw}{w - 1} - \frac{dw}{w}$$

provides the invariant $J = (v - 1)(w - 1)/w$. Equating this invariant to an arbitrary constant and substituting the expressions for v and w, one obtains the nonlinear superposition formula (3.2.17):

$$\frac{(x - x_2)(x_3 - x_1)}{(x_1 - x)(x_2 - x_3)} = C.$$

Example 6.7.4. The system of two homogeneous linear equations

$$\frac{dx}{dt} = a_{11}(t)x + a_{12}(t)y, \qquad \frac{dy}{dt} = a_{21}(t)x + a_{22}(t)y \qquad (6.7.27)$$

has form (6.7.5) with the following coefficients:

$$T_1 = a_{11}(t), \quad T_2 = a_{12}(t), \quad T_3 = a_{21}(t), \quad T_4 = a_{22}(t),$$

$$\xi_1 = (x, 0), \quad \xi_2 = (y, 0), \quad \xi_3 = (0, x), \quad \xi_4 = (0, y).$$

Hence, the Vessiot-Guldberg-Lie algebra has dimension four and is spanned by

$$X_1 = x \frac{\partial}{\partial x}, \quad X_2 = y \frac{\partial}{\partial x}, \quad X_3 = x \frac{\partial}{\partial y}, \quad X_4 = y \frac{\partial}{\partial y}. \qquad (6.7.28)$$

Estimation (6.7.10) is written $2m \geq 4$ and shows that one needs at least two ($m = 2$) particular solutions. The calculations show that the three-point representation of operators (6.7.28) indeed provides two invariants:

$$J_1 = \frac{xy_2 - x_2y}{x_1y_2 - x_2y_1}, \quad J_2 = \frac{x_1y - xy_1}{x_1y_2 - x_2y_1}, \qquad (6.7.29)$$

such that the general solution is expressible in form (6.7.11) via two particular solutions, (x_1, y_1) and (x_2, y_2). The latter are presupposed to be linearly independent, and hence $x_1y_2 - x_2y_1 \neq 0$. The explicit formula (6.7.9) for the general solution is obtained by solving the equations $J_1 = C_1, J_2 = C_2$ with respect to x and y and provides the linear superposition:

$$x = C_1x_1 + C_2x_2, \quad y = C_1y_1 + C_2y_2.$$

Example 6.7.5. In the case of two non-homogeneous linear equations

$$\frac{dx}{dt} = a_{11}(t)x + a_{12}(t)y + b_1(t), \quad \frac{dy}{dt} = a_{21}(t)x + a_{22}(t)y + b_2(t), \quad (6.7.30)$$

one has to add to the coefficients T_α and ξ_α of the previous example the following:

$$T_5 = b_1(t), \quad \xi_5 = (1, 0); \quad T_6 = b_2(t), \quad \xi_6 = (0, 1).$$

Hence, the algebra L_4 of Example 6.7.4 extends to the algebra L_6 spanned by

$$X_1 = x\frac{\partial}{\partial x}, \quad X_2 = y\frac{\partial}{\partial x}, \quad X_3 = x\frac{\partial}{\partial y}, \quad X_4 = y\frac{\partial}{\partial y}, \quad X_5 = \frac{\partial}{\partial x}, \quad X_6 = \frac{\partial}{\partial y}.$$

Representation (6.7.9) is provided by the linear superposition formula:

$$x = x_1 + C_1(x_2 - x_1) + C_2(x_3 - x_1), \quad y = y_1 + C_1(y_2 - y_1) + C_2(y_3 - y_1). \quad (6.7.31)$$

Example 6.7.6. Lie considered the following system of linear equations:

$$\frac{dx}{dt} = a(t)y + b_1(t), \quad \frac{dy}{dt} = -a(t)x + b_2(t), \qquad (6.7.32)$$

which, unlike the general system (6.7.30), requires only two particular solutions. Lie's reasoning is based on the fact that operators (6.7.6),

$$X_1 = \frac{\partial}{\partial x}, \quad X_2 = \frac{\partial}{\partial y}, \quad X_3 = y\frac{\partial}{\partial x} - x\frac{\partial}{\partial y}, \qquad (6.7.33)$$

generate the group of the rotation and two translations in the plane. Since this group conserves all distances, any three solutions (x_1, y_1), (x_2, y_2), (x, y) are connected by the relations

$$(x - x_1)^2 + (y - y_1)^2 = K_1, \quad (x - x_2)^2 + (y - y_2)^2 = K_2. \qquad (6.7.34)$$

Let us dwell on this example whose discussion discloses the advantages to be gained from the use of invariants of $m + 1$ points under Vessiot-Guldberg-Lie algebras. Estimation (6.7.10), $2m \geq 3$, determines the minimum $m = 2$ of necessary particular solutions. Consequently, we take the three-point representation of operators (6.7.33):

$$V_1 = \frac{\partial}{\partial x} + \frac{\partial}{\partial x_1} + \frac{\partial}{\partial x_2}, \quad V_2 = \frac{\partial}{\partial y} + \frac{\partial}{\partial y_1} + \frac{\partial}{\partial y_2},$$

$$V_3 = y\frac{\partial}{\partial x} - x\frac{\partial}{\partial y} + y_1\frac{\partial}{\partial x_1} - x_1\frac{\partial}{\partial y_1} + y_2\frac{\partial}{\partial x_2} - x_2\frac{\partial}{\partial y_2}.$$

A basis of the common invariants for V_1 and V_2 is given by

$$u_1 = x - x_1, \quad v_1 = y - y_1, \quad u_2 = x - x_2, \quad v_2 = y - y_2.$$

Restricting the action of V_3 to these invariants, one obtains the infinitesimal simultaneous rotation of the vectors $\boldsymbol{u} = (u_1, u_2)$ and $\boldsymbol{v} = (v_1, v_2)$:

$$\tilde{V}_3 = v_1\frac{\partial}{\partial u_1} - u_1\frac{\partial}{\partial v_1} + v_2\frac{\partial}{\partial u_2} - u_2\frac{\partial}{\partial v_2}.$$

The reckoning shows that basic invariants of this rotation are the magnitudes $|\boldsymbol{u}|$ and $|\boldsymbol{v}|$ and the scalar product $\boldsymbol{u} \cdot \boldsymbol{v}$ of the vectors $\boldsymbol{u}$ and $\boldsymbol{v}$. Returning to the original variables, one ultimately arrives at the following basic invariants of three points for the Vessiot-Guldberg-Lie algebra (6.7.33):

$$\psi_1 = (x - x_1)^2 + (y - y_1)^2, \quad \psi_2 = (x - x_2)^2 + (y - y_2)^2,$$

$$\psi_3 = (x - x_1)(x - x_2) + (y - y_1)(y - y_2).$$

Hence, the general nonlinear superposition (6.7.11), involving two particular solutions, (x_1, y_1) and (x_2, y_2), has the form:

$$J_1(\psi_1, \psi_2, \psi_3) = K_1, \quad J_2(\psi_1, \psi_2, \psi_3) = K_2, \quad K_i = \text{const.}, \tag{6.7.35}$$

where J_1 and J_2 are arbitrary functions of three variables such that their Jacobian with respect to x, y does not vanish identically, the latter condition meaning that Eqs. (6.7.35) can be solved with respect to x and y. Letting, e.g. $J_1 = \psi_1$ and $J_2 = \psi_2$, one obtains Lie's representation (6.7.34) of the general solution. Another simple nonlinear superposition is obtained by letting $J_1 = \psi_1$ and $J_2 = \psi_3$:

$$(x - x_1)^2 + (y - y_1)^2 = K_1, \quad (x - x_1)(x - x_2) + (y - y_1)(y - y_2) = K_3. \tag{6.7.36}$$

Representations (6.7.34) and (6.7.36) of the general solution provide two different (i.e., functionally independent) nonlinear superpositions.

Thus, the general solution of system (6.7.32) can be represented as the linear superposition (6.7.31) of *three* particular solutions or, alternatively, as a nonlinear superposition (6.7.35) of *two* particular solutions.

Example 6.7.7. Theorem 6.7.1 associates with any Lie algebra a system of differential equations admitting a superposition of solutions. Consider, as an illustrative example, the three-dimensional algebra spanned by

$$X_1 = \frac{\partial}{\partial x}, \quad X_2 = 2x\frac{\partial}{\partial x} + y\frac{\partial}{\partial y}, \quad X_3 = x^2\frac{\partial}{\partial x} + xy\frac{\partial}{\partial y}. \qquad (6.7.37)$$

This algebra is a three-dimensional subalgebra of the eight-dimensional Lie algebra of the projective group on the plane. Accordingly, the first equation of the associated system (6.7.5),

$$\frac{dx}{dt} = T_1(t) + 2T_2(t)x + T_3(t)x^2, \quad \frac{dy}{dt} = T_2(t)y + T_3(t)xy, \qquad (6.7.38)$$

is the Riccati equation (6.7.24) with $P = T_1, Q = 2T_2, R = T_3$. Operators (6.7.37) span the Vessiot-Guldberg-Lie algebra L_3 for system (6.7.38). Estimation (6.7.10), $2m \geq 3$, determines the minimum $m = 2$ of necessary particular solutions. Consequently, we take the three-point representation of operators (6.7.37):

$$V_1 = \frac{\partial}{\partial x} + \frac{\partial}{\partial x_1} + \frac{\partial}{\partial x_2},$$

$$V_2 = 2x\frac{\partial}{\partial x} + y\frac{\partial}{\partial y} + 2x_1\frac{\partial}{\partial x_1} + y_1\frac{\partial}{\partial y_1} + 2x_2\frac{\partial}{\partial x_2} + y_2\frac{\partial}{\partial y_2},$$

$$V_3 = x^2\frac{\partial}{\partial x} + xy\frac{\partial}{\partial y} + x_1^2\frac{\partial}{\partial x_1} + x_1y_1\frac{\partial}{\partial y_1} + x_2^2\frac{\partial}{\partial x_2} + x_2y_2\frac{\partial}{\partial y_2}.$$

Operator V_1 provides five invariants, $y, y_1, y_2, z_1 = x_1 - x, z_2 = x_2 - x_1$. Restricting V_2 to these invariants, one obtains the dilation generator

$$\tilde{V}_2 = 2z_1\frac{\partial}{\partial z_1} + 2z_2\frac{\partial}{\partial z_2} + y\frac{\partial}{\partial y} + y_1\frac{\partial}{\partial y_1} + y_2\frac{\partial}{\partial y_2}.$$

Its independent invariants are

$$u_1 = \frac{z_2}{z_1}, \quad u_2 = \frac{y^2}{x_1 - x}, \quad u_3 = \frac{y_1^2}{x_1 - x}, \quad u_4 = \frac{y_2^2}{x_1 - x}.$$

Hence, a basis of the common invariants of V_1 and V_2 :

$$u_1 = \frac{x_2 - x_1}{x_1 - x}, \quad u_2 = \frac{y^2}{x_1 - x}, \quad u_3 = \frac{y_1^2}{x_1 - x}, \quad u_4 = \frac{y_2^2}{x_1 - x}.$$

It remains to find the restriction $\tilde{V}_3$ of V_3 to the above invariants:

$$\tilde{V}_3 = V_3(u_1)\frac{\partial}{\partial u_1} + \cdots + V_3(u_4)\frac{\partial}{\partial u_4}.$$

The reckoning shows that

$$V_3(u_1) = \frac{(x_2 - x_1)(x - x_2)}{x - x_1} \equiv (x_1 - x)(1 + u_1)u_1,$$

$$V_3(u_2) = -y^2 \equiv -(x_1 - x)u_2, \quad V_3(u_3) = y_1^2 \equiv (x_1 - x)u_3,$$

$$V_3(u_4) = \frac{x + x_1 - 2x_2}{x - x_1} y_2^2 \equiv (x_1 - x)(1 + 2u_1)u_4.$$

Hence,

$$\tilde{V}_3 = (x_1 - x)\left((1 + u_1)u_1 \frac{\partial}{\partial u_1} - u_2 \frac{\partial}{\partial u_2} + u_3 \frac{\partial}{\partial u_3} + (1 + 2u_1)u_4 \frac{\partial}{\partial u_4} \right).$$

Consequently, the equation $\tilde{V}_3(\psi(u_1, \ldots, u_4)) = 0$ is equivalent to

$$(1 + u_1)u_1 \frac{\partial \psi}{\partial u_1} - u_2 \frac{\partial \psi}{\partial u_2} + u_3 \frac{\partial \psi}{\partial u_3} + (1 + 2u_1)u_4 \frac{\partial \psi}{\partial u_4} = 0,$$

whence, by solving the characteristic system

$$\frac{du_1}{(1 + u_1)u_1} = -\frac{du_2}{u_2} = \frac{du_3}{u_3} = \frac{du_4}{(1 + 2u_1)u_4},$$

one obtains the following three independent invariants:

$$\psi_1 = u_2 u_3 \equiv \frac{y^2 y_1^2}{(x_1 - x)^2}, \quad \psi_2 = \frac{u_1 u_2}{1 + u_1} \equiv \frac{(x_2 - x_1)y^2}{(x_1 - x)(x_2 - x)},$$

$$\psi_3 = \frac{u_4}{(1 + u_1)u_1} \equiv \frac{(x_1 - x)y_2^2}{(x_2 - x_1)(x_2 - x)}.$$

Hence, the general nonlinear superposition (6.7.11), involving two particular solutions, (x_1, y_1) and (x_2, y_2), is written

$$J_1(\psi_1, \psi_2, \psi_3) = C_1, \quad J_2(\psi_1, \psi_2, \psi_3) = C_2, \tag{6.7.39}$$

where J_1 and J_2 are arbitrary functions of three variables such that their Jacobian with respect to x, y does not vanish identically (cf. Example 6.7.6). Letting, e.g. $J_1 = \sqrt{\psi_1}$ and $J_2 = \sqrt{\psi_2 \psi_3}$, i.e., specifying (6.7.39) in the form

$$\frac{yy_1}{x_1 - x} = C_1, \quad \frac{yy_2}{x_2 - x} = C_2,$$

one arrives at the following representation of the general solution via two particular solutions:

$$x = \frac{C_1 x_1 y_2 - C_2 x_2 y_1}{C_1 y_2 - C_2 y_1}, \quad y = \frac{C_1 C_2 (x_2 - x_1)}{C_1 y_2 - C_2 y_1}.$$

Recall that the general solution of the single Riccati equation (6.7.24) requires *three* particular solutions. It is remarkable that when the same Riccati equation is integrated in the system of two coupled equations (6.7.38), one needs to know its *two* particular solutions only!

Example 6.7.8. The following system of two nonlinear equations:

$$\frac{dx}{dt} = xy^2 - \frac{x}{2t}, \quad \frac{dy}{dt} = x^2y - \frac{y}{2t}, \tag{6.7.40}$$

arises in nonlinear optics (see further Section 7.2.6). It has form (6.7.5) with the two-dimensional Vessiot-Guldberg-Lie algebra L_2 spanned by

$$X_1 = xy^2\frac{\partial}{\partial x} + x^2y\frac{\partial}{\partial y}, \quad X_2 = x\frac{\partial}{\partial x} + y\frac{\partial}{\partial y}.$$

Estimation (6.7.10) is written $m \geq 1$, hence one particular solution may be enough to express the general solution. To verify that this is indeed possible, let us find the invariants $J(x, y; x_1, x_2)$ of the two-point representation of the above operators:

$$V_1 = xy^2\frac{\partial}{\partial x} + x^2y\frac{\partial}{\partial y} + x_1y_1^2\frac{\partial}{\partial x_1} + x_1^2y_1\frac{\partial}{\partial y_1},$$

$$V_2 = x\frac{\partial}{\partial x} + y\frac{\partial}{\partial y} + x_1\frac{\partial}{\partial x_1} + y_1\frac{\partial}{\partial y_1}.$$

Since $[X_1, X_2] = -2X_1$, and hence X_1 spans an ideal of L_2, it is convenient to begin the calculations with the operator V_1. The first and the last equations of the characteristic system for $V_1(J) = 0$,

$$\frac{dx}{y^2x} = \frac{dy}{x^2y} = \frac{dx_1}{y_1^2x_1} = \frac{dy_1}{x_1^2y_1},$$

provide the invariants $z = x^2 - y^2$ and $z_1 = x_1^2 - y_1^2$, respectively. It remains to solve one equation, e.g.

$$\frac{dx}{x(x^2 - z)} = \frac{dx_1}{x_1(x_1^2 - z_1)},$$

where z and z_1 should be regarded as constants. The integration yields a third invariant,

$$z_2 = \frac{1}{z}\ln\frac{x}{\sqrt{x^2 - z}} - \frac{1}{z_1}\ln\frac{x_1}{\sqrt{x_1^2 - z_1}}$$

or, upon substituting the values of z and z_1,

$$z_2 = \frac{\ln x - \ln y}{x^2 - y^2} - \frac{\ln x_1 - \ln y_1}{x_1^2 - y_1^2}.$$

Now, the restriction of the operator V_2 to the invariants z, z_1, z_2,

$$\tilde{V}_2 = 2\left(z\frac{\partial}{\partial z} + z_1\frac{\partial}{\partial z_1} - z_2\frac{\partial}{\partial z_2}\right),$$

readily yields two independent invariants, e.g.

$$J_1 = \frac{z}{z_1}, \quad J_2 = z\,z_2.$$

Substituting here the values of z, z_1, z_2 and equating the resulting expressions for J_i to arbitrary constants C_i, we arrive at the following nonlinear superposition (6.7.11) for system (6.7.40):

$$J_1 \equiv \frac{x^2 - y^2}{x_1^2 - y_1^2} = C_1, \quad J_2 \equiv \ln x - \ln y - \frac{x^2 - y^2}{x_1^2 - y_1^2}\left(\ln x_1 - \ln y_1\right) = C_2.$$

6.7.4 Integration of systems using nonlinear superposition

The Vessiot-Guldberg-Lie algebra furnishes a theoretical basis for a new general integration theory for systems of ordinary differential equations admitting a nonlinear superposition [21].

To illustrate the approach, let us consider the simple case of systems of two coupled equations:

$$\frac{dx}{dt} = T_1(t)\xi_1^1(x, y) + T_2(t)\xi_2^1(x, y),$$

$$\frac{dy}{dt} = T_1(t)\xi_1^2(x, y) + T_2(t)\xi_2^2(x, y),$$
(6.7.41)

with a two-dimensional Vessiot-Guldberg-Lie algebra L_2 spanned by

$$X_1 = \xi_1^1(x, y)\frac{\partial}{\partial x} + \xi_1^2(x, y)\frac{\partial}{\partial y}, \quad X_2 = \xi_2^1(x, y)\frac{\partial}{\partial x} + \xi_2^2(x, y)\frac{\partial}{\partial y}. \quad (6.7.42)$$

To solve system (6.7.41), it suffices to transform the basic operators (6.7.42) and the corresponding equations (6.7.41) to the standard forms given in Table 6.7.1 and obtained in accordance with Section 6.5.3.

Table 6.7.1 Standard forms of operators (6.7.42) and systems (6.7.41)

	Vessiot-Guldberg-Lie algebra	Canonical forms of (6.7.41)
I	$X_1 = \dfrac{\partial}{\partial x}, \; X_2 = \dfrac{\partial}{\partial y}$	$\dfrac{dx}{dt} = T_1(t), \; \dfrac{dy}{dt} = T_2(t)$
II	$X_1 = \dfrac{\partial}{\partial y}, \; X_2 = x\dfrac{\partial}{\partial y}$	$\dfrac{dx}{dt} = 0, \; \dfrac{dy}{dt} = T_1(t) + T_2(t)x$
III	$X_1 = \dfrac{\partial}{\partial y}, \; X_2 = x\dfrac{\partial}{\partial x} + y\dfrac{\partial}{\partial y}$	$\dfrac{dx}{dt} = T_2(t)x, \; \dfrac{dy}{dt} = T_1(t) + T_2(t)y$
IV	$X_1 = \dfrac{\partial}{\partial y}, \; X_2 = y\dfrac{\partial}{\partial y}$	$\dfrac{dx}{dt} = 0, \; \dfrac{dy}{dt} = T_1(t) + T_2(t)y$

Example 6.7.9. Let us apply our method of integration to the system of equations (6.7.40) from Example 6.7.8:

$$\frac{dx}{dt} = xy^2 - \frac{x}{2t}, \quad \frac{dy}{dt} = x^2 y - \frac{y}{2t}. \tag{6.7.40}$$

Its Vessiot-Guldberg-Lie algebra is the two-dimensional algebra spanned by

$$X_1 = xy^2 \frac{\partial}{\partial x} + x^2 y \frac{\partial}{\partial y}, \quad X_2 = x \frac{\partial}{\partial x} + y \frac{\partial}{\partial y}. \tag{6.7.43}$$

We have

$$[X_1, \ X_2] = -2X_1, \quad \xi_1 \eta_2 - \eta_1 \xi_2 = xy(y^2 - x^2) \neq 0.$$

Hence, operators (6.7.43) span an L_2 of type III in the classification of Table 6.7.1.

Consequently, we can transform operators (6.7.43), and hence system (6.7.40) to the standard form III of Table 6.7.1. We first find canonical variables $\tilde{x}, \tilde{y}$ for the first operator in (6.7.43) by solving the equations

$$X_1(\tilde{x}) = 0, \quad X_1(\tilde{y}) = 1.$$

Whence

$$\tilde{x} = x^2 - y^2, \quad \tilde{y} = \frac{\ln y - \ln x}{x^2 - y^2}. \tag{6.7.44}$$

One can verify that variables (6.7.44) are, in fact, the canonical variables required for our algebra L_2. Indeed, operators (6.7.43) are written in the form of type III of Table 6.7.1:

$$X_1 = \frac{\partial}{\partial \tilde{y}}, \quad X_2 = 2\left(\tilde{x} \frac{\partial}{\partial \tilde{x}} + \tilde{y} \frac{\partial}{\partial \tilde{y}} \right).$$

Hence, rewriting Eqs. (6.7.40) in the new variables (6.7.44), one obtains

$$\frac{d\tilde{x}}{dt} = -\frac{\tilde{x}}{t}, \quad \frac{d\tilde{y}}{dt} = 1 + \frac{\tilde{y}}{t}. \tag{6.7.45}$$

Integration of Eqs. (6.7.45) yields

$$\tilde{x} = \frac{C_1}{t}, \quad \tilde{y} = C_2 t + t \ln t. \tag{6.7.46}$$

Now we solve Eqs. (6.7.44) with respect to x and y :

$$x = \sqrt{\frac{\tilde{x}}{1 - e^{2\tilde{x}\tilde{y}}}}, \quad y = \sqrt{\frac{\tilde{x}}{e^{-2\tilde{x}\tilde{y}} - 1}},$$

substitute here solutions (6.7.46) and finally arrive at the following general solution to the system of Eqs. (6.7.40):

$$x = \sqrt{\frac{k}{t(1 - \zeta^2)}}, \quad y = \zeta \sqrt{\frac{k}{t(1 - \zeta^2)}}. \tag{6.7.47}$$

Here, $\zeta = C t^k$, where C and k are arbitrary constants.

Problems to Chapter 6

6.1. Check if the equation $y'' - y'^2 + xy = 0$ admits the group of dilations with the generator

$$X = x\frac{\partial}{\partial x} + y\frac{\partial}{\partial y}.$$

6.2. Find the most general equations of the first and second orders admitting the groups of dilations with the generators

$$\text{(i) } X = x\frac{\partial}{\partial x} + y\frac{\partial}{\partial y}, \quad \text{(ii) } X = y\frac{\partial}{\partial y}, \quad \text{(iii) } X = x\frac{\partial}{\partial x}.$$

6.3. Check that the third-order equation

$$\mu^2 \mu''' = \nu(2\mu\mu'' - \mu'^2), \quad \nu = \text{const.} \neq 0, \tag{P6.1}$$

where $\mu = \mu(x)$, admits the operators

$$X_1 = \frac{\partial}{\partial x}, \quad X_2 = x\frac{\partial}{\partial x} + \mu\frac{\partial}{\partial \mu}$$

and, using these symmetries, reduce Eq. (P6.1) to a first-order equation.

6.4. Calculate the infinitesimal symmetries of the second-order equations

$$\text{(i) } y'' + \frac{y'}{x} - e^y = 0 \quad \text{and} \quad \text{(ii) } y'' - \frac{y'}{x} + e^y = 0. \tag{P6.2}$$

6.5. Test for linearization the following second-order equations:

$$\text{(i) } y'' = yy'^2 - xy'^3, \quad \text{(ii) } y'' + 3yy' + y^3 = 0, \tag{P6.3}$$

$$\text{(iii) } y'' = 2\left(\frac{y'^2}{y} - \frac{xy'}{1+x^2}\right), \quad \text{(iv) } y'' = \frac{y'}{y^2} - \frac{1}{xy},$$

$$\text{(v) } y'' + \frac{y'}{x} - e^y = 0, \quad \text{(vi) } y'' + \left(1 - \frac{xy'}{y}\right)^3 = 0.$$

6.6. Consider Eq. (6.5.6),

$$y'' = \frac{y'}{y^2} - \frac{1}{xy},$$

and the second operator in (6.5.7) admitted by this equation:

$$X_2 = x\frac{\partial}{\partial x} + \frac{y}{2}\frac{\partial}{\partial y}.$$

(i) Find canonical variables t, u such that X_2 becomes $X_2 = \partial/\partial t$.

(ii) Rewrite Eq. (6.5.6) in the variables t, u setting $u = u(t)$.

(iii) Reduce the order of the resulting second-order equation $u'' = \phi(u, u')$ by using the substitution $u' = p(u)$.

6.7. Integrate the equation

$$y'' + \frac{k}{(1 + \omega^2 x^2)^2} y = 0, \qquad k, \ \omega = \text{const.} \neq 0,$$

using the following two symmetries of this equation:

$$X_1 = (1 + \omega^2 x^2)\frac{\partial}{\partial x} + \omega^2 xy\frac{\partial}{\partial y}, \quad X_2 = y\frac{\partial}{\partial y}.$$

Write the solution for the arbitrary parameters k and ω, and for the particular case $k = 3\omega^2$.

6.8. Let $X_1, \ldots, X_r$ span a vector space L_r, and let $[X_i, X_j] \in L_r$ for all i, j. Prove that $[X, Y] \in L_r$ for any $X, Y \in L_r$ (cf. Definition 6.5.2).

6.9. Prove that transformation (6.5.27) maps Eq. (6.5.26) to $u'' = 0$.

6.10. Find the general form of the first-order equations admitting the operator

$$X = \sqrt{2}\,x\frac{\partial}{\partial x} + y\frac{\partial}{\partial y}$$

and integrate them.

6.11. Find the general form of the first-order double homogeneous equations, i.e., equations $y' = f(x, y)$ admitting two independent operators

$$X_1 = y\frac{\partial}{\partial y}, \quad X_2 = x\frac{\partial}{\partial x}.$$

6.12. Find the general form of the second-order double homogeneous equations $y'' = f(x, y, y')$.

6.13. Check that function (6.5.35) satisfies Eq. (6.5.34).

6.14. Integrate the equation $y'' = xy' - 4y$ by first finding its particular polynomial solution and then applying the methods of Section 6.5.5.

6.15. Verify that the change of variables $t = y$, $u = x^2 + y^2$ maps the equation $xy'' = y' + y'^3$ to the linear equation $u'' = 0$ and check that $x^2 + y^2 + Ay + B = 0$ provides the implicit solution to the equation $xy'' = y' + y'^3$ (see Example 6.5.11).

6.16. Verify that the non-linear equation

$$y'' + 2\left(y' - \frac{y}{x}\right)^3 = 0$$

admits the algebra L_2 spanned by

$$X_1 = x^2 \frac{\partial}{\partial x} + xy \frac{\partial}{\partial y}, \quad X_2 = xy \frac{\partial}{\partial x} + y^2 \frac{\partial}{\partial y}$$

and integrate the equation using this algebra. Clarify whether the L_2 given above is the maximal Lie algebra admitted by the equation in question.

6.17. Solve the following initial value problems:

(i) $y'' + 2\left(y' - \frac{y}{x}\right)^3 = 0, \quad y(1) = 0, \quad y'(1) = 1,$

(ii) $y'' + 2\left(y' - \frac{y}{x}\right)^3 = 0, \quad y(1) = y'(1) = 0,$

(iii) $y'' + 2\left(y' - \frac{y}{x}\right)^3 = 0, \quad y\left(\frac{1}{2}\right) = \frac{1}{2}, \quad y'\left(\frac{1}{2}\right) = \frac{3}{4},$

(iv) $y'' + 2\left(y' - \frac{y}{x}\right)^3 = 0, \quad y(2) = \sqrt{2}, \quad y'(2) = \frac{1}{2\sqrt{2}},$

(v) $y'' + 2\left(y' - \frac{y}{x}\right)^3 = 0, \quad y(2) = 3\sqrt{2}, \quad y'(2) = \frac{7}{2\sqrt{2}}.$

6.18. Find the general solution of the nonlinear equation (6.6.53):

$$y''' - \left(\frac{6}{y}y' + \frac{3}{x}\right)y'' + 6\left(\frac{y'^3}{y^2} + \frac{y'^2}{xy} + \frac{y'}{x^2} + \frac{y}{x^3}\right) = 0.$$

6.19. Find the general solution of the nonlinear equation (6.6.74):

$$y''' + \frac{1}{y'}\left[-3y''^2 - xy'^5\right] = 0.$$

Chapter 7

Nonlinear partial differential equations

This chapter contains simple applications of Lie groups for finding exact solutions to nonlinear partial differential equations. Basic conservation theorems are also presented here.

Additional reading: L.V. Ovsyannikov [32], W.F. Ames [1], N.H. Ibragimov [14], [18], [19], [20], P.J. Olver [31], G.W. Bluman and S. Kumei [2], B.J. Cantwell [3].

7.1 Symmetries

The definitions of one-parameter groups and a symmetry groups for partial differential equations are the same as that for ordinary differential equations. For example, in the case of two independent variables t, x and one dependent variable u, transformations (6.2.1) are replaced by invertible transformations of the variables t, x, u :

$$\bar{t} = f(t, x, u, a), \quad \bar{x} = g(t, x, u, a), \quad \bar{u} = h(t, x, u, a), \tag{7.1.1}$$

and the main group property (6.2.4)—(6.2.5) is replaced by the following equations:

$$\begin{aligned}
\bar{\bar{t}} &\equiv f(\bar{t}, \bar{x}, \bar{u}, b) = f(t, x, u, a + b), \\
\bar{\bar{x}} &\equiv g(\bar{t}, \bar{x}, \bar{u}, b) = g(t, x, u, a + b), \\
\bar{\bar{u}} &\equiv h(\bar{t}, \bar{x}, \bar{u}, b) = h(t, x, u, a + b).
\end{aligned} \tag{7.1.2}$$

Thus, we use the following definition.

Definition 7.1.1. A set G of invertible transformations (7.1.1) is called a one-parameter group of transformations in the space of variables t, x, u if G contains the identity transformation $\bar{t} = t$, $\bar{x} = x$, $\bar{u} = u$, the inverse to its transformations and obeys the group property (7.1.2).

7.1.1 Definition and calculation of symmetry groups

Let us formulate the definition of a symmetry group for partial differential equations by considering, e.g. evolutionary equations of the second order:

$$u_t = F(t, x, u, u_x, u_{xx}), \quad \partial F/\partial u_{xx} \neq 0. \tag{7.1.3}$$

Definition 7.1.2. A one-parameter group G of transformations (7.1.1) is said to be admitted by Eq. (7.1.3) if Eq. (7.1.3) has the same form in the new variables $\bar{t}, \bar{x}, \bar{u}$, i.e.,

$$\bar{u}_{\bar{t}} = F(\bar{t}, \bar{x}, \bar{u}, \bar{u}_{\bar{x}}, \bar{u}_{\bar{x}\bar{x}}), \tag{7.1.4}$$

where the function F is the same as in Eq. (7.1.3). Then G is called a *symmetry group* of Eq. (7.1.3).

The construction of the symmetry group G is equivalent to determination of its *infinitesimal transformations*

$$\bar{t} \approx t + a\tau(t, x, u), \quad \bar{x} \approx x + a\xi(t, x, u), \quad \bar{u} \approx u + a\eta(t, x, u) \tag{7.1.5}$$

obtained from (7.1.1) by expanding into Taylor series with respect to the group parameter a and keeping only the terms linear in a. The infinitesimal transformation (7.1.5) provides the generator of the group G, i.e., the differential operator

$$X = \tau(t, x, u)\frac{\partial}{\partial t} + \xi(t, x, u)\frac{\partial}{\partial x} + \eta(t, x, u)\frac{\partial}{\partial u} \tag{7.1.6}$$

acting on any differentiable function $J(t, x, u)$ as follows:

$$X(J) = \tau(t, x, u)\frac{\partial J}{\partial t} + \xi(t, x, u)\frac{\partial J}{\partial x} + \eta(t, x, u)\frac{\partial J}{\partial u}.$$

Generator (7.1.6) is called an operator admitted by Eq. (7.1.3) or an *infinitesimal symmetry* for Eq. (7.1.3).

The group transformations (7.1.1) corresponding to generator (7.1.6) are found by solving the *Lie equations*

$$\frac{d\bar{t}}{da} = \tau(\bar{t}, \bar{x}, \bar{u}), \quad \frac{d\bar{x}}{da} = \xi(\bar{t}, \bar{x}, \bar{u}), \quad \frac{d\bar{u}}{da} = \eta(\bar{t}, \bar{x}, \bar{u}), \tag{7.1.7}$$

with the initial conditions:

$$\bar{t}\big|_{a=0} = t, \quad \bar{x}\big|_{a=0} = x, \quad \bar{u}\big|_{a=0} = u.$$

Let us turn to Eq. (7.1.4). The quantities $\bar{u}_{\bar{t}}$, $\bar{u}_{\bar{x}}$ and $\bar{u}_{\bar{x}\bar{x}}$ involved in (7.1.4) can be obtained by means of the usual rule of change of derivatives treating Eqs. (7.1.1) as a change of variables. Then, expanding the resulting expressions for $\bar{u}_{\bar{t}}$, $\bar{u}_{\bar{x}}$, $\bar{u}_{\bar{x}\bar{x}}$ into Taylor series with respect to the parameter a, one can obtain the infinitesimal form of these expressions:

$$\bar{u}_{\bar{t}} \approx u_t + a \, \zeta_0(t, x, u, u_t, u_x), \quad \bar{u}_{\bar{x}} \approx u_x + a \, \zeta_1(t, x, u, u_t, u_x),$$

$$\bar{u}_{\bar{x}\bar{x}} \approx u_{xx} + a \, \zeta_2(t, x, u, u_t, u_x, u_{tx}, u_{xx}),$$

(7.1.8)

where $\zeta_0, \zeta_1, \zeta_2$ are given by the following *prolongation formulae*:

$$\zeta_0 = D_t(\eta) - u_t D_t(\tau) - u_x D_t(\xi),$$

$$\zeta_1 = D_x(\eta) - u_t D_x(\tau) - u_x D_x(\xi),$$

(7.1.9)

$$\zeta_2 = D_x(\zeta_1) - u_{tx} D_x(\tau) - u_{xx} D_x(\xi).$$

Here, D_t and D_x denote the total differentiations with respect to t and x:

$$D_t = \frac{\partial}{\partial t} + u_t \frac{\partial}{\partial u} + u_{tt} \frac{\partial}{\partial u_t} + u_{tx} \frac{\partial}{\partial u_x},$$

$$D_x = \frac{\partial}{\partial x} + u_x \frac{\partial}{\partial u} + u_{tx} \frac{\partial}{\partial u_t} + u_{xx} \frac{\partial}{\partial u_x}.$$

Substitution of (7.1.5) and (7.1.8) into Eq. (7.1.4) yields

$$\bar{u}_{\bar{t}} - F(\bar{t}, \bar{x}, \bar{u}, \bar{u}_{\bar{x}}, \bar{u}_{\bar{x}\bar{x}}) \approx u_t - F(t, x, u, u_x, u_{xx})$$

$$+ a\left(\zeta_0 - \frac{\partial F}{\partial u_{xx}}\zeta_2 - \frac{\partial F}{\partial u_x}\zeta_1 - \frac{\partial F}{\partial u}\eta - \frac{\partial F}{\partial x}\xi - \frac{\partial F}{\partial t}\tau\right).$$

Therefore, by virtue of Eq. (7.1.3), Eq. (7.1.4) yields

$$\zeta_0 - \frac{\partial F}{\partial u_{xx}}\zeta_2 - \frac{\partial F}{\partial u_x}\zeta_1 - \frac{\partial F}{\partial u}\eta - \frac{\partial F}{\partial x}\xi - \frac{\partial F}{\partial t}\tau = 0,$$

(7.1.10)

where u_t is replaced by $F(t, x, u, u_x, u_{xx})$ in $\zeta_0, \zeta_1, \zeta_2$.

Equation (7.1.10) defines all infinitesimal symmetries of Eq. (7.1.3) and therefore it is called the *determining equation*. Conventionally, it is written in the compact form

$$X\left[u_t - F(t, x, u, u_x, u_{xx})\right] = 0,$$

(7.1.11)

where X denotes the *prolongation* of operator (7.1.6) to the first and second order derivatives:

$$X = \tau\frac{\partial}{\partial t} + \xi\frac{\partial}{\partial x} + \eta\frac{\partial}{\partial u} + \zeta_0\frac{\partial}{\partial u_t} + \zeta_1\frac{\partial}{\partial u_x} + \zeta_2\frac{\partial}{\partial u_{xx}}.$$

The determining equation (7.1.10) (or its equivalent (7.1.11)) is a linear homogeneous partial differential equation of the second order for unknown functions $\tau(t, x, u)$, $\xi(t, x, u)$, $\eta(t, x, u)$. In consequence, the set of all solutions to the determining equation is a vector space L. Furthermore, the determining equation possesses the following significant and less evident property. The vector space L is a *Lie algebra*, i.e., it is closed with respect to the *commutator*. In other words, L contains, together with any operators X_1, X_2, their commutator $[X_1, X_2]$ defined by

$$[X_1, X_2] = X_1 X_2 - X_2 X_1.$$

In particular, if $L = L_r$ is finite-dimensional and has a basis $X_1, \ldots, X_r$, then

$$[X_\alpha, X_\beta] = c_{\alpha\beta}^\gamma X_\gamma,$$

where $c_{\alpha\beta}^\gamma$ are constant coefficients known as the *structure constants* of L_r.

The determining equation (7.1.10) should be satisfied identically with respect to $t, x, u, u_x, u_{xx}, u_{tx}$ treated as six independent variables. Consequently, the determining equation decomposes into a system of several equations. As a rule, this is an over-determined system since it contains more equations than three unknown functions τ, ξ and η. Therefore, in practical applications, the determining equation can be explicitly solved. The following preparatory lemma due to S. Lie simplifies the calculations.

Lemma 7.1.1. For Eq. (7.1.3), the symmetry transformations (7.1.1) have the form

$$\bar{t} = f(t, a), \quad \bar{x} = g(t, x, u, a), \quad \bar{u} = h(t, x, u, a). \tag{7.1.12}$$

It means that one can search the infinitesimal symmetries in the form

$$X = \tau(t)\frac{\partial}{\partial t} + \xi(t, x, u)\frac{\partial}{\partial x} + \eta(t, x, u)\frac{\partial}{\partial u}. \tag{7.1.13}$$

Proof. Let us single out in the determining equation (7.1.10) the terms containing u_{tx}. The prolongation formulae (7.1.9) show that u_{tx} is contained only in ζ_2, namely, in the term $u_{tx}D_x(\tau)$. Since Eq. (7.1.10) holds identically in $t, x, u, u_x, u_{xx}, u_{tx}$, we have $D_x(\tau) \equiv \tau_x + u_x\tau_u = 0$. Hence, $\tau_x = \tau_u = 0$, i.e., $\tau = \tau(t)$, and operator (7.1.6) has form (7.1.13).

Example 7.1.1. Let us calculate the symmetries of the following nonlinear equation, known as the Burgers equation:

$$u_t = u_{xx} + uu_x. \tag{7.1.14}$$

According to Lemma 7.1.1, we search the infinitesimal symmetries in form (7.1.13). For operators (7.1.13), the prolongation formulae (7.1.9) yield

$$\zeta_0 = D_t(\eta) - u_x D_t(\xi) - \tau'(t)u_t, \quad \zeta_1 = D_x(\eta) - u_x D_x(\xi),$$
$$\zeta_2 = D_x(\zeta_1) - u_{xx}D_x(\xi) \equiv D_x^2(\eta) - u_x D_x^2(\xi) - 2u_{xx}D_x(\xi). \tag{7.1.15}$$

In our example, the determining equation (7.1.10) has the form

$$\zeta_0 - \zeta_2 - u\zeta_1 - \eta u_x = 0, \tag{7.1.16}$$

where ζ_0, ζ_1 and ζ_2 are given by (7.1.15). Let us single out and annul the terms with u_{xx}. Bearing in mind that u_t has to be replaced by $u_{xx} + uu_x$ and substituting into ζ_2 the expressions

$$
\begin{aligned}
D_x^2(\xi) &= D_x(\xi_x + \xi_u u_x) = \xi_u u_{xx} + \xi_{uu} u_x^2 + 2\xi_{xu} u_x + \xi_{xx}, \\
D_x^2(\eta) &= D_x(\eta_x + \eta_u u_x) = \eta_u u_{xx} + \eta_{uu} u_x^2 + 2\eta_{xu} u_x + \eta_{xx},
\end{aligned} \tag{7.1.17}
$$

we arrive at the following equation:

$$
2\xi_u u_x + 2\xi_x - \tau'(t) = 0.
$$

It splits into two equations, namely $\xi_u = 0$ and $2\xi_x - \tau'(t) = 0$. The first equation shows that ξ depends only on t, x, and integration of the second equation yields

$$
\xi = \frac{1}{2}\tau'(t)\, x + p(t). \tag{7.1.18}
$$

It follows from (7.1.18) that $D_x^2(\xi) = 0$. Now the determining equation (7.1.16) reduces to the form

$$
u_x^2 \eta_{uu} + \left[\frac{1}{2}\tau'(t)u + \frac{1}{2}\tau''(t)x + p'(t) + 2\eta_{xu} + \eta \right] u_x + u\eta_x + \eta_{xx} - \eta_t = 0
$$

and splits into three equations:

$$
\begin{aligned}
\eta_{uu} = 0, \quad u\eta_x + \eta_{xx} - \eta_t = 0, \\
\frac{1}{2}\tau'(t)u + \frac{1}{2}\tau''(t)x + p'(t) + 2\eta_{xu} + \eta = 0.
\end{aligned} \tag{7.1.19}
$$

The first equation of (7.1.19) yields $\eta = \sigma(t, x)u + \mu(t, x)$, and the third equation of (7.1.19) becomes

$$
\left(\frac{1}{2}\tau'(t) + \sigma \right) u + \frac{1}{2}\tau''(t)x + p'(t) + 2\sigma_x + \mu = 0,
$$

whence

$$
\sigma = -\frac{1}{2}\tau'(t), \quad \mu = -\frac{1}{2}\tau''(t)x - p'(t).
$$

Thus, we have

$$
\eta = -\frac{1}{2}\tau'(t)\, u - \frac{1}{2}\tau''(t)x - p'(t). \tag{7.1.20}
$$

Finally, substitution of (7.1.20) into the second equation of (7.1.19) yields

$$
\frac{1}{2}\tau'''(t)x + p''(t) = 0,
$$

whence $\tau'''(t) = 0$, $p''(t) = 0$, and hence,

$$
\tau(t) = C_1 t^2 + 2C_2 t + C_3, \quad p(t) = C_4 t + C_5.
$$

Invoking (7.1.18) and (7.1.20), we ultimately arrive at the following general solution of the determining equation (7.1.16):

$$\tau = C_1 t^2 + 2C_2 t + C_3, \ \xi = C_1 tx + C_2 x + C_4 t + C_5, \ \eta = -(C_1 t + C_2)u - C_1 x - C_4.$$

It contains five arbitrary constants C_i. It means that the infinitesimal symmetries of the Burgers equation (7.1.14) form the five-dimensional Lie algebra spanned by the following linearly independent operators:

$$X_1 = \frac{\partial}{\partial t}, \quad X_2 = \frac{\partial}{\partial x}, \quad X_3 = 2t\frac{\partial}{\partial t} + x\frac{\partial}{\partial x} - u\frac{\partial}{\partial u},$$

$$X_4 = t\frac{\partial}{\partial x} - \frac{\partial}{\partial u}, \quad X_5 = t^2\frac{\partial}{\partial t} + tx\frac{\partial}{\partial x} - (x + tu)\frac{\partial}{\partial u}. \tag{7.1.21}$$

7.1.2 Group transformations of solutions

Any symmetry transformation of a differential equation carries over any solution of the differential equation into its solution. It means that, just like in the case of ordinary differential equations, the solutions of a partial differential equation are permuted among themselves under the action of a symmetry group. The solutions may also be individually unaltered, then they are called *invariant solutions*. Accordingly, group analysis provides two basic ways for construction of exact solutions: *group transformations* of known solutions and construction of *invariant solutions*.

In this section, the method of groups transformations of solutions is illustrated by the linear heat equation $u_t - u_{xx} = 0$. An example of a nonlinear equation, namely the Burgers equation, is considered in Section 7.2.2.

The method is based on the fact that a symmetry group transforms any solution of the equation in question into a solution of the same equation. Namely, let (7.1.1) be a symmetry transformation group of Eq. (7.1.3), and let a function

$$u = \Phi(t, x)$$

solve Eq. (7.1.3). Since (7.1.1) is a symmetry transformation, the above solution can be also written in the new variables:

$$\bar{u} = \Phi(\bar{t}, \bar{x}).$$

Replacing here $\bar{u}, \bar{t}, \bar{x}$ from (7.1.1), we get

$$h(t, x, u, a) = \Phi\big(f(t, x, u, a), g(t, x, u, a)\big). \tag{7.1.22}$$

Having solved Eq. (7.1.22) with respect to u, one obtains a one-parameter family of new solutions to Eq. (7.1.3).

The infinitesimal symmetries of the linear heat equation

$$u_t - u_{xx} = 0 \tag{7.1.23}$$

comprise the infinite-dimensional algebra with the generator

$$X_\tau = \tau \frac{\partial}{\partial u},$$

where $\tau = \tau(t, x)$ is an arbitrary solution of Eq. (7.1.23), and the 6-dimensional Lie algebra spanned by (see Problem 7.2)

$$X_1 = \frac{\partial}{\partial t}, \quad X_2 = \frac{\partial}{\partial x}, \quad X_3 = 2t\frac{\partial}{\partial t} + x\frac{\partial}{\partial x}, \quad X_4 = u\frac{\partial}{\partial u},$$

$$X_5 = 2t\frac{\partial}{\partial x} - xu\frac{\partial}{\partial u}, \quad X_6 = t^2\frac{\partial}{\partial t} + tx\frac{\partial}{\partial x} - \frac{1}{4}(2t + x^2)u\frac{\partial}{\partial u}. \tag{7.1.24}$$

Consider, e.g. the operator X_5. It generates the *heat representation of the Galilean transformation*:

$$\bar{t} = t, \quad \bar{x} = x + 2at, \quad \bar{u} = u\,e^{-(ax+a^2t)}. \tag{7.1.25}$$

Any solution $u = \Phi(t, x)$ of the heat equation can be converted into a new solution by transformation (7.1.25). Since the heat equation is invariant under this transformation, we write it in the form $\bar{u}_{\bar{t}} - \bar{u}_{\bar{x}\bar{x}} = 0$ using the variables $\bar{t}, \bar{x}, \bar{u}$, and take the solution in the same variables:

$$\bar{u} = \Phi(\bar{t}, \bar{x}).$$

Then we substitute here expressions (7.1.25) for $\bar{t}, \bar{x}, \bar{u}$, and upon solving the resulting equation

$$u\,e^{-(ax+a^2t)} = \Phi(t, x + 2at)$$

for u, we obtains the new solution

$$u = e^{ax+a^2t}\Phi(t, x + 2at) \tag{7.1.26}$$

involving the parameter a.

Exercise 7.1.1. Obtain the solutions of the heat equation by transformation (7.1.25) and by formula (7.1.26) applied to the following two simple solutions:

$$\text{(i) } u = 1, \qquad \text{(ii) } u = x.$$

Solution. (i) Inserting $\bar{u} = 1$ into (7.1.25), i.e., letting

$$u\,e^{-(ax+a^2t)} = 1,$$

one obtains the following new solution (7.1.26):

$$u = e^{ax+a^2t}.$$

The same result is obtained from (7.1.26) by letting $\Phi(t, x + 2at) = 1$.

(ii) Inserting $\bar{u} = \bar{x}$ into (7.1.25), i.e., letting

$$u\,e^{-(ax+a^2t)} = x + 2at,$$

one obtains the following new solution (7.1.26):

$$u = (x + 2at)\,e^{ax+a^2t}.$$

The same result is obtained from (7.1.26) by letting $\Phi(t, x + 2at) = x + 2at$.

7.2 Group invariant solutions

7.2.1 Introduction

If a group transformation maps a solution into itself, we arrive at what is called a *self-similar* or *group invariant solution*. Consider, e.g. evolutionary equations (7.1.3). Given an infinitesimal symmetry (7.1.6) of Eq. (7.1.3), the invariant solutions under the one-parameter group generated by X are obtained as follows. One calculates two independent *invariants* $J_1 = \lambda(t, x)$ and $J_2 = \mu(t, x, u)$ by solving the equation

$$X(J) \equiv \tau(t, x, u)\frac{\partial J}{\partial t} + \xi(t, x, u)\frac{\partial J}{\partial x} + \eta(t, x, u)\frac{\partial J}{\partial u} = 0,$$

or its characteristic system:

$$\frac{dt}{\tau(t, x, u)} = \frac{dx}{\xi(t, x, u)} = \frac{du}{\eta(t, x, u)}. \qquad (7.2.1)$$

Then one designates one of the invariants as a function of the other, e.g.

$$\mu = \phi(\lambda), \qquad (7.2.2)$$

and solves Eq. (7.2.2) with respect to u. Finally, one substitutes the expression for u into Eq. (7.1.3) and obtains an ordinary differential equation for the unknown function $\phi(\lambda)$ of one variable. This procedure reduces the number of independent variables by one.

Exercise 7.2.1. Find the invariant solution of the heat equation (7.1.23) under group (7.1.25) with generator (7.1.24):

$$X = 2t\frac{\partial}{\partial x} - xu\frac{\partial}{\partial u}.$$

Solution. There are two independent invariants for X. One of them is t, while the other is obtained from the characteristic equation

$$\frac{dx}{2t} + \frac{du}{xu} = 0, \quad \text{or} \quad \frac{xdx}{2t} + \frac{du}{u} = 0.$$

Integration of the latter equation yields the invariant

$$J = u\, e^{\frac{x^2}{4t}}.$$

Consequently, one seeks the invariant solution in the form $J = \phi(t)$, or

$$u = \phi(t)\, e^{-\frac{x^2}{4t}}.$$

Now, substitute this expression into the heat equation. We have

$$u_t = \left(\phi' + \frac{x^2}{4t^2}\,\phi\right) e^{-\frac{x^2}{4t}}, \quad u_x = -\frac{x}{2t}\,\phi\, e^{-\frac{x^2}{4t}}, \quad u_{xx} = \left(\frac{x^2}{4t^2} - \frac{1}{2t}\right)\phi\, e^{-\frac{x^2}{4t}},$$

and the second-order partial differential equation $u_t - u_{xx} = 0$ reduces to the first-order ordinary differential equation:

$$\frac{d\phi}{dt} + \frac{\phi}{2t} = 0.$$

It follows that $\phi(t) = C/\sqrt{t}$, $C = \text{const}$. Hence, the invariant solution is

$$u = \frac{C}{\sqrt{t}}\, e^{-\frac{x^2}{4t}}.$$

7.2.2 The Burgers equation

We know from Section 7.1.1, Example 7.1.1, that the Burgers equation (7.1.14), $u_t = u_{xx} + u u_x$, has five infinitesimal symmetries (7.1.21):

$$X_1 = \frac{\partial}{\partial t}, \quad X_2 = \frac{\partial}{\partial x}, \quad X_3 = 2t\frac{\partial}{\partial t} + x\frac{\partial}{\partial x} - u\frac{\partial}{\partial u},$$

$$X_4 = t\frac{\partial}{\partial x} - \frac{\partial}{\partial u}, \quad X_5 = t^2\frac{\partial}{\partial t} + tx\frac{\partial}{\partial x} - (x + tu)\frac{\partial}{\partial u}. \tag{7.2.3}$$

Consider the generator X_5 from (7.2.3). The Lie equations have the form

$$\frac{d\bar{t}}{da} = \bar{t}^2, \quad \frac{d\bar{x}}{da} = \bar{t}\bar{x}, \quad \frac{d\bar{u}}{da} = -(\bar{x} + \bar{t}\bar{u}).$$

Integrating them and using the condition $a = 0 : \bar{t} = t, \bar{x} = x, \bar{u} = u$ we get

$$\bar{t} = \frac{t}{1 - at}, \quad \bar{x} = \frac{x}{1 - at} \quad \bar{u} = (1 - at)\, u - ax. \tag{7.2.4}$$

Substituting transformation (7.2.4) into Eq. (7.1.22), one maps any known solution $u = \Phi(t, x)$ of the Burgers equation to the following one-parameter set of new solutions:

$$u = \frac{ax}{1 - at} + \frac{1}{1 - at}\,\Phi\!\left(\frac{t}{1 - at}, \frac{x}{1 - at}\right). \tag{7.2.5}$$

Example 7.2.1. One can obtain many examples, by choosing as an initial solution $u = \Phi(t,x)$, any invariant solution. Let us take, e.g. the invariant solution under the space translation generated by X_2 from (7.2.3). In this case the invariants are $\lambda = t$ and $\mu = u$, and Eq. (7.2.2) is written $u = \phi(t)$. Substitution into the Burgers equation yields the obvious constant solution $u = k$. It is mapped by (7.2.5) to the following one-parameter set of solutions:

$$u = \frac{k + ax}{1 - at}.$$

Example 7.2.2. One of the physically significant types of solutions is obtained by assuming the invariance under the time translation group generated by X_1. This assumption provides the stationary solution

$$u = \Phi(x)$$

for which the Burgers equation yields

$$\Phi'' + \Phi\Phi' = 0. \tag{7.2.6}$$

Integrate it once:

$$\Phi' + \frac{\Phi^2}{2} = C_1,$$

and integrate again by setting $C_1 = 0, C_1 = \nu^2 > 0, C_1 = -\omega^2 < 0$ to obtain:

$$\Phi(x) = \frac{2}{x + C},$$

$$\Phi(x) = \nu\, \text{th}\left(C + \frac{\nu}{2}x\right), \tag{7.2.7}$$

$$\Phi(x) = \omega \tan\left(C - \frac{\omega}{2}x\right).$$

The Galilean transformation $\bar{t} = t, \bar{x} = x + at, \bar{u} = u - a$ generated by X_4 maps the stationary solutions (7.2.7) to travelling waves $u = f(x - ct)$.

Example 7.2.3. If one applies transformation (7.2.5) to the stationary solutions (7.2.7), one obtains the following new non-stationary solutions:

$$u = \frac{ax}{1 - at} + \frac{2}{x + C(1 - at)},$$

$$u = \frac{1}{1 - at}\left[ax + \nu\, \text{th}\left(C + \frac{\nu x}{2(1 - at)}\right)\right], \tag{7.2.8}$$

$$u = \frac{1}{1 - at}\left[ax + \omega \tan\left(C - \frac{\omega x}{2(1 - at)}\right)\right].$$

Example 7.2.4. Let us find the invariant solutions under the projective group generated by X_5. The characteristic system

$$\frac{dt}{t^2} = \frac{dx}{tx} = -\frac{du}{x + tu}$$

provides the invariants $\lambda = x/t$ and $\mu = x + tu$. Hence, the general expression (7.2.2) for the invariant solutions takes the form

$$u = -\frac{x}{t} + \frac{1}{t}\Phi(\lambda), \quad \lambda = \frac{x}{t}. \tag{7.2.9}$$

Substituting this expression into the Burgers equation (7.1.14), one obtains for $\Phi(\lambda)$ precisely Eq. (7.2.6). Hence, its general solution is obtained from (7.2.7) where x is replaced by λ. The corresponding invariant solutions are obtained by substituting into (7.2.9) the resulting expressions for $\Phi(\lambda)$. For example, using for $\Phi(\lambda)$ the second formula of (7.2.7) by letting there $\nu = \pi$, one obtains the solution

$$u = -\frac{x}{t} + \frac{\pi}{t}\operatorname{th}\left(C + \frac{\pi x}{2t}\right). \tag{7.2.10}$$

It is important in nonlinear acoustics and was derived by R.V. Khokhlov in 1961 by physical reasoning.

Example 7.2.5. The invariant solutions under the group of dilations gener-ated by X_3 lead to what is called in physics *similarity solutions* because of their connection with the dimensional analysis. The characteristic system

$$\frac{dt}{2t} = \frac{dx}{x} = -\frac{du}{u}$$

provides the following invariants: $\lambda = x/\sqrt{t}, \mu = \sqrt{t}\, u$. Consequently, one seeks the invariant solutions in the form

$$u = \frac{1}{\sqrt{t}}\Phi(\lambda), \quad \lambda = \frac{x}{\sqrt{t}},$$

and arrives at the following equation for the similarity solutions of the Burgers equation:

$$\Phi'' + \Phi\Phi' + \frac{1}{2}(\lambda\Phi' + \Phi) = 0. \tag{7.2.11}$$

Integrating once, one has

$$\Phi' + \frac{1}{2}(\Phi^2 + \lambda\Phi) = C.$$

Letting $C = 0$, one obtains the solution (found in physics by O.V. Rudenko)

$$u = \frac{2}{\sqrt{\pi t}}\frac{e^{-x^2/(4t)}}{B + \operatorname{erf}(x/(2\sqrt{t}))},$$

where B is an arbitrary constant and

$$\operatorname{erf}(z) = \frac{2}{\sqrt{\pi}}\int_0^z e^{-s^2}ds$$

is the error function.

7.2.3   A nonlinear boundary-value problem

Consider the following nonlinear equation

$$\Delta u = e^u, \tag{7.2.12}$$

where $u = u(x, y)$ and $\Delta u = u_{xx} + u_{yy}$ is the Laplacian with two independent variables. Equation (7.2.12) admits the operator

$$X = \xi \frac{\partial}{\partial x} + \eta \frac{\partial}{\partial y} - 2\xi_x \frac{\partial}{\partial u}, \tag{7.2.13}$$

where $\xi(x, y)$ and $\eta(x, y)$ are arbitrary solutions of the Cauchy-Riemann system

$$\xi_x - \eta_y = 0, \quad \xi_y + \eta_x = 0. \tag{7.2.14}$$

Consequently, one can express the general solution of the nonlinear equation (7.2.12) via the solution of the Laplace equation

$$\Delta v = 0 \tag{7.2.15}$$

in the form

$$u = \ln\left(2\frac{v_x^2 + v_y^2}{v^2}\right). \tag{7.2.16}$$

In other words, the nonlinear equation (7.2.12) is mapped to the linear equation (7.2.15) by transformation (7.2.16). However, this transformation is not particularly useful in dealing with concrete problems as it is clear from the following example taken from [16].

Consider the following boundary-value problem in the circle of radius $r = 1$:

$$\Delta u = e^u, \quad u\big|_{r=1} = 0, \tag{7.2.17}$$

where $r = \sqrt{x^2 + y^2}$. The general solution (7.2.16) is not suitable for solving our problem since it leads to the nonlinear boundary-value problem

$$\Delta v = 0, \quad (v_x^2 + v_y^2 - \frac{1}{2}v^2)\big|_{r=1} = 0.$$

In order to solve problem (7.2.17) it is convenient to use the polar coordinates

$$x = r\cos\varphi, \quad y = r\sin\varphi. \tag{7.2.18}$$

In these coordinates Eq. (7.2.12) is written

$$u_{rr} + \frac{1}{r}u_r + \frac{1}{r^2}u_{\varphi\varphi} = e^u. \tag{7.2.19}$$

Since the differential equation and the boundary condition in problem (7.2.17) are invariant with respect to the rotation group, we will seek the solution depending only on the variable r. Then Eq. (7.2.19) is written

$$u_{rr} + \frac{1}{r}u_r = e^u. \tag{7.2.20}$$

We will assume that the function $u(r)$ is bounded at the "singular" point $r = 0$ and formulate the boundary conditions of problem (7.2.17) in the following form:

$$u(1) = 0, \quad u(0) < \infty. \tag{7.2.21}$$

One can integrate Eq. (7.2.20) by means of Lie's method. Namely, it has two infinitesimal symmetries

$$X_1 = r\frac{\partial}{\partial r} - 2\frac{\partial}{\partial u}, \quad X_2 = r\ln r\frac{\partial}{\partial r} - 2(1 + \ln r)\frac{\partial}{\partial u}. \tag{7.2.22}$$

We reduce the first operator to the form $X_1 = \partial/\partial t$ by the change of variables

$$t = \ln r, \quad z = u + 2\ln r.$$

Equation (7.2.20) is written in these variables as follows

$$\frac{d^2 z}{dt^2} = e^z. \tag{7.2.23}$$

Integration by means of the standard substitution $dz/dt = p(z)$ yields

$$\int \frac{dz}{\sqrt{C_1 + 2e^z}} = t + C_2. \tag{7.2.24}$$

Evaluating the integral in (7.2.24) one can verify that the condition $u(0) < \infty$ is not satisfied if $C_1 \leqslant 0$. Therefore, we calculate the integral for $C_1 > 0$. For the sake of convenience we set $C_1 = \lambda^2$ and $C_2 = \ln c$. Then we evaluate the integral, rewrite the result in the old variables and obtain the following solution:

$$u = \ln \frac{2\lambda^2 (cr)^\lambda}{r^2[1 - (cr)^\lambda]^2}. \tag{7.2.25}$$

It follows that

$$u \approx (\lambda - 2)\ln r \quad (r \to 0),$$

and the condition $u(0) < \infty$ entails that $\lambda = 2$. Furthermore, the boundary condition $u(1) = 0$ takes the form

$$8c^2 = (1 - c^2)^2,$$

whence

$$c^2 = 5 \pm 2\sqrt{6}. \tag{7.2.26}$$

Hence, problem (7.2.17) has two solutions

$$u = \ln(8c^2) - \ln(1 - c^2 r^2)^2 \tag{7.2.27}$$

with

$$c^2 = 5 - 2\sqrt{6}$$

and

$$c^2 = 5 + 2\sqrt{6},$$

respectively. The first solution corresponding to the case $c^2 = 5 - 2\sqrt{6}$ is bounded everywhere in the circle

$$x^2 + y^2 \leq 1,$$

whereas the second solution corresponding to $c^2 = 5 + 2\sqrt{6}$ is unbounded on the circle

$$x^2 + y^2 = r_*^2$$

with the radius $r_* = 1/c \approx 0.33$ (see Fig.7.1 and Fig.7.2).

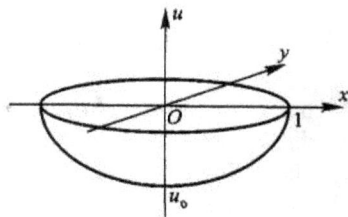

Figure 7.1: Solution (7.2.27) with $c^2 = 5 - 2\sqrt{6}$ is bounded.

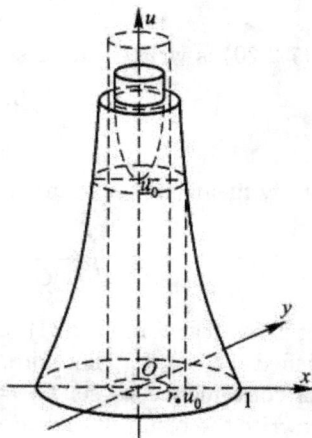

Figure 7.2: Unbounded solution, $c^2 = 5 + 2\sqrt{6}$, $u(0) = u(r_0) = u_0 \approx 4.37$; $r_* = 1/c$, $r_0 = \sqrt{2}/c$.

7.2.4   Invariant solutions for an irrigation system

Consider the nonlinear partial differential equation (2.3.39),

$$C(\psi)\psi_t = [K(\psi)\psi_x]_x + [K(\psi)(\psi_z - 1)]_z - S(\psi),$$

modelling soil water motion in an irrigation system.

The infinitesimal symmetries of Eq. (2.3.39) with arbitrary coefficients form a Lie algebra called the *principal Lie algebra* L_P for Eq. (2.3.39). It is a three-dimensional algebra spanned by

$$X_1 = \frac{\partial}{\partial t}, \quad X_2 = \frac{\partial}{\partial x}, \quad X_3 = \frac{\partial}{\partial z}.$$

There are 29 particular types of the coefficients $C(\psi), K(\psi), S(\psi)$ when an extension of the algebra L_P occurs. Let us consider here one of cases when L_P extends by three operators. Namely, consider the equation ($M = $ const.)

$$\frac{4}{Me^{4\psi} - 1}\psi_t = \left(e^{-4\psi}\psi_x\right)_x + \left(e^{-4\psi}\psi_z\right)_z + 4e^{-4\psi}\psi_z + M - e^{-4\psi}. \quad (7.2.28)$$

Equation (7.2.28) admits a six-dimensional Lie algebra L_6 obtained by adding to the basis X_1, X_2, X_3 of L_P the following three operators:

$$X_4 = t\frac{\partial}{\partial t} - \frac{1}{4}(Me^{4\psi} - 1)\frac{\partial}{\partial \psi},$$

$$X_5 = \sin x \ e^{-z}\frac{\partial}{\partial x} - \cos x \ e^{-z}\frac{\partial}{\partial z} + \frac{1}{2}\cos x \ e^{-z}(Me^{4\psi} - 1)\frac{\partial}{\partial \psi},$$

$$X_6 = \cos x \ e^{-z}\frac{\partial}{\partial x} + \sin x \ e^{-z}\frac{\partial}{\partial z} - \frac{1}{2}\sin x \ e^{-z}(Me^{4\psi} - 1)\frac{\partial}{\partial \psi}.$$

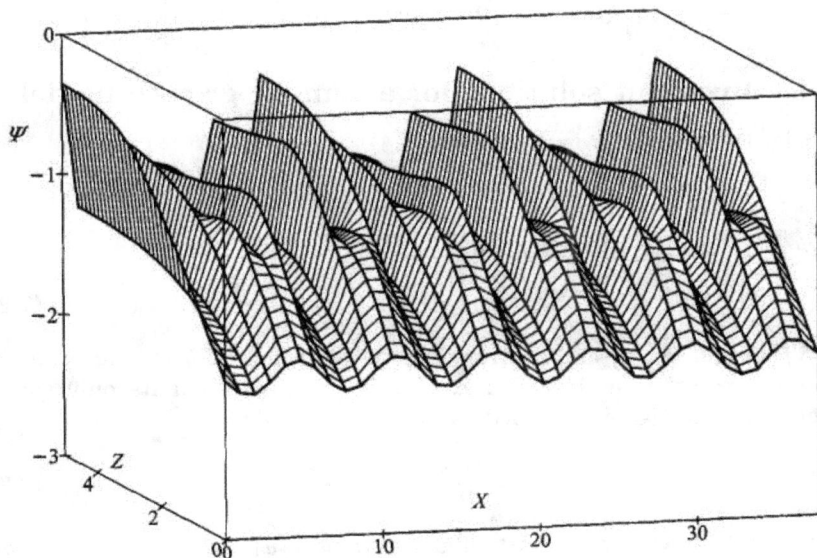

Figure 7.3: Plot of solution (7.2.30), $M = 4, l_1 = -2, l_2 = -4, t = 0.01$.

Let us find the invariant solutions based on the two-dimensional subalgebra $L_2 \subset L_6$ spanned by X_4, X_5. Invariants $J(t, x, z, \psi)$ of L_2 are defined by the system of linear partial differential equations

$$X_4(J) = 0, \quad X_5(J) = 0. \tag{7.2.29}$$

This system provides the following basis of invariants:

$$v = te^{2z}\left(e^{-4\psi} - M\right), \quad \lambda = e^z \sin x.$$

The invariant solutions have the form

$$te^{2z}\left(e^{-4\psi} - M\right) = \Phi(\lambda),$$

whence, upon solving for ψ:

$$\psi = -\frac{1}{4}\ln\left|M + \frac{e^{-2z}}{t}\Phi(\lambda)\right|.$$

Substitution into Eq. (7.2.28) yields

$$\Phi''(\lambda) = 4,$$

whence

$$\Phi(\lambda) = 2\lambda^2 + l_1\lambda + l_2, \quad l_1,\ l_2 = \text{const.}$$

Thus, the invariant solution is given by

$$\psi = -\frac{1}{4}\ln\left|M + \frac{e^{-2z}}{t}\left(2e^{2z}\sin^2 x + l_1 e^z \sin x + l_2\right)\right|. \tag{7.2.30}$$

7.2.5   Invariant solutions for a tumour growth model

Consider the tumour growth model (2.5.4):

$$u_t = f(u) - (uc_x)_x, \quad c_t = -g(c, u),$$

where $f(u)$ and $g(c, u)$ satisfy the conditions

$$f(u) > 0, \quad g_c(c, u) > 0, \quad g_u(c, u) > 0. \tag{7.2.31}$$

If $f(u)$ and $g(c, u)$ are arbitrary functions, system (2.5.4) is invariant only under the translations in t and x. In other words, it admits only the two-dimensional Lie algebra spanned by

$$X_1 = \frac{\partial}{\partial t}, \quad X_2 = \frac{\partial}{\partial x}. \tag{7.2.32}$$

There are, however, many particular functions $f(u)$ and $g(c, u)$ when system (2.5.4) has more symmetries[1]. For example, if

$$f = \alpha u, \quad g = G(ue^{-c}),$$

---

[1]N.H. Ibragimov and N. Säfström, *Communications in Nonlinear Science and Numerical Simulation*, vol. 9(1), 61-68, 2004.

where α is an arbitrary constant and G an arbitrary function, the corresponding system

$$u_t = \alpha u - (uc_x)_x, \quad c_t = -G(ue^{-c})$$

has, along with X_1, X_2, the additional symmetry

$$X_3 = \frac{\partial}{\partial c} + u\frac{\partial}{\partial u}.$$

Let us take, e.g. $G(ue^{-c}) = ue^{-c}$ and consider the system

$$u_t = \alpha u - (uc_x)_x, \quad c_t = -ue^{-c}, \quad \alpha = \text{const.} \tag{7.2.33}$$

Let us find the invariant solutions under the one-parameter group generated by the operator

$$X_1 + X_3 = \frac{\partial}{\partial t} + \frac{\partial}{\partial c} + u\frac{\partial}{\partial u}.$$

The equation $(X_1 + X_3)J = 0$ gives three independent invariants:

$$x, \quad \psi_1 = c - t, \quad \psi_2 = ue^{-t}.$$

The corresponding invariant solutions are defined by

$$c = t + \psi_1(x), \quad u = e^t\psi_2(x). \tag{7.2.34}$$

We have

$$u_t = e^t\psi_2(x), \quad u_x = e^t\psi_2'(x),$$
$$c_t = 1, \quad c_x = \psi_1'(x), \quad c_{xx} = \psi_1''(x). \tag{7.2.35}$$

Substitution of (7.2.34) and (7.2.35) into the first equation of (7.2.33) yields

$$e^t\psi_2(x) = \alpha e^t\psi_2(x) - e^t\psi_1'(x)\psi_2'(x) - e^t\psi_2(x)\psi_1''(x),$$

or

$$(1 - \alpha)\psi_2(x) + \psi_1'(x)\psi_2'(x) + \psi_2(x)\psi_1''(x) = 0. \tag{7.2.36}$$

The second equation of (7.2.33) yields

$$1 = -\psi_2(x)e^{-\psi_1(x)},$$

whence

$$\psi_2(x) = -e^{\psi_1(x)}. \tag{7.2.37}$$

Now Eq. (7.2.36) becomes

$$\psi_1''(x) + \psi_1'^2 + (1 - \alpha) = 0.$$

It follows that

$$\psi_1(x) = \ln|A_2(x + A_1)| \quad \text{for } \alpha = 1, \tag{7.2.38}$$

$$\psi_1(x) = x\sqrt{\alpha - 1} + \ln\left|A_2\left(1 \pm e^{2\sqrt{\alpha-1}(A_1-x)}\right)\right| \quad \text{for } \alpha > 1, \qquad (7.2.39)$$

$$\psi_1(x) = \ln\left|A_2 \cos\left(\sqrt{1-\alpha}(A_1 - x)\right)\right| \quad \text{for } \alpha < 1, \qquad (7.2.40)$$

where A_1 and A_2 are arbitrary constants.

Substituting (7.2.37)~(7.2.40) into (7.2.34), we obtain the following three different invariant solutions to system (7.2.33):

$$c(t,x) = t + \ln|A_2(x + A_1)|,$$
$$u(t,x) = -e^t|A_2(x + A_1)| \quad \text{if } \alpha = 1; \qquad (7.2.41)$$

$$c(t,x) = t + x\sqrt{\alpha - 1} + \ln\left|A_2\left(1 \pm e^{2\sqrt{\alpha-1}(A_1-x)}\right)\right|,$$
$$u(t,x) = -e^{t+x\sqrt{\alpha-1}}\left|A_2\left(1 \pm e^{2\sqrt{\alpha-1}(A_1-x)}\right)\right| \quad \text{if } \alpha > 1; \qquad (7.2.42)$$

and

$$c(t,x) = t + \ln\left|A_2 \cos\left(\sqrt{1-\alpha}(A_1 - x)\right)\right|,$$
$$u(t,x) = -e^t\left|A_2 \cos\left(\sqrt{1-\alpha}(A_1 - x)\right)\right| \quad \text{if } \alpha < 1. \qquad (7.2.43)$$

Conditions (7.2.31) select, however, solution (7.2.43) with $\alpha < 0$. Indeed, only in this case the functions $f(u) = \alpha u$ and $g(c,u) = ue^{-c}$ satisfy conditions (7.2.31):

$$f(u) = \alpha u > 0, \quad g_c(c,u) = -ue^{-c} > 0, \quad g_u(c,u) = e^{-c} > 0$$

and hence solution (7.2.43) with $\alpha < 0$ is relevant to model (7.2.33).

7.2.6 An example from nonlinear optics

The phenomena of the wave front correction for optical radiations in laser systems are simulated by nonlinear equations called the system of phase-conjugated reflection equations, known also as wave front reversal. A simplified model obtained by considering steady-state waves is described, upon choosing particular parameters of a medium, by the following system from nonlinear optics (see [21], Section 11.2.7):

$$\left(\frac{\partial}{\partial z} - i\Delta\right)E_1 = |E_2|^2 E_1, \quad \left(\frac{\partial}{\partial z} + i\Delta\right)E_2 = |E_1|^2 E_2, \qquad (7.2.44)$$

where Δ is the Laplace operator in the (x,y) plane, E_1 and E_2 are complex amplitudes of incident and phase conjugated (amplified) light waves, respectively.

Equations (7.2.44) are invariant under the translations of x, y, z, rotations in the (x,y) plane and appropriate dilations of the dependent and independent variables. One can find an additional symmetry group using an analogy of the left-hand sides of (7.2.44) and the heat equation. Namely, we denote by E_α^* the complex conjugate quantities to E_α and look for an infinitesimal symmetry of system (7.2.44) in the following particular form:

$$X = 2z\frac{\partial}{\partial x} + \sum_{\alpha=1}^{2}\left(f_\alpha(x,z)E_\alpha\frac{\partial}{\partial E_\alpha} + g_\alpha(x,z)E_\alpha^*\frac{\partial}{\partial E_\alpha^*}\right),$$

obtained as an analog of generator (7.1.24),

$$X = 2t\frac{\partial}{\partial x} - xu\frac{\partial}{\partial u}$$

of the Galilean transformation for the heat equation. Substitution of the operator taken into the above specific form in the determining equations yields

$$X = 2z\frac{\partial}{\partial x} + ix\left(E_1\frac{\partial}{\partial E_1} - E_2\frac{\partial}{\partial E_2} - E_1^*\frac{\partial}{\partial E_1^*} + E_2^*\frac{\partial}{\partial E_2^*}\right). \qquad (7.2.45)$$

Consider the two-dimensional Lie algebra spanned by operator (7.2.45) and the generator $\partial/\partial y$ of the translations in y. The basis of invariants is

$$z, \quad u_1 = E_1 e^{-ix^2/(4z)}, \quad u_2 = E_2 e^{ix^2/(4z)},$$

and the complex conjugates for u_1 and u_2. Hence, we have the following general form of the invariant solutions:

$$E_1 = u_1(z)e^{ix^2/(4z)}, \quad E_2 = u_2(z)e^{-ix^2/(4z)}. \qquad (7.2.46)$$

For simplicity sake, we consider the case of real functions u_1 and u_2. Then the substitution of expressions (7.2.46) into Eqs. (7.2.44) yields

$$\frac{du_1}{dz} = u_2^2 u_1 - \frac{u_1}{2z}, \quad \frac{du_2}{dz} = u_1^2 u_2 - \frac{u_2}{2z}. \qquad (7.2.47)$$

Equations (7.2.47) have been solved in Section 6.7.4, Example 6.7.9, where u_1, u_2 and z are denoted by x, y and t, respectively. Making the corresponding replacement in Eqs. (6.7.47), we have the following general solution to Eqs. (7.2.47):

$$u_1 = \sqrt{\frac{k}{z(1-\zeta^2)}}, \quad u_2 = \zeta\sqrt{\frac{k}{z(1-\zeta^2)}}, \qquad (7.2.48)$$

where $\zeta = Cz^k$, $C, k = \text{const.}$

Finally, we substitute (7.2.48) into (7.2.46) and obtain the following invariant solution of system (7.2.44):

$$E_1 = \sqrt{\frac{k}{z(1-\zeta^2)}}\, e^{ix^2/(4z)}, \quad E_2 = \zeta\sqrt{\frac{k}{z(1-\zeta^2)}}\, e^{-ix^2/(4z)},$$

where $\zeta = Cz^k$ contains two arbitrary constants, C and k.

7.3   Invariance and conservation laws

We discuss in this section a general method for constructing conservation laws for differential equations obtained from the variational principle. The method is based on two *conservation theorems*. First of them is Noether's theorem [30] associating conservation laws with *invariance of variational integrals*. The second theorem, formulated in Section 7.3.6, generalizes Noether's theorem and associates conservation laws with *invariance of the extremal values of variational integrals*. It has been proved in [13].

The conservation theorems discussed here have numerous applications. Some of them are included as illustrative examples. Many other applications in mechanics, physics and engineering sciences are collected in [18]-[20].

7.3.1   Introduction

Conservation laws provide one of the basic principles in formulating and investigating mathematical models.

In everyday life, one encounters a great variety of evident conservation laws, e.g. traditional customs, or natural laws such as alternations day-night, summer-winter, invariable position of stars, etc. Some conservation laws may not be clearly evident. One of such "hidden conservation laws" is what I jokingly call the *conservation of the number of problems*. It states that "*Each individual has a fixed number of problems. If somebody helps him to solve one of his problems, a new problem immediately replaces the solved one.*" In my numerous observations I have seen its manifestation over and over.

Our concern is, however, on mathematical conservation laws. The concept of a conservation law is motivated by the conservation of such quantities as energy, linear and angular momenta, etc. that arise in classical mechanics. These quantities are conserved in the sense that they are constant on each trajectory of a given dynamical system (see Section 1.5.1). Namely, a function $T = T(t, q, v)$ of time t, the position coordinates $q = (q^1, \ldots, q^s)$ and the velocity $v = (v^1, \ldots, v^s)$ is called a conserved quantity if it satisfies the equation

$$D_t(T) = 0 \qquad (7.3.1)$$

on each trajectory $q = q(t)$ of the dynamical system in question. Since

$$D_t(T) \equiv \frac{\partial T}{\partial t} + v^\alpha \frac{\partial T}{\partial q^\alpha} + \dot{v}^\alpha \frac{\partial T}{\partial v^\alpha} \qquad (7.3.2)$$

is the total derivative with respect to time, we can formulate the definition as follows. Let $q = q(t)$ be a given trajectory and $v = \dot{q}(t)$ the corresponding velocity. Then the conservation equation (7.3.1) means that the function $T(t) = T(t, q(t), v(t))$ satisfies the equation $dT(t)/dt = 0$. In other words, the conserved quantity $T(t, q, v)$ is constant on each trajectory. Therefore, T is also called a *constant of motion*.

For example, a free motion of a single particle with the mass m is described by the equation $m\dot{v} = 0$. The energy $E = m|v|^2/2$ of the particle is a constant of the free motion. Indeed, its total derivative $D_t(E) = m\dot{v} \cdot v$ vanishes on any trajectory due to the equation of motion $m\dot{v} = 0$.

The extension of the conservation law (7.3.1) to continuous systems leads to the following definition valid for any number $n \geq 1$ of independent variables. Consider the partial differential equations, e.g. of the second order:

$$F(x, u, u_{(1)}, u_{(2)}) = 0, \tag{7.3.3}$$

where $x = (x^1, \ldots, x^n)$ are the independent variables, $u = (u^1, \ldots, u^m)$ are the dependent variables, and $u_{(1)} = \{u_i^\alpha\}$ and $u_{(2)} = \{u_{ij}^\alpha\}$ denote the first and second order derivatives, respectively (see the notation in Section 1.4.3).

Definition 7.3.1. A vector field $C(x, u, u_{(1)})$ with n components,

$$C = (C^1, \ldots, C^n) \tag{7.3.4}$$

is called a *conserved vector* if it satisfies the equation

$$\operatorname{div} C \equiv D_i(C^i) = 0 \tag{7.3.5}$$

on each solution $u = u(x)$ of Eq. (7.3.3). Equation (7.3.5) is termed a *conservation law* for Eq. (7.3.3).

Let us assume that one of the independent variables is time, e.g. $x^n = t$. Then, Eq. (7.3.5) implies existence of a function $T(t, u, u_{(1)})$ which is a constant of motion. Namely, the following statement is valid.

Lemma 7.3.1. Let the conservation law (7.3.5) hold. Then the integral

$$T(t) = \int_{\mathbb{R}^{n-1}} C^n(x, u(x), u_{(1)}) dx^1 \cdots dx^{n-1} \tag{7.3.6}$$

is constant along any solution $u = u(x)$ to Eq. (7.3.3), i.e., satisfies Eq. (7.3.1), provided that the components of the conserved vector C decrease rapidly and vanish at infinity. Accordingly, C^n is termed the *density of the conservation law* (7.3.5).

Proof. Let Ω be an $(n-1)$-dimensional tube domain in the space of all independent variables $(x^1, \ldots, x^{n-1}, t)$ defined as follows:

$$\Omega : \quad \sum_{i=1}^{n-1} (x^i)^2 = r^2, \quad t_1 \leq t \leq t_2,$$

where r and t_1, t_2 with $t_1 < t_2$ are arbitrary constants. Let S be the boundary of Ω and let ν be the unit outward normal to the surface S. Applying the divergence theorem to the tube domain Ω and using Eq. (7.3.5) one obtains

$$\int_S C \cdot \nu \, d\sigma = \int_\Omega \operatorname{div} C = 0. \tag{7.3.7}$$

Invoking that the components C^i of the vector C are rapidly vanishing functions at infinity, we can let $r \to \infty$ and neglect the integral over the cylindrical part of the surface S in the left-hand side of Eq. (7.3.7). It remains to evaluate the integrals over the lower base K_1 (where $t = t_1$) and the upper base K_2 (where $t = t_2$) of the tube domain Ω. To this end, we note that $C \cdot \nu = -C^n|_{t=t_1}$ and $C \cdot \nu = C^n|_{t=t_2}$ at K_1 and K_2, respectively. Furthermore, when $r \to \infty$, both K_1 and K_2 coincide with the $(n-1)$-dimensional space $\mathbb{R}^{n-1}$ of the spatial variables $(x^1, \ldots, x^{n-1})$. Therefore, the integral in the left-hand side of Eq. (7.3.7) at $r \to \infty$ becomes

$$\int_{\mathbb{R}^{n-1}} (C^n|_{t=t_2} - C^n|_{t=t_1})\, dx^1 \cdots dx^{n-1},$$

and hence Eq. (7.3.7) yields that

$$\int_{\mathbb{R}^{n-1}} C^n dx^1 \cdots dx^{n-1}\bigg|_{t=t_1} = \int_{\mathbb{R}^{n-1}} C^n dx^1 \cdots dx^{n-1}\bigg|_{t=t_2}.$$

Since the instants t_1 and t_2 are arbitrary, we conclude that integral (7.3.6) is constant along any solution $u = u(x)$ to Eq. (7.3.3), thus completing the proof.

Definition 7.3.2. If the divergence $D_i(C^i)$ of a vector field $C(x, u, u_{(1)})$ vanishes not only for solutions of Eq. (7.3.3), but for any function $u(x)$, then C is called a *trivial conserved vector*.

Remark 7.3.1. Vector fields $C = (C^1(x), C^2(x), C^3(x))$ in $\mathbb{R}^3$ satisfying the equation div $C = 0$ identically in a certain domain are called *solenoidal vectors*.[2] Thus, a three-dimensional vector $C(x, u, u_{(1)})$ is a trivial conserved vector if and only if it is solenoidal for any function $u = u(x)$. Since we deal with arbitrary dimensions, we will use in what follows our nomenclature a *trivial conserved vector*.

Two conserved vectors are considered to be identical if one is obtained from the other by adding a trivial conserved vector. Furthermore, it is clear from the linearity of the conservation equation (7.3.5) that if Eq. (7.3.3) has several conserved vectors, $C_1, \ldots, C_r$, their linear combination $C = k_1 C_1 + \cdots + k_1 C_r$ with constant coefficients is also a conserved vector for Eq. (7.3.3). We summarize.

Definition 7.3.3. Let $C_1, \ldots, C_r$ be conserved vectors for Eq. (7.3.3). They are said to be *linearly dependent* if there exist constants $k_1, \ldots, k_r$, not all zero, such that the linear combination $k_1 C_1 + \cdots + k_1 C_r$ is a trivial conserved vector, and *linearly independent* otherwise.

---

[2]Constant vector fields provide examples of solenoidal vectors. Furthermore, vectors of the form $C = \text{curl}\, A$ are solenoidal (see Problem 7.9). Moreover, it is known in vector analysis that all solenoidal vector fields $C = (C^1(x), C^2(x), C^3(x))$ can be presented in the form curl A. (see examples in Problems 7.11 and 7.12).

7.3.2 Preliminaries

Consider the variational integral (1.5.3),

$$\int_V L(x, u, u_{(1)})dx, \tag{7.3.8}$$

and the Euler-Lagrange equations (1.5.4):

$$\frac{\delta L}{\delta u^\alpha} \equiv \frac{\partial L}{\partial u^\alpha} - D_i\left(\frac{\partial L}{\partial u_i^\alpha}\right) = 0, \quad \alpha = 1, \ldots, m, \tag{7.3.9}$$

where the Lagrangian L involves the independent variables $x = (x^1, \ldots, x^n)$, the dependent variables $u = (u^1, \ldots, u^m)$ and the first-order derivatives $u_{(1)} = \{u_i^\alpha\}$ of u with respect to x.

Let G be a one-parameter group of transformations

$$\bar{x}^i = f^i(x, u, a), \quad \bar{u}^\alpha = \varphi^\alpha(x, u, a) \tag{7.3.10}$$

with the generator

$$X = \xi^i(x, u)\frac{\partial}{\partial x^i} + \eta^\alpha(x, u)\frac{\partial}{\partial u^\alpha}. \tag{7.3.11}$$

Definition 7.3.4. Integral (7.3.8) is said to be invariant under the group G if the following equation holds for any domain V and *any function* $u(x)$:

$$\int_{\bar{V}} L(\bar{x}, \bar{u}, \bar{u}_{(1)})d\bar{x} = \int_V L(x, u, u_{(1)})dx. \tag{7.3.12}$$

Here, $\bar{V} \subset \mathbb{R}^n$ is a domain obtained from V by transformation (7.3.10).

The following statements make Noether's theorem transparent. Proofs of the lemmas are given, in a more general formulation, in [21], Chapter 8.

Lemma 7.3.2. Integral (7.3.8) is invariant under the group G if and only if the following equation holds:

$$X(L) + LD_i(\xi^i) = 0, \tag{7.3.13}$$

where X denotes the first prolongation of generator (7.3.11), hence,

$$X(L) = \xi^i\frac{\partial L}{\partial x^i} + \eta^\alpha\frac{\partial L}{\partial u^\alpha} + \zeta_i^\alpha\frac{\partial L}{\partial u_i^\alpha}, \quad \zeta_i^\alpha = D_i(\eta^\alpha) - u_j^\alpha D_i(\xi^j).$$

Lemma 7.3.3. The following equation holds for any function $L(x, u, u_{(1)})$:

$$X(L) + LD_i(\xi^i) \equiv W^\alpha\frac{\delta L}{\delta u^\alpha} + D_i(C^i), \tag{7.3.14}$$

where

$$W^\alpha = \eta^\alpha - \xi^j u_j^\alpha, \quad \alpha = 1, \ldots, m, \tag{7.3.15}$$

and

$$C^i = \xi^i L + (\eta^\alpha - \xi^j u_j^\alpha)\frac{\partial L}{\partial u_i^\alpha}, \quad i = 1, \ldots, n. \tag{7.3.16}$$

Lemma 7.3.4. A function $F(x, u, u_{(1)})$ is the divergence of a certain vector field $H = (H^1, \ldots, H^n)$ if and only if the variational derivative of F vanishes:

$$F = D_i(H^i) \quad \text{iff} \quad \frac{\delta F}{\delta u^\alpha} = 0, \ \alpha = 1, \ldots, m. \tag{7.3.17}$$

7.3.3   Noether's theorem

Theorem 7.3.1. Let the variational integral (7.3.8) be invariant under the group with generator (7.3.11). In other words, let the invariance test (7.3.13) be satisfied. Then the vector field $C = (C^1, \ldots, C^n)$ defined by

$$C^i = \xi^i L + (\eta^\alpha - \xi^j u_j^\alpha) \frac{\partial L}{\partial u_i^\alpha}, \quad i = 1, \ldots, n, \tag{7.3.18}$$

is a conserved vector for Eqs. (7.3.9), i.e., C satisfies the conservation law (7.3.5) $D_i(C^i) = 0$.

Proof. The statement follows from Lemmas 7.3.2 and 7.3.3.

Remark 7.3.2. The invariance of the variational integral (7.3.8) manifestly implies the invariance of the Euler-Lagrange equations (7.3.9) under the group G. Hence, Noether's theorem gives a constructive way of determining conservation laws using a known symmetry group G of the Euler-Lagrange equations, provided that G has the *additional property to leave invariant the variational integral*.

Corollary 7.3.1. Lemma 7.3.4 shows that one can add to the Lagrangian any divergence type function F. Consequently, the invariance condition (7.3.13) can be replaced by the following *divergence condition*:

$$X(L) + LD_i(\xi^i) = D_i(B^i). \tag{7.3.19}$$

Then one obtains, instead of (7.3.18), the following conserved vector:

$$C^i = \xi^i L + (\eta^\alpha - \xi^j u_j^\alpha) \frac{\partial L}{\partial u_i^\alpha} - B^i. \tag{7.3.20}$$

7.3.4   Higher-order Lagrangians

Mathematical physics provides various mathematical models described by Lagrangians of the first and second order. For example, we had a second-order Lagrangian (2.6.27),

$$L = \frac{1}{2} [\rho u_t^2 - \mu(u_{xx} + u_{yy})^2],$$

in the problem of vibrating plates. Therefore, it is useful to have, along with the conservation formula (7.3.18), the similar formula for equations described by second-order Lagrangians.

Consider, in the previous notation, a Lagrangian $L(x, u, u_{(1)}, u_{(2)})$ involving second-order derivatives. Then Noether's theorem states that the invariance of the variational integral leads to the conservation law (7.3.5), $D_i(C^i) = 0$, where the Euler-Lagrange equations have the form

$$\frac{\delta L}{\delta u^\alpha} \equiv \frac{\partial L}{\partial u^\alpha} - D_i\left(\frac{\partial L}{\partial u_i^\alpha}\right) + D_i D_k\left(\frac{\partial L}{\partial u_{ik}^\alpha}\right) = 0 \qquad (7.3.21)$$

and the conserved vector (7.3.18) is modified as follows:

$$C^i = L\xi^i + W^\alpha\left[\frac{\partial L}{\partial u_i^\alpha} - D_k\left(\frac{\partial L}{\partial u_{ik}^\alpha}\right)\right] + D_k(W^\alpha)\frac{\partial L}{\partial u_{ik}^\alpha}. \qquad (7.3.22)$$

Here W^α is again defined by (7.3.15), $W^\alpha = \eta^\alpha - \xi^j u_j^\alpha$.

Remark 7.3.1 on conservation laws under the divergence condition is applicable in the case of higher-order Lagrangians as well.

7.3.5 Conservation theorems for ODEs

Let us adapt the conservation theorem to systems of ordinary differential equations. We will slightly change the notation used above in discussing dynamical systems, e.g. in Eq. (7.3.2). Namely, we will assume again that the independent variable is time t, but we will denote the dependent variables (e.g. coordinates of particles) by $x = (x^1, \ldots, x^n)$. The velocities are denoted by $v = (v^1, \ldots, v^n)$, where $v^i = \dot{x}^i \equiv dx^i/dt$.

We will consider Lagrangians of the form

$$L(t, x, v) \qquad (7.3.23)$$

and the corresponding Euler-Lagrange equations

$$\frac{\delta L}{\delta x^i} \equiv \frac{\partial L}{\partial x^i} - D_t\left(\frac{\partial L}{\partial v^i}\right) = 0, \quad i = 1, \ldots, n, \qquad (7.3.24)$$

where D_t is the total differentiation in t (cf. Eq. (7.3.2)):

$$D_t = \frac{\partial}{\partial t} + v^i\frac{\partial}{\partial x^i} + \dot{v}^i\frac{\partial}{\partial v^i}.$$

Let G be a group of transformations

$$\bar{t} = \varphi(t, x, a), \quad \bar{x}^i = \psi^i(t, x, a), \qquad (7.3.25)$$

with a generator

$$X = \xi(t, x)\frac{\partial}{\partial t} + \eta^i(t, x)\frac{\partial}{\partial x^i}. \qquad (7.3.26)$$

In this notation, the infinitesimal invariance test (7.3.13) for the variational integral with a Lagrangian (7.3.23) is written:

$$X(L) + LD_t(\xi) = 0. \qquad (7.3.27)$$

Then Theorem 7.3.1 can be formulated as follows.

Theorem 7.3.2. Let the invariance test (7.3.27) be satisfied. Then

$$T = \xi L + (\eta^i - \xi v^i)\frac{\partial L}{\partial v^i} \tag{7.3.28}$$

is a conserved quantity, i.e., satisfies the *conservation law*

$$D_t(T) \equiv \frac{\partial T}{\partial t} + v^i\frac{\partial T}{\partial x^i} + \dot{v}^i\frac{\partial T}{\partial v^i} = 0 \tag{7.3.29}$$

for all solutions $x(t)$ of Eqs. (7.3.24).

Furthermore, the divergence condition (7.3.19) and formula (7.3.20) for the corresponding conservation vector are given in the following statement.

Theorem 7.3.3. Let the following divergence condition be satisfied:

$$X(L) + LD_t(\xi) = D_t(B). \tag{7.3.30}$$

Then

$$T = \xi L + (\eta^i - \xi v^i)\frac{\partial L}{\partial v^i} - B \tag{7.3.31}$$

is a conserved quantity for Eqs. (7.3.24).

7.3.6   Generalization of Noether's theorem

Theorem 7.3.1 and Remark 7.3.2 show that the invariance of the variational integral under the action of the group G admitted by the Euler-Lagrange equations furnishes a *sufficient condition* for (7.3.18) to be a conserved vector. Likewise, Corollary 7.3.1 gives a sufficient condition for (7.3.20) to be a conserved vector. Examples show that the invariance and the divergence property are not *necessary conditions* (see Section 7.3.7 and Section 7.3.9). Therefore, it was desirable to obtain the necessary and sufficient condition for (7.3.20) to be a conserved vector.

The solution to this problem was given in [13] (see also [21]). The result is formulated in the following theorem. The term *extremal values of the variational integral* used in the theorem refers to the values of the integral (7.3.8) on the solutions $u(x)$ of the Euler-Lagrange equations (7.3.9).

Theorem 7.3.4. Let the Euler-Lagrange equations (7.3.9) admit a continuous group G with generator (7.3.11). Then vector (7.3.18),

$$C^i = \xi^i L + (\eta^\alpha - \xi^j u_j^\alpha)\frac{\partial L}{\partial u_i^\alpha},$$

provides a conservation law for Eqs. (7.3.9) if and only if the extremal values of integral (7.3.8) are invariant under G. The infinitesimal test for the invariance of the extremal values of integral (7.3.8) is

$$X(L) + LD_i(\xi^i) = F^\alpha\frac{\delta L}{\delta u^\alpha}, \tag{7.3.32}$$

where $F^\alpha = F^\alpha(x, u, u_{(1)})$ are different from W^α defined by (7.3.15).

Remark 7.3.3. If $F^\alpha = W^\alpha$, then (7.3.18) defines a trivial conservation law. Indeed, Eq. (7.3.32) with $F^\alpha = W^\alpha$ and identity (7.3.14) yield that $D_i(C^i)$ vanishes identically. Hence, C^i is a trivial conserved vector.

Corollary 7.3.2. Equation (7.3.32) for the invariance of extremal values can be replaced by the *divergence condition on extremal values*:

$$X(L) + LD_i(\xi^i) = F^\alpha \frac{\delta L}{\delta u^\alpha} + D_i(B^i), \quad F^\alpha \neq W^\alpha. \tag{7.3.33}$$

Then one obtains, instead of (7.3.18), the following conserved vector:

$$C^i = \xi^i L + (\eta^\alpha - \xi^j u_j^\alpha) \frac{\partial L}{\partial u_i^\alpha} - B^i. \tag{7.3.34}$$

In the case of ordinary differential equations we use the notation of Section 7.3.5 and write Eqs. (7.3.33)—(7.3.34) as follows:

$$X(L) + LD_t(\xi) = F^k \frac{\delta L}{\delta x^k} + D_t(B), \quad F^k \neq W^k, \tag{7.3.35}$$

$$T = \xi L + (\eta^i - \xi v^i) \frac{\partial L}{\partial v^i} - B. \tag{7.3.36}$$

7.3.7 Examples from classical mechanics

The motion of a single particle with a constant mass m in a potential field $U(t, x)$ is described by the Lagrangian

$$L = \frac{m}{2} |v|^2 - U(t, x), \quad |v|^2 = \sum_{i=1}^{3} (v^i)^2, \tag{7.3.37}$$

where $x = (x^1, x^2, x^3)$ is the position vector of the particle, $v = (v^1, v^2, v^3)$ is the velocity of the particle. The Euler-Lagrange equations (7.3.24) yield

$$m \frac{d^2 x^i}{dt^2} = -\frac{\partial U}{\partial x^i}, \quad i = 1, 2, 3. \tag{7.3.38}$$

The infinitesimal transformations of the group of transformations (7.3.25) are often written in mechanics in the following form:

$$\bar{t} = t + \delta t, \quad \bar{x} = x + \delta x, \tag{7.3.39}$$

where $\delta t = a\xi$, $\delta x^k = a\eta^k$ with the group parameter a.

Example 7.3.1. *Free motion of a particle.* The free motion corresponds to $U = 0$. Then Eqs. (7.3.38) yield the equation of free motion:

$$m \frac{d^2 x}{dt^2} = 0, \tag{7.3.40}$$

with the Lagrangian $L = \frac{m}{2}|v|^2$. Equation (7.3.40) admits the Galilean group comprising the translation in time, translations and rotations in the space coordinates, and the Galilean transformation. Their generators are

$$X_0 = \frac{\partial}{\partial t}, \quad X_i = \frac{\partial}{\partial x^i}, \quad X_{ij} = x^j\frac{\partial}{\partial x^i} - x^i\frac{\partial}{\partial x^j}, \quad Y_i = t\frac{\partial}{\partial x^i}, \qquad (7.3.41)$$

where $i, j = 1, 2, 3$. Let us find the corresponding conservation laws.

(i) *Time translation*. Its generator is X_0. The conserved quantity obtained by substituting the coordinates $\xi = 1$ and $\eta^1 = \eta^2 = \eta^3 = 0$ of X_0 into Eqs. (7.3.28) and denoting $E = -T$, is the energy:

$$E = \frac{m}{2}|v|^2.$$

(ii) *Space translations*. Consider the x^1-translation generated by X_1 with the coordinates $\xi = 0$, $\eta^1 = 1$, $\eta^2 = \eta^3 = 0$. Equation (7.3.28) yields, upon setting $T = p^1$, the conserved quantity $p^1 = mv^1$. The use of all space translations with the generators X_i furnishes us with the vector valued conservation quantity, namely the linear momentum

$$p = mv.$$

(iii) *Rotations*. Consider the rotation around the x^3-axis. Its generator X_{12} has the coordinates $\xi = 0$, $\eta^1 = x^2$, $\eta^2 = -x^1$, $\eta^3 = 0$. Substitution into (7.3.28) yields the conserved quantity $M_3 = m(x^2v^1 - x^1v^2)$. Using all rotations with the generators X_{12}, X_{13}, X_{23}, we conclude that the invariance under rotations leads to conservation of the angular momentum (2.2.9):

$$M = m(x \times v).$$

(iv) *Galilean transformations*. Their generators Y_i, unlike X_0, X_i, X_{ij}, do not satisfy the invariance test (7.3.13). Namely, the extended action of Y_1:

$$Y_1 = t\frac{\partial}{\partial x^1} + \frac{\partial}{\partial v^1}$$

yields $Y_1(L) + LD_t(\xi) = mv^1 \equiv D_t(mx^1)$. Hence, the divergence condition (7.3.30) is satisfied and one can apply Theorem 7.3.3. Then Eq. (7.3.31) with $B = mx^1$ gives the conserved quantity $T = m(tv^1 - x^1)$. Using all operators Y_i and denoting T by Q, one obtains the vector valued conserved quantity

$$Q = m(tv - x).$$

Conservation of the vector Q is known, in the case of a system of particles, as the *center-of-mass theorem*.

Example 7.3.2. *Kepler's problem*. The two-body problem (e.g., the sun and a planet) is known as Kepler's problem (see Section 2.2.3). It has the Lagrangian

$$L = \frac{m}{2}|v|^2 - \frac{\mu}{r}, \quad \text{where} \quad r = |x|, \; \mu = \text{const.} \qquad (7.3.42)$$

The equations of motion (7.3.38) are written

$$m\frac{d^2 x^k}{dt^2} = \mu\frac{x^k}{r^3}, \quad k = 1, 2, 3, \tag{7.3.43}$$

or in the vector form:

$$m\frac{d^2 \boldsymbol{x}}{dt^2} = \mu\frac{\boldsymbol{x}}{r^3}.$$

Equations (7.3.43) admit five generators of form (7.3.26):

$$X_0 = \frac{\partial}{\partial t}, \quad X_{ij} = x^j\frac{\partial}{\partial x^i} - x^i\frac{\partial}{\partial x^j}, \quad Z = 3t\frac{\partial}{\partial t} + 2x^i\frac{\partial}{\partial x^i}. \tag{7.3.44}$$

The generators X_0 and X_{ij} of translation in time and rotations lead again to the conservation of energy E and angular momentum $\boldsymbol{M}$, respectively:

$$E = \frac{m}{2}|\boldsymbol{v}|^2 + \frac{\mu}{r}, \quad \boldsymbol{M} = m(\boldsymbol{x} \times \boldsymbol{v}).$$

The operator Z does not lead to conservation laws (see Problem 7.13).

Example 7.3.3. Note that the three infinitesimal rotations corresponding to the operators X_{ij} from (7.3.44) are written in form (7.3.39) as follows:

$$\delta t = 0, \quad \delta \boldsymbol{x} = \boldsymbol{x} \times \boldsymbol{a}, \tag{7.3.45}$$

where $\boldsymbol{a} = (a^1, a^2, a^3)$. Equations (7.3.43) admit also the following generalization of (7.3.45) (see [14], Section 25.1):

$$\delta t = 0, \quad \delta \boldsymbol{x} = [\boldsymbol{x} \times (\boldsymbol{v} \times \boldsymbol{a})] + [(\boldsymbol{x} \times \boldsymbol{v}) \times \boldsymbol{a}]. \tag{7.3.46}$$

The infinitesimal transformation (7.3.46) corresponds to the generators

$$\widehat{X}_i = \left(2x^i v^k - x^k v^i - (\boldsymbol{x} \cdot \boldsymbol{v})\delta_i^k\right)\frac{\partial}{\partial x^k}, \quad i = 1, 2, 3. \tag{7.3.47}$$

Let us consider the first operator of (7.3.47):

$$\widehat{X}_1 = -\left(x^2 v^2 + x^3 v^3\right)\frac{\partial}{\partial x^1} + \left(2x^1 v^2 - x^2 v^1\right)\frac{\partial}{\partial x^2} + \left(2x^1 v^3 - x^3 v^1\right)\frac{\partial}{\partial x^3}.$$

The reckoning shows that the action of the first prolongation of $\widehat{X}_1$ (see Problem 7.14) on Lagrangian (7.3.42) has the form:

$$\widehat{X}_1(L) = -W^k\frac{\delta L}{\delta x^k} + D_t\left(-\frac{2\mu}{r}x^1\right). \tag{7.3.48}$$

It follows from (7.3.48) that $\widehat{X}_1$, and hence all symmetries (7.3.47) meet neither the invariance test (7.3.27) nor the divergence condition (7.3.30). Consequently, Noether's theorem does not apply to symmetries (7.3.47). On the other hand,

(7.3.48) shows that the divergence condition (7.3.35) on extremal values is satisfied with $F^k = -W^k$. Hence, (7.3.36) yields the conserved quantity

$$T_1 = 2m\left(x^1\left[(v^2)^2 + (v^3)^2\right] - x^2 v^1 v^2 - x^3 v^1 v^3\right) + \frac{2\mu}{r}\,x^1.$$

Making similar calculations for other two operators (7.3.47) and denoting T_i by $2A^i$, we arrive at the conservation of the Laplace vector (2.2.10):

$$A = [v \times M] + \mu\frac{x}{r}\,.$$

Note that conservation of the Laplace vector yields that planets move on elliptic orbits (see [21], Section 9.7.4). Thus, symmetries (7.3.47) are responsible for Kepler's first law.

7.3.8   Derivation of Einstein's formula for energy

The geometric essence of the *special relativity*, formulated by A. Einstein in 1905 as a new physical theory, is that the three-dimensional Euclidean space and the Galilean group are replaced by a four-dimensional space-time (called the Minkowski space) and the Lorentz group, respectively.

The Lorentz group has the following generators:

$$X_0 = \frac{\partial}{\partial t}\,, \ X_i = \frac{\partial}{\partial x^i}\,, \ X_{ij} = x^j\frac{\partial}{\partial x^i} - x^i\frac{\partial}{\partial x^j}\,, \ X_{0i} = t\frac{\partial}{\partial x^i} + \frac{1}{c^2}x^i\frac{\partial}{\partial t}\,, \quad (7.3.49)$$

where $i, j = 1, 2, 3$, and c is the light velocity. Generators (7.3.41) of the Galilean group are obtained from (7.3.49) by letting $c \to \infty$.

The requirement of the invariance under the Lorentz group leads to the following *relativistic Lagrangian* for a free particle with mass m :

$$L = -mc^2\sqrt{1 - \beta^2}, \quad \text{where} \quad \beta^2 = \frac{|v|^2}{c^2}\,. \qquad (7.3.50)$$

Let us apply Noether's theorem to Lagrangian (7.3.50) taking, e.g. the time translation with the generator X_0. The coordinates of X_0 are $\xi = 1$ and $\eta^i = 0$. Substituting them into formula (7.3.28) and setting $E = -T$, one arrives at Einstein's formula for the *relativistic energy*:

$$E = \frac{mc^2}{\sqrt{1 - \beta^2}} \approx mc^2 + \frac{1}{2}m|v|^2.$$

Likewise, one can obtain all other relativistic conservation laws by using generators (7.3.49) of the Lorentz group (see [21], Section 9.7.5).

7.3.9   Conservation laws for the Dirac equations

Consider the Dirac equation (2.3.32),

$$\gamma^k \frac{\partial \psi}{\partial x^k} + m\psi = 0, \tag{7.3.51}$$

together with the conjugate equation

$$\frac{\partial \tilde{\psi}}{\partial x^k} \gamma^k - m\tilde{\psi} = 0. \tag{7.3.52}$$

Here $\tilde{\psi}$ is the row vector defined by

$$\tilde{\psi} = \overline{\psi}^T \gamma^4, \tag{7.3.53}$$

where $\overline{\psi}$ is the complex-conjugate to ψ and T denotes transposition.

Equations (7.3.51)~(7.3.52) can be obtained from the Lagrangian

$$L = \frac{1}{2} \left[\tilde{\psi} \left(\gamma^k \frac{\partial \psi}{\partial x^k} + m\psi \right) - \left(\frac{\partial \tilde{\psi}}{\partial x^k} \gamma^k - m\tilde{\psi} \right) \psi \right]. \tag{7.3.54}$$

Indeed, we have

$$\frac{\delta L}{\delta \psi} = -\left(\frac{\partial \tilde{\psi}}{\partial x^k} \gamma^k - m\tilde{\psi} \right), \qquad \frac{\delta L}{\delta \tilde{\psi}} = \gamma^k \frac{\partial \psi}{\partial x^k} + m\psi.$$

The Dirac equations (7.3.51)~(7.3.52) provide examples for illustrating both Noether's theorem and Theorem 7.3.4.

Example 7.3.4. The simplest example for illustration of Noether's theorem is provided by the usual linear superposition written as the group of transformations $\psi' = \psi + a\varphi(x)$, $\tilde{\psi}' = \tilde{\psi} + a\tilde{\varphi}(x)$ with the generator

$$X_\varphi = \varphi^k(x) \frac{\partial}{\partial \psi^k} + \tilde{\varphi}_k(x) \frac{\partial}{\partial \tilde{\psi}_k}. \tag{7.3.55}$$

Here the vectors $\varphi(x)$ and $\tilde{\varphi}(x)$ are related by (7.3.53) and have the components $\varphi^k(x)$ and $\tilde{\varphi}_k(x)$, respectively. They solve Eqs. (7.3.51)—(7.3.52):

$$\gamma^k \frac{\partial \varphi}{\partial x^k} + m\varphi = 0, \qquad \frac{\partial \tilde{\varphi}}{\partial x^k} \gamma^k - m\tilde{\varphi} = 0. \tag{7.3.56}$$

Let us check if Lagrangian (7.3.54) and generator (7.3.55) obey either the invariance test (7.3.13) or the divergence condition (7.3.19). Since $\xi = 0$, the left-hand side of (7.3.13) reduces to $X_\varphi(L)$ which is equal to

$$X_\varphi(L) = \frac{1}{2} \left[\tilde{\varphi} \left(\gamma^k \frac{\partial \psi}{\partial x^k} + m\psi \right) - \left(\frac{\partial \tilde{\psi}}{\partial x^k} \gamma^k - m\tilde{\psi} \right) \varphi \right]$$

and does not vanish identically. Hence, the invariance test (7.3.13) is not satisfied. Let us check the divergence condition (7.3.19), i.e., check if the above value of $X_\varphi(L)$ is the divergence. This can be done by means of Lemma 7.3.4. The reckoning yields:

$$\frac{\delta}{\delta\psi}X_\varphi(L) = -\frac{1}{2}\left(\frac{\partial\tilde{\varphi}}{\partial x^k}\gamma^k - m\tilde{\varphi}\right), \qquad \frac{\delta}{\delta\tilde{\psi}}X_\varphi(L) = \frac{1}{2}\left(\gamma^k\frac{\partial\varphi}{\partial x^k} + m\varphi\right).$$

These expressions vanish due to Eqs. (7.3.56). Therefore, according to Lemma 7.3.4, $X_\varphi(L)$ is the divergence, i.e., condition (7.3.19) is satisfied. One can verify that $X_\varphi(L) = D_k(B^k)$ with

$$B^k = -\frac{1}{2}\left(\tilde{\psi}\gamma^k\varphi - \tilde{\varphi}\gamma^k\psi\right).$$

Let us find quantities (7.3.18). We have

$$\varphi\frac{\partial L}{\partial\psi_{,k}} + \tilde{\varphi}\frac{\partial L}{\partial\tilde{\psi}_{,k}} = \frac{1}{2}\left(\tilde{\psi}\gamma^k\varphi - \tilde{\varphi}\gamma^k\psi\right),$$

where we used the notation $D_k(\psi) = \psi_{,k}$, $D_k(\tilde{\psi}) = \tilde{\psi}_{,k}$. Substituting these quantities and the above expression for B^k into (7.3.20), we arrive at the conserved vector

$$C_\varphi^k = \tilde{\psi}\gamma^k\varphi(x) - \tilde{\varphi}(x)\gamma^k\psi.$$

Problems to Chapter 7

7.1. Find the transformations of the dilation group with the generator Z from (7.3.44):

$$Z = 3t\frac{\partial}{\partial t} + 2x^i\frac{\partial}{\partial x^i}.$$

7.2. Find all infinitesimal symmetries for:

(i) one-dimensional heat equation $u_t = u_{xx}$,

(ii) two-dimensional heat equation $u_t = u_{xx} + u_{yy}$,

(iii) three-dimensional heat equation $u_t = u_{xx} + u_{yy} + u_{zz}$.

7.3. Find the invariant solutions of the one-dimensional heat equation $u_t = u_{xx}$ obtained by using the infinitesimal symmetry (see Problem 7.2(i))

$$X = X_1 + kX_4 = \frac{\partial}{\partial t} + ku\frac{\partial}{\partial u}, \qquad k = \text{const.}$$

7.4. Investigate the invariant solutions of the two-dimensional heat equation $u_t = u_{xx} + u_{yy}$ obtained by using:

(i) one infinitesimal symmetry (cf. Problem 7.3)

$$X = \frac{\partial}{\partial t} + ku\frac{\partial}{\partial u}, \quad k = \text{const.},$$

(ii) the two-dimensional Lie algebra spanned by (see Problem 7.2(ii))

$$X = \frac{\partial}{\partial t} + ku\frac{\partial}{\partial u}, \quad Y = X_6 = y\frac{\partial}{\partial x} - x\frac{\partial}{\partial y}.$$

7.5. The motion of a planet around the sun is governed by the system of differential equations (2.2.8):

$$m\frac{d^2\boldsymbol{x}}{dt^2} = \frac{\alpha}{r^3}\boldsymbol{x}, \quad \alpha = \text{const.}$$

Show that the energy of the planet defined by

$$E = \frac{m}{2}|\boldsymbol{v}|^2 + \frac{\alpha}{r}, \quad \text{where} \quad |\boldsymbol{v}|^2 = \sum_{i=1}^{3}(v^i)^2,$$

is a constant of motion, i.e., $dE/dt = 0$.

7.6. The motion of a particle with mass m in an arbitrary central potential field $U = U(r)$ is described by the Lagrangian $L = \frac{m}{2}|\boldsymbol{v}|^2 - U(r)$. For this Lagrangian, find the equations of motion, that is, the Euler–Lagrange equations.

7.7. The action integral with the Lagrangian of the previous problem, $L = \frac{m}{2}|\boldsymbol{v}|^2 - U(r)$, is invariant under the translation in time t with the generator $X_1 = \partial/\partial t$ and the rotations of the spatial variables x^1, x^2, x^3 with the generators

$$X_{12} = x^2\frac{\partial}{\partial x^1} - x^1\frac{\partial}{\partial x^2}, \quad X_{13} = x^3\frac{\partial}{\partial x^1} - x^1\frac{\partial}{\partial x^3}, \quad X_{23} = x^3\frac{\partial}{\partial x^2} - x^2\frac{\partial}{\partial x^3}.$$

Calculate the corresponding conservation laws given by Noether's theorem and compare with Problem 2.4.

7.8. Consider the central potential field of the form

$$U(r) = \frac{k}{r^2}, \quad k = \text{const.},$$

and the corresponding equations of motion of a particle:

$$m\frac{d^2x^i}{dt^2} = 2k\frac{x^i}{r^4}, \quad i = 1, 2, 3.$$

These equations admit, along with the time translation and the rotations of the spatial variables (see Problem 7.7), the dilation and the projective transformation generated by

$$X_5 = 2t\frac{\partial}{\partial t} + x^1\frac{\partial}{\partial x^1} + x^2\frac{\partial}{\partial x^2} + x^3\frac{\partial}{\partial x^3}$$

and

$$X_6 = t^2\frac{\partial}{\partial t} + t\left(x^1\frac{\partial}{\partial x^1} + x^2\frac{\partial}{\partial x^2} + x^3\frac{\partial}{\partial x^3}\right),$$

respectively. Examine the applicability of the Noether theorem to the symmetries X_5, X_6 and find the corresponding conservation laws T_5, T_6.

7.9. Let a be a three-dimensional vector field. Verify that curl a is solenoidal, i.e., div curl $a = 0$ (see the property 9 in (1.3.15)).

7.10. Let $x = (x, y, z)$. Show that

$$\operatorname{div}\left(\frac{x}{r^3}\right) = 0.$$

Generalize this property to higher dimensions. Namely, let $x = (x^1, \ldots, x^n)$ and $r = \sqrt{(x^1)^2 + \cdots + (x^n)^2}$. Find s such that

$$\operatorname{div}\left(\frac{x}{r^s}\right) \equiv \sum_{i=1}^{n}\frac{\partial}{\partial x^i}\left(\frac{x}{r^s}\right) = 0.$$

7.11. Check that the vector

$$B = \left(\frac{x}{x^2 + y^2}, \frac{y}{x^2 + y^2}, 0\right)$$

is solenoidal and find a vector A such that $B = \operatorname{curl} A$.

7.12. Return to Problem 7.10 and find a vector A such that

$$\frac{x}{r^3} = \operatorname{curl} A.$$

7.13. Examine the generator Z from Problem 7.1 for applicability of the conservation theorems, i.e., check properties (7.3.30) and (7.3.33).

7.14. Find the first prolongation (i.e., extension to $v^k = dx^k/dt$) of the following operator from Section 7.3.7:

$$\widehat{X}_1 = -\left(x^2v^2 + x^3v^3\right)\frac{\partial}{\partial x^1} + \left(2x^1v^2 - x^2v^1\right)\frac{\partial}{\partial x^2} + \left(2x^1v^3 - x^3v^1\right)\frac{\partial}{\partial x^3}.$$

7.15. Transform the Black-Scholes equation (2.4.15),

$$u_t + \frac{1}{2}A^2x^2u_{xx} + Bxu_x - Cu = 0,$$

into an equation with constant coefficients by the change of the variable x into $y = \ln|x|$.

7.16. Solve the following over-determined system (four equations for two dependent variables $\xi(x,y)$ and $\eta(x,y)$) used in Section 6.5.1, Example 6.5.2:

$$\xi_y = 0, \quad 3(\eta_y + \eta) - 2\xi_x = 0, \quad \eta_x = 0, \quad \xi_{xx} = 0.$$

Chapter 8

Generalized functions or distributions

This chapter provides an easy to follow introduction to basic concepts and methods of the distribution theory with emphasis on useful tools. Furthermore, the invariance principle used in Chapter 9 for calculating fundamental solutions invites a reconstruction of the theory of group invariant solutions and an extension of Lie's infinitesimal technique to the space of distributions. This is done in Section 8.4.

Additional reading: L. Schwartz [34], I.M. Gel'fand and G.E. Shilov [8], N.H. Ibragimov [17].

8.1 Introduction of generalized functions

Modern developments in applied mathematics, in particular investigations of nonlinear problems in fluid mechanics, necessitated discontinuous solutions for differential equations. Therefore, S.L. Sobolev introduced in the 1930s the so-called *generalized solutions*. Moreover, even earlier the observation was made that discontinuous solutions with certain singularities play a major part in tackling problems of mathematical physics. J. Hadamard in the 1920s termed these solutions *elementary;* at present they are mostly referred to as *fundamental solutions*. It was the endeavour to perceive the mathematical nature of these solutions that gave birth to the modern theory of *generalized functions* (S.L. Sobolev, 1936) or *distributions* (L. Schwartz, 1950). The most useful generalized function is Dirac's δ-function. It was introduced in theoretical physics by P.A.M. Dirac in the 1930s and became one of efficient tools in the general theory of differential equations.

8.1.1   Heuristic considerations

Recall that the definition of a differentiable function implies that its differentiation results in a classical (usual) function. One might still attempt to differentiate non-differentiable functions generalizing the notion of differentiation. This is one of possible ways for introducing generalized functions. The crucial idea in this approach is to transfer differentiation from a non-differentiable function to a differentiable one. An appropriate tool for implementing this idea is provided by integration by parts for functions of a single variable and the divergence theorem (1.3.18) for functions of several variables.

Let us consider the case of a single variable x. Provided that functions $u(x)$ and $\varphi(x)$ are continuously differentiable in $a \leq x \leq b$, formula (1.2.12) of integration by parts yields

$$\int_a^b \varphi \mathrm{d}u = (u\varphi)\Big|_a^b - \int_a^b u\mathrm{d}\varphi.$$

Furthermore, assuming that the function $\varphi(x)$ vanishes outside of a bounded interval[1] of the x-axis and designating differentiation with respect to x by D one obtains

$$\int_{-\infty}^{+\infty} \varphi D(u)\mathrm{d}x = -\int_{-\infty}^{+\infty} uD(\varphi)\mathrm{d}x.$$

For the sake of convenience one interprets the integrals involved here as scalar products $(\,,\,)$ of the corresponding functions. Then, the above equation is written in the form:

$$(Du, \varphi) = -(u, D\varphi). \tag{8.1.1}$$

Reading the above equation from left to right one can interpret it merely as a means of transfer of the differentiation D from a differentiable function $u(x)$ to another function of this type, $\varphi(x)$. However, one arrives at a non-trivial conclusion reading it otherwise and assuming that only the function φ is continuously differentiable. The function u is subject to the condition of convergence of the integral in the right-hand side of Eq. (8.1.1).

The above approach is a decisive step for introducing a new differentiation. Namely, let u be a function that might be non-differentiable in the classical sense but satisfying the convergence condition of the integral in the right-hand side of Eq. (8.1.1). A *generalized derivative* of the function u is such a "function" Du (in a generalized sense) that satisfies Eq. (8.1.1) for any continuously differentiable function φ with a bounded support. It is evident that the generalized differentiation can be iterated till the function φ is differentiable. For instance, the generalized derivative D^2u of the second order is obtained as follows:

$$(D^2u, \varphi) = -(Du, D\varphi) = (u, D^2\varphi).$$

---

[1]The smallest closed interval, outside of which the function $\varphi(x)$ vanishes, is termed the *support* of this function and is denoted by supp(φ). If the interval supp(φ) is bounded, then $\varphi(x)$ is referred to as a function with a bounded (compact) support.

In order to be able to deal with derivatives of any order let us assume that the function φ is infinitely differentiable. The set of infinitely differentiable functions with a bounded support is denoted by C_0^∞. The functions $\varphi \in C_0^\infty$ are referred to as *test functions*.

To understand the nature of the generalized derivative Du one should take into account that for a fixed u the expression (Du, φ) maps every function $\varphi \in C_0^\infty$ into a number equal to $(u, D\varphi)$. This operation is linear:

$$(Du, c_1\varphi_1 + c_2\varphi_2) = c_1(Du, \varphi_1) + c_2(Du, \varphi_2); \quad c_1, c_2 = \text{const.},$$

and continuous:

$$(Du, \varphi_k) \to (Du, \varphi) \quad \text{whenever} \quad \varphi_k \to \varphi \text{ in } C_0^\infty.$$

Thus, the generalized derivative Du is a *linear continuous functional* over the space C_0^∞. These heuristic considerations can be also applied to functions in several variables $x = (x^1, \dots, x^n) \in \mathbb{R}^n$ and entail the following definition.

8.1.2 Definition and examples of distributions

Definition 8.1.1. A *generalized function* or *distribution* is a linear continuous functional f over the space C_0^∞ of C^∞ functions $\varphi(x) = \varphi(x^1, \dots, x^n)$ with compact support. In other words, f maps any function $\varphi \in C_0^\infty$ into a number denoted by (f, φ). This operation is *linear*:

$$(f, \varphi_1 + \varphi_2) = (f, \varphi_1) + (f, \varphi_2), \quad (f, c\varphi) = c(f, \varphi); \quad c = \text{const.}, \quad (8.1.2)$$

and *continuous*:

$$\text{if } \varphi_k \to \varphi \text{ in } C_0^\infty, \quad \text{then } (f, \varphi_k) \to (f, \varphi). \quad (8.1.3)$$

The convergence $\varphi_k \to \varphi$ in C_0^∞ means that the following conditions are satisfied:

1. $\varphi_k, \varphi \in C_0^\infty$,

2. the supports of all members of the sequence φ_k are contained in one and the same bounded closed subset of $\mathbb{R}^n$,

3. the functions $\varphi_k(x)$ converge uniformly to $\varphi(x)$ together with all their derivatives.

Example 8.1.1. Let a function $f(x)$ be locally integrable, i.e., integrable in any bounded domain of $\mathbb{R}^n$. The integral

$$(f, \varphi) = \int_{\mathbb{R}^n} f(x)\varphi(x)\mathrm{d}x \quad (8.1.4)$$

defines a generalized function. The generalized functions of this type are said to be *regular* and the others are termed *singular*.

Example 8.1.2. The *Heaviside function* $\theta(x)$ in one variable x is

$$\theta(x) = \begin{cases} 0, x < 0, \\ 1, x > 0. \end{cases} \tag{8.1.5}$$

It determines the regular generalized function (8.1.4) defined as follows:

$$(\theta, \varphi) = \int_0^{+\infty} \varphi(x)\mathrm{d}x. \tag{8.1.6}$$

Likewise, the Heaviside function $\theta(x - x_0)$:

$$\theta(x - x_0) = \begin{cases} 0, x < x_0, \\ 1, x > x_0. \end{cases}$$

defines the regular generalized function by the integral formula

$$(\theta(x - x_0), \varphi(x)) = \int_{x_0}^{+\infty} \varphi(x)\mathrm{d}x.$$

Example 8.1.3. *Dirac's δ-function* is the simplest but extremely important singular generalized function. It is denoted by δ or $\delta(x)$ and is defined by the formula:

$$(\delta, \varphi) = \varphi(0). \tag{8.1.7}$$

This formula is also written as

$$(\delta(x), \varphi(x)) = \varphi(0),$$

where $x \in \mathbb{R}^n$. Likewise, $\delta(x - x_0)$ is defined by the formula:

$$(\delta(x - x_0), \varphi(x)) = \varphi(x_0).$$

The alternative notation for $\delta(x - a)$ is $\delta_{(a)}$. It is often referred to as the δ-function at the point a.

8.1.3 Representations of the δ-function as a limit

The derivation of the following useful representations of the δ-function can be found, e.g. in [8], Chapter I, Section 2.5.

Theorem 8.1.1. Consider the case of a single variable x. The following equations hold:

$$\delta(x) = \lim_{\varepsilon \to 0} \frac{\varepsilon}{\pi(x^2 + \varepsilon^2)}, \tag{8.1.8}$$

$$\delta(x) = \lim_{\nu \to \infty} \frac{\sin(\nu x)}{\pi x}, \tag{8.1.9}$$

$$\delta(x) = \lim_{\nu \to \infty} \frac{1}{2\pi} \int_{-\nu}^{\nu} e^{i\xi x} d\xi, \tag{8.1.10}$$

$$\delta(x) = \lim_{t \to +0} \frac{1}{2\sqrt{\pi t}} e^{-\frac{x^2}{4t}}, \tag{8.1.11}$$

where the symbol $t \to +0$ means that t tends to zero assuming positive values. The convergence $f_n \to \delta$ of distributions is defined by

$$(f_n, \varphi) \to (\delta, \varphi).$$

We will use also the extension of (8.1.11) to the case of several variables when $x = (x^1, \ldots, x^n)$. Denoting $r = \sqrt{(x^1)^2 + \cdots + (x^n)^2}$, we have

$$\lim_{t \to +0} \frac{1}{(2\sqrt{\pi t})^n} e^{-\frac{r^2}{4t}} = \delta(x). \tag{8.1.12}$$

8.2 Operations with distributions

8.2.1 Multiplication by a function

Multiplication of a distribution f by a C^∞ function $\alpha(x)$ is defined by the action of the product αf on test functions φ as follows:

$$(\alpha f, \varphi) = (f, \alpha\varphi). \tag{8.2.1}$$

The right-hand side of this equation is well defined since $\alpha\varphi \in C_0^\infty$.

Example 8.2.1. It follows from definition (8.1.7) of the δ-function and the multiplication rule (8.2.1) that

$$\alpha(x)\delta = \alpha(0)\delta. \tag{8.2.2}$$

Indeed,

$$(\alpha(x)\delta, \varphi) = (\delta, \alpha\varphi) = \alpha(0)\varphi(0) = \alpha(0)(\delta, \varphi) = (\alpha(0)\delta, \varphi).$$

8.2.2 Differentiation

Differentiation of a distribution f is defined by the equation

$$\left(D_i f, \varphi\right) = -\left(f, D_i \varphi\right). \tag{8.2.3}$$

Here the total differentiation D_i coincides with the partial differentiation, $D_i = \partial/\partial x^i$, since φ depends on the independent variables x^i only. This definition manifests that any generalized function is infinitely differentiable, the higher-order derivatives being obtained by successive application of formula (8.2.3). For instance, $(D_j D_i f, \varphi) = -(D_i f, D_j \varphi) = (f, D_i D_j \varphi)$. It follows in particular that $D_j D_i f = D_i D_j f$.

Example 8.2.2. The derivative of the Heaviside function is the δ-function:

$$\theta'(x) = \delta(x). \tag{8.2.4}$$

Indeed, formulae (8.1.1) and (8.1.6) yield

$$(\theta', \varphi) = -(\theta, \varphi') = -\int_0^\infty \varphi'(x)\mathrm{d}x = -\varphi\Big|_0^\infty = \varphi(0) = (\delta, \varphi).$$

8.2.3 Direct product of distributions

Let $f(x)$ and $g(t)$ be two distributions acting on functions of n variables $x = (x^1, \ldots, x^n)$ and m variables $t = (t^1, \ldots, t^m)$, respectively. Let $\varphi(x,t)$ be a test function depending on $n+m$ variables $(x^1, \ldots, x^n, t^1, \ldots, t^m)$.

Definition 8.2.1. The *direct product* $f(x) \otimes g(t)$ of $f(x)$ and $g(t)$ is the distribution acting on test functions $\varphi(x,t)$ as follows:

$$\left(f(x) \otimes g(t), \varphi(x,t)\right) = \left(f(x), \left(g(t), \varphi(x,t)\right)\right). \tag{8.2.5}$$

One can readily derive from Definition 8.2.1 the following properties of the direct product (see, e.g. [8], Chapter 1, §5, or [34], Chapter 3):

$$\left(f(x) \otimes g(t), \varphi(x)\psi(t)\right) = (f(x), \varphi(x))(g(t), \psi(t)), \tag{8.2.6}$$

$$f(x) \otimes g(t) = g(t) \otimes f(x), \tag{8.2.7}$$

$$\delta(x) \otimes \delta(t) = \delta(x,t), \tag{8.2.8}$$

where $\delta(x), \delta(t)$ and $\delta(x,t)$ are Dirac's δ-functions of the respective variables.

8.2.4 Convolution

The *convolution* $(f * g)(x)$ of two usual functions $f(x)$ and $g(x)$ in $\mathbb{R}^n$ is a function defined by the following integral:

$$(f * g)(x) = \int_{\mathbb{R}^n} f(y)g(x - y)\mathrm{d}y. \qquad (8.2.9)$$

In order to define the convolution of distributions let us begin with regular distributions. Namely, we consider action (8.1.4) of convolution (8.2.9) on test functions and change the order of integration. We have

$$(f * g, \varphi) = \int (f * g)(x)\varphi(x)\mathrm{d}x = \int \varphi(x) \left[\int f(y)g(x - y)\mathrm{d}y \right] \mathrm{d}x$$

$$= \int f(y)\mathrm{d}y \int g(x - y)\varphi(x)\mathrm{d}x = \int f(y) \left[\int g(z)\varphi(z + y)\mathrm{d}z \right] \mathrm{d}y.$$

Denoting z in the last integral by x, we conclude that the convolution $f * g$ of regular distributions f and g is a distribution acting on test functions as follows:

$$(f * g, \varphi) = \int f(y) \left[\int g(x)\varphi(x + y)\mathrm{d}x \right] \mathrm{d}y. \qquad (8.2.10)$$

Equation (8.2.10) can be written in the form

$$(f * g, \varphi) = (f(y), (g(x), \varphi(x + y))).$$

This preliminary consideration entails the following definition.

Definition 8.2.2. Let f and g be any distributions such that at least one of them has a compact support. Their convolution $f * g$ is a distribution determined by

$$(f * g, \varphi) = (f(y), (g(x), \varphi(x + y))). \qquad (8.2.11)$$

Using definition (8.2.5) of the direct product, Eq. (8.2.11) can be written

$$(f * g, \varphi) = (f(y) \otimes g(x), \varphi(x + y)). \qquad (8.2.12)$$

Remark 8.2.1. Even though $\varphi(x)$ has a compact support, $\varphi(x + y)$ will not have this property in the space of the variables x, y. Consequently, Eq. (8.2.12) may have no meaning for arbitrary distributions f and g. The convolution is well defined for distributions obeying certain conditions. The compactness of the support of f or g is one of such conditions. We will not discuss here the general restrictions since Definition 8.2.2 is sufficient for our purposes. For the definition of the *support* of distributions, see, e.g. [34], Ch. II, §1.3, or [8], §1.4.

The following properties of the convolution are used in the next chapter.

Theorem 8.2.1. Convolution (8.2.11) is commutative:

$$f * g = g * f. \tag{8.2.13}$$

Proof. The statement follows from Eq. (8.2.12) and commutativity (8.2.7) of the direct product.

Theorem 8.2.2. The convolution with the δ-function exists for any distribution f and satisfies the following equation:

$$f * \delta = f. \tag{8.2.14}$$

Theorem 8.2.3. Let the distributions f and g have compact supports. Then the differentiation $D_i = \partial/\partial x^i$ of the convolution satisfies the following equations:

$$D_i(f * g) = (D_i f) * g = f * (D_i g). \tag{8.2.15}$$

8.3   The distribution $\Delta(r^{2-n})$

8.3.1   The mean value over the sphere

Let us introduce the notation and make preliminary calculations. Let

$$x = (x^1, \ldots, x^n) \in \mathbb{R}^n, \quad |x| = \sqrt{(x^1)^2 + \cdots + (x^n)^2}.$$

The mean value $\overline{\varphi}(r)$ of a function $\varphi(x)$ over the sphere Ω_r of radius r and center 0 is defined by

$$\overline{\varphi}(r) = \frac{1}{S_r} \int_{\Omega_r} \varphi(x) dS. \tag{8.3.1}$$

The sphere Ω_r is the set of the points x such that $|x| = r$, and S_r is the surface area of the sphere Ω_r. Since Ω_r is similar to the unit sphere and has the dimension $n - 1$, we have (cf. Section 1.1.3):

$$S_r = r^{n-1} \omega_n, \tag{8.3.2}$$

where $\omega_n = 2\sqrt{\pi^n}/\Gamma(n/2)$ is the surface area of the unit sphere in the n-dimensional space. It follows from Eq. (8.3.1) that $\overline{\varphi}(r)$ has one and the same value at each point $x \in \Omega_r$ independently of the position of x on the sphere Ω_r. Therefore, $\overline{\varphi}(r)$ is said to be a *spherically symmetric function*.

8.3.2   Solution of the Laplace equation $\Delta v(r) = 0$

Lemma 8.3.1. Let v be any spherically symmetric function, i.e., $v = v(r)$. Then the Laplace equation

$$\Delta v \equiv \sum_{i=1}^{n} v_{ii} = 0,$$

where $v_i = D_i(v) = \partial v / \partial x^i$, $v_{ii} = D_i^2(v)$, is written:

$$v'' + \frac{n-1}{r} v' = 0, \qquad (8.3.3)$$

where $v' = dv/dr$.

Proof. The statement follows from the equations

$$D_i(v) \equiv \frac{\partial v}{\partial x^i} = v' \frac{x^i}{r}, \quad D_i^2(v) = v'' \frac{(x^i)^2}{r^2} + v' \left[\frac{1}{r} - \frac{(x^i)^2}{r^3} \right].$$

Remark 8.3.1. One can prove that the mean value of the Laplacian of a function $\varphi(x)$ is identical with the Laplacian of the mean value of $\varphi(x)$, i.e.,

$$\overline{\Delta\varphi}(r) = \Delta\overline{\varphi}(r) = \overline{\varphi}'' + \frac{n-1}{r} \overline{\varphi}'. \qquad (8.3.4)$$

Let us find all spherically symmetric solutions $v = v(r)$ of the Laplace equation, i.e., integrate the ordinary differential equation (8.3.3). We have

$$v'' + \frac{n-1}{r} v' = \frac{1}{r} \left[rv'' + (n-1) v' \right]$$

$$= \frac{1}{r} \left[(rv')' + (n-2) v' \right] = \frac{1}{r} \left[rv' + (n-2) v \right]'.$$

Hence, Eq. (8.3.3) is written $[rv' + (n-2) v]' = 0$ and yields

$$rv' + (n-2) v = C. \qquad (8.3.5)$$

If $n > 2$, we set $C = (n-2)C_1$ and rewrite Eq. (8.3.5) in the separable form $rv' + (n-2)(v - C_1) = 0$. If $n = 2$, Eq. (8.3.5) has the separable form $rv' = C$. If $n = 1$, Eq. (8.3.5) is written $v'' = 0$. The integration is simple in all cases and yields the following.

Theorem 8.3.1. The general solution to Eq. (8.3.3) has the form

$$v = C_1 + C_2 r^{2-n}, \quad \text{if} \quad n \neq 2; \qquad (8.3.6)$$

$$v = C_1 + C_2 \ln r, \quad \text{if} \quad n = 2. \qquad (8.3.7)$$

8.3.3 Evaluation of the distribution $\Delta(r^{2-n})$

Theorem 8.3.1 shows that the fundamental system of solutions for Eq. (8.3.3) comprises the trivial constant solution, e.g. $v = 1$ and the solutions r^{2-n} and $\ln r$ for $n > 2$ and $n = 2$, respectively. The functions r^{2-n} and $\ln r$ have a singularity at $r = 0$, i.e., at the origin $x = 0$, and do not have classical derivatives at the singular point. In consequence, the Laplacian of these functions annuls at $x \neq 0$ and provides distributions near the singular point $x = 0$. We will evaluate here the distribution $\Delta(r^{2-n})$ when $n > 2$.

Theorem 8.3.2. Let $n > 2$. Then $\Delta(r^{2-n})$ is the distribution given by

$$\Delta(r^{2-n}) = (2 - n)\omega_n \,\delta(x), \qquad (8.3.8)$$

where $\delta(x)$ is Dirac's δ-function defined by Eq. (8.1.7) and ω_n is the surface area of the unit sphere.

Proof. It is manifest that the function r^{2-n} is locally integrable. Therefore, invoking the differentiation rule (8.2.3) and definition (8.1.4) of distributions determined by locally integrable functions, and then applying the equation

$$\int_{\mathbb{R}^n} f(x)dx = \int_0^\infty dr \int_{\Omega_r} f \, dS \qquad (8.3.9)$$

we obtain

$$(\Delta(r^{2-n}), \varphi) = (r^{2-n}, \Delta\varphi) = \int_{\mathbb{R}^n} r^{2-n} \Delta\varphi \, dx = \int_0^\infty dr \int_{\Omega_r} r^{2-n} \Delta\varphi \, dS.$$

Equations (8.3.1) and (8.3.2) yield

$$\int_{\Omega_r} r^{2-n} \Delta\varphi \, dS = r^{2-n} S_r \overline{\Delta\varphi}(r) = \omega_n r^{2-n} r^{n-1} \overline{\Delta\varphi}(r) = \omega_n r \overline{\Delta\varphi}(r).$$

Collecting the above equations and using Eq. (8.3.4), we have

$$(\Delta(r^{2-n}), \varphi) = \omega_n \int_0^\infty r \overline{\Delta\varphi}(r)dr = \omega_n \int_0^\infty [r\overline{\varphi}'' + (n-1)\overline{\varphi}']dr. \qquad (8.3.10)$$

Now, we evaluate the integral in the right-hand side of Eq. (8.3.10):

$$\int_0^\infty [r\overline{\varphi}'' + (n-1)\overline{\varphi}']dr = \int_0^\infty [(r\overline{\varphi}')' + (n-2)\overline{\varphi}']dr = \left[r\overline{\varphi}' + (n-2)\overline{\varphi}\right]_0^\infty.$$

Since $\varphi \in C_0^\infty$, it vanishes at the infinity. Therefore,

$$\left[r\overline{\varphi}' + (n-2)\overline{\varphi}\right]_0^\infty = -(n-2)\overline{\varphi}(0) = -(n-2)\varphi(0).$$

Substituting this expression into Eq. (8.3.10), we obtain

$$(\Delta(r^{2-n}), \varphi) = (2-n)\omega_n\varphi(0) = ((2-n)\omega_n\delta(x), \varphi(x)),$$

thus proving Eq. (8.3.8).

Remark 8.3.2. When $n = 2$, the following equation is valid instead of Eq. (8.3.8):

$$\Delta(\ln r) = \omega_2 \,\delta(x). \qquad (8.3.11)$$

Remark 8.3.3. Equation (8.3.8) holds also in the case $n = 1$. Thus, Eqs. (8.3.8) and (8.3.11) are encapsulated in the following general formula:

$$\Delta(r^{2-n}) = (2 - n)\frac{2\pi^{n/2}}{\Gamma(n/2)}\,\delta(x), \quad n \neq 2,$$

$$\Delta(\ln r) = \frac{2\pi^{n/2}}{\Gamma(n/2)}\,\delta(x), \quad n = 2. \tag{8.3.12}$$

The following particular cases are significant for applications:

$$\Delta(|x|) = 2\,\delta(x), \quad n = 1,$$

$$\Delta(\ln r) = 2\pi\delta(x), \quad n = 2, \tag{8.3.13}$$

$$\Delta(r^{-1}) = -4\pi\delta(x), \quad n = 3.$$

8.4 Transformations of distributions

The present section contains an extension of Lie's infinitesimal technique to the space of distributions. The results are employed in the next chapter.

8.4.1 Motivation by linear transformations

Consider, for the simplicity sake, the one-dimensional case and begin with the widely known transformation termed the *shift* of distributions. It corresponds to the translation $\bar{x} = x - a$ of the independent variable and is defined in accordance with the following transformation law of regular distributions (8.1.4). Let $f(x)$ be a locally integrable function of the single variable x. The shift $f(x - a)$ of the corresponding regular distribution is given by the formula of change of variables:

$$\Big(f(x - a), \varphi(x)\Big) = \int f(x - a)\varphi(x)\mathrm{d}x = \int f(\bar{x})\varphi(\bar{x} + a)\mathrm{d}\bar{x}.$$

Using here notation (8.1.4), one has

$$\Big(f(x - a), \varphi(x)\Big) = \Big(f(\bar{x}), \varphi(\bar{x} + a)\Big) \tag{8.4.1}$$

or, denoting $\bar{x}$ by x again, one obtains the following shift formula:

$$\Big(f(x - a), \varphi(x)\Big) = \Big(f(x), \varphi(x + a)\Big). \tag{8.4.2}$$

Equation (8.4.2) is used as the definition of the shift for arbitrary distributions. For example, the shift $\delta(x - a)$ of the δ-function has the form

$$\Big(\delta(x - a), \varphi(x)\Big) = \Big(\delta(x), \varphi(x + a)\Big) = \varphi(a). \tag{8.4.3}$$

An arbitrary linear transformation of generalized functions is defined in a similar way. Namely, let $f(x)$ be a locally integrable function in $\mathbb{R}^n$, and let

$$\overline{x} = Ax - a$$

be the general linear transformation, where $a = (a^1, \ldots, a^n)$ is an n-dimensional vector and $A = (a_{ij})$ is an arbitrary $n \times n$ matrix such that $\det A \neq 0$. Then, the change of variables in integral (8.1.4) yields

$$\left(f(Ax - a), \varphi(x) \right) = \int f(Ax - a)\varphi(x)dx = |\det A|^{-1} \int f(\overline{x})\varphi[A^{-1}(\overline{x} + a)]d\overline{x}.$$

This equation defines the linear transformation of arbitrary distributions:

$$\left(f(Ax - a), \varphi(x) \right) = \left(|\det A|^{-1} f(\overline{x}), \varphi[A^{-1}(\overline{x} + a)] \right), \tag{8.4.4}$$

or, upon denoting $\overline{x}$ by x :

$$\left(f(Ax - a), \varphi(x) \right) = \left(|\det A|^{-1} f(x), \varphi[A^{-1}(x + a)] \right). \tag{8.4.5}$$

8.4.2 Change of variables in the δ-function

Let a function $p(x)$ of one variable x be such that $p(0) = 0$, $p'(0) \neq 0$. Then

$$\delta(p(x)) = \frac{1}{p'(0)}\delta(x). \tag{8.4.6}$$

If $p(x)$ has several zeros, e.g. at points $a_1, \ldots, a_s$ and if $p'(a_\sigma) \neq 0$ for all $\sigma = 1, 2, \ldots, s$, then (see, e.g. [4], Ch. VI, §3.3, Eq. (2)):

$$\delta(p(x)) = \sum_{\sigma=1}^{s} \frac{1}{p'(a_\sigma)} \delta(x - a_\sigma). \tag{8.4.7}$$

Example 8.4.1. Application of (8.4.7) to $\delta(x^2 - a^2)$, $a \neq 0$, yields

$$\delta(x^2 - a^2) = \frac{1}{2a}\left[\delta(x - a) - \delta(x + a) \right]. \tag{8.4.8}$$

In the case of several variables $x = (x^1, \ldots, x^n)$, Eq. (1.2.52) yields

$$\delta(p(x)) = \frac{1}{|J(0)|} \delta(x), \quad J(0) = \det\left(\frac{\partial p^i}{\partial x^j} \right)_{x=0}. \tag{8.4.9}$$

8.4.3 Arbitrary group transformations

Consider an arbitrary transformation T of points $x \in \mathbb{R}^n$ into $\overline{x} \in \mathbb{R}^n$:

$$\overline{x} = T x. \tag{8.4.10}$$

The transformation T is written in coordinates in the form

$$\bar{x}^i = \bar{x}^i(x), \quad i = 1, \ldots, n,$$

and its Jacobian is

$$J = \det\left(\frac{\partial \bar{x}^i}{\partial x^j}\right). \tag{8.4.11}$$

We assume that $J \neq 0$ and that $\bar{x}^i(x)$ are infinitely differentiable.

Definition 8.4.1. The transformation $f \to \bar{f}$ of an arbitrary distribution f into a distribution $\bar{f}$ associated with (8.4.10) is defined by the following equation:

$$\Big(f(Tx), \varphi(x)\Big) = \Big(\bar{f}(\bar{x}), \varphi(T^{-1}\bar{x})\Big). \tag{8.4.12}$$

Definition 8.4.1 is motivated by the linear transformations. Namely, the shift formula (8.4.1) has form (8.4.12) with $Tx = x - a$ and $\bar{f} = f$. Likewise, transformation (8.4.4) has form (8.4.12) with $Tx = Ax - a$ and $\bar{f} = |\det A|^{-1} f$.

In order to obtain the transformation of distributions associated with (8.4.10), we repeat the reasoning used in the case of linear transformations. Namely, let $f(x)$ be a locally integrable function in $\mathbb{R}^n$. The formula of change of variables in integrals yields

$$\Big(f(Tx), \varphi(x)\Big) = \int f(Tx)\varphi(x)dx = \int J^{-1}f(\bar{x})\,\varphi(T^{-1}\bar{x})d\bar{x}.$$

This motivates the following equation for *arbitrary distributions:*

$$\Big(f(Tx), \varphi(x)\Big) = \Big(J^{-1}f(\bar{x}),\, \varphi(T^{-1}\bar{x})\Big), \tag{8.4.13}$$

where J is Jacobian (8.4.11). Equation (8.4.13) can be also written, upon denoting $\bar{x}$ in its right-hand side by x again, as follows:

$$\Big(f(Tx), \varphi(x)\Big) = \Big(J^{-1}f(x), \varphi(T^{-1}x)\Big). \tag{8.4.14}$$

Comparison of Eq. (8.4.13) with Eq. (8.4.12) defines the following transformation of arbitrary distributions:

$$\bar{f} = J^{-1}f. \tag{8.4.15}$$

Example 8.4.2. For the scaling transformation $\bar{x} = xe^a$ in $\mathbb{R}^n$, formula (8.4.15) yields

$$\bar{f} = e^{-na} f. \tag{8.4.16}$$

Example 8.4.3. Consider the two-dimensional case and take the rotation group

$$\bar{x} = x\cos a + y\sin a, \quad \bar{y} = y\cos a - x\sin a.$$

Formula (8.4.15) yields

$$\bar{f} = f. \tag{8.4.17}$$

8.4.4 Infinitesimal transformation of distributions

Consider a one-parameter group G of transformations in $\mathbb{R}^n$:

$$\overline{x} = T_a(x), \qquad\qquad (8.4.18)$$

with the infinitesimal transformation

$$\overline{x}^i \approx x^i + a\xi^i(x), \quad i = 1, \ldots, n. \qquad\qquad (8.4.19)$$

According to Eq. (8.4.15), we extend the action of the group to distributions f as follows:

$$\overline{x} = T_a(x), \quad \overline{f} = J^{-1} f. \qquad\qquad (8.4.20)$$

Here, Jacobian (8.4.11), $J = \det(\partial \overline{x}^i / \partial x^j)$, is positive when the group parameter a assumes small values.

Expanding transformations (8.4.20) into Taylor's series in a near $a = 0$ and taking into account that $J = 1$ at $a = 0$ one obtains, along with the infinitesimal transformation of x given by (8.4.19), the following infinitesimal transformation of f :

$$\overline{f} \approx f - a f \left[\frac{\mathrm{d}J}{\mathrm{d}a}\Big|_{a=0} \right].$$

Applying the rule for differentiating determinants (see Section 1.3.6) to Jacobian (8.4.11), one obtains the following equation (see Problem 8.7):

$$\frac{\mathrm{d}J}{\mathrm{d}a}\Big|_{a=0} = D_i(\xi^i), \qquad\qquad (8.4.21)$$

where

$$D_i(\xi^i) = \sum_{i=1}^{n} \frac{\partial \xi^i}{\partial x^i} .$$

Thus, the infinitesimal transformation (8.4.19) of the independent variables is accompanied by the following infinitesimal transformation of distributions:

$$\overline{f} \approx f - a D_i(\xi^i) f. \qquad\qquad (8.4.22)$$

In particular, letting $f(x) = \delta(x)$ and invoking (8.2.2), one obtains the infinitesimal transformation of the δ-function:

$$\overline{\delta} \approx \delta - a \left[D_i(\xi^i)\big|_{x=0} \right] \delta. \qquad\qquad (8.4.23)$$

Problems to Chapter 8

8.1. Let x be a single variable. Find the action of the derivatives of the δ-function on test functions. Namely, evaluate

    (i) $(\delta'(x), \varphi(x))$,   (ii) $(\delta''(x), \varphi(x))$,   (iii) $(\delta'(x - x_0), \varphi(x))$.

8.2. Evaluate the following generalized functions:

(i) $\lim\limits_{\varepsilon \to 0} \dfrac{2\varepsilon x}{\pi(x^2 + \varepsilon^2)^2}$,   (ii) $\lim\limits_{\nu \to \infty} \displaystyle\int_{-\nu}^{\nu} i\xi e^{i\xi x} d\xi$,   (iii) $\lim\limits_{\nu \to \infty} \displaystyle\int_{-\nu}^{\nu} \xi^2 e^{i\xi x} d\xi$.

8.3. Let $\alpha(x)$ be a C^∞ function and f be any distribution. Prove that
$$(\alpha(x)f)' = \alpha'(x)f + \alpha(x)f'.$$

8.4. Let $\alpha(x)$ be any C^∞ function. Prove that
$$(\alpha(x)\theta(x))' = \alpha(0)\delta(x) + \theta(x)\alpha'(x).$$

8.5. Prove Eq. (8.2.6), $\big(f(x) \otimes g(t), \varphi(x)\psi(t)\big) = \big(f(x), \varphi(x)\big)\big(g(t), \psi(t)\big)$.

8.6. Prove Eq. (8.2.8), $\delta(x) \otimes \delta(t) = \delta(x, t)$.

8.7. Prove the equation $f * g = g * f$ for convolution (8.2.10) of regular distributions.

8.8. Prove properties (8.2.14) and (8.2.15) of the convolution.

8.9. Verify Eq. (8.3.4) for $n = 2$ and $n = 3$.

8.10. Evaluate the distribution $\Delta(\ln r)$ when $n = 2$ and prove Eq. (8.3.11).

8.11. Discuss Eq. (8.3.9).

8.12. Prove Eq. (8.4.21).

8.13. Let $\alpha(x)$ be any C^∞ function. Prove that
$$\alpha(x)\delta'(x) = \alpha(0)\delta'(x) - \alpha'(0)\delta(x).$$

8.14. Consider a change of the variable x given by $y = p(x)$ and assume that $p(0) = 0$, $p'(0) \neq 0$. Prove Eq. (8.4.6), $\delta(p(x)) = p'(0)^{-1}\,\delta(x)$.

8.15. Let the function $p(x)$ has several zeros, e.g. at points $a_1, \ldots, a_s$, and let $p'(a_\sigma) \neq 0$ for all $\sigma = 1, 2, \ldots, s$. Prove Eq. (8.4.7):
$$\delta(p(x)) = \sum_{\sigma=1}^{s} \frac{1}{p'(a_\sigma)}\, \delta(x - a_\sigma).$$

8.16. Derive Eq. (8.4.8), $\delta(x^2 - a^2) = [\delta(x - a) - \delta(x + a)]/(2a)$.

Chapter 9

Invariance principle and fundamental solutions

This chapter contains a new group theoretic approach to initial value problems based on the *invariance principle*. The method is efficient for solving linear equations both with constant and variable coefficients.

Additional reading: N.H. Ibragimov [20], Chapter 3.

9.1   Introduction

Lie group methods are usually regarded to be not particularly useful for solving the *initial value problem*. This is due to the fact that arbitrary initial conditions break the symmetry group of a differential equation in question. However, it is shown in [15] that the fundamental solutions of the classical equations of mathematical physics are in fact group-invariant solutions. This observation, amplified by the formulation of the *invariance principle* in initial value problems, led to the development of a systematic group theoretic approach to the fundamental solutions [17]. This new approach combines the philosophy of Lie symmetries with the theory of distributions.

The key point of the success of the group theoretic approach to fundamental solutions is that the initial value problem determining the fundamental solution, unlike the general problem with arbitrary boundary or initial conditions, inherits certain symmetries of the differential equation. In consequence, one can find the fundamental solutions by searching them as invariant solutions under the inherited symmetry group.

The fundamental solutions for elliptic and parabolic equations are usual functions and can be obtained by means of the classical Lie theory (see Sections 9.2.3 and 9.2.4). For hyperbolic equations, however, the fundamental solutions are distributions (see Section 9.4). Therefore, one needs differential equations with distributions presented in Section 9.4.2.

The new method, unlike the Fourier transform method, is independent on a choice of coordinates and is applicable not only to linear equations with *constant coefficients* but also to equations with *variable coefficients*.

9.2   The invariance principle

9.2.1   Formulation of the invariance principle

A general principle formulated in [15] (see also [21]) and called an *invariance principle*, adjusts Lie group theory to tackling boundary value problems, in particular, an initial value or the Cauchy problem. This principle states that if a differential equation admits a Lie group G and if a boundary/initial value problem is invariant under a subgroup $H \subset G$, then one should seek the solution to the problem among H-invariant solutions of the differential equation in question. The invariance principle applicable both to linear and nonlinear equations, but in what follows, it is employed to derive fundamental solutions for linear partial differential equations. Accordingly, the invariance principle is formulated here only for linear equations.

Definition 9.2.1. Let L be a linear partial differential operator with independent variables $x = (x^1, \ldots, x^n)$. Let the differential equation $L(u) = f(x)$ admit a group G. Then the boundary value problem

$$L(u) = f(x), \quad u\big|_S = h(x) \tag{9.2.1}$$

is said to be invariant under a subgroup $H \subset G$ if
   1) the manifold S is invariant under H,
   2) the boundary (initial) condition $u\big|_S = h(x)$ is invariant under the group $\widetilde{H}$ induced on S by H, i.e., $\widetilde{H}$ is the action of H restricted to S.

The invariance principle: If the boundary (initial) value problem (9.2.1) is invariant under the group H, one should seek the solution of the problem among the functions invariant under H.

9.2.2   Fundamental solution of linear equations with constant coefficients

Definition 9.2.2. A fundamental solution for a linear differential operator L with constant coefficients is a generalized function $\mathcal{E}$ such that

$$L\mathcal{E} = \delta. \tag{9.2.2}$$

The characteristic property of fundamental solutions is given by the following statement.

Theorem 9.2.1. Let $\mathcal{E}$ be a fundamental solution of a linear differential operator L with constant coefficients. Then the function

$$u = \mathcal{E} * f \qquad (9.2.3)$$

solves the non-homogeneous equation

$$Lu = f. \qquad (9.2.4)$$

Proof. Using properties (8.2.11)—(8.2.15) of the convolution, we have

$$Lu = L(\mathcal{E} * f) = (L\mathcal{E}) * f = \delta * f = f.$$

9.2.3 Application to the Laplace equation

Consider the Laplace equation with an arbitrary number $n \geq 3$ of variables $x = (x^1, \ldots, x^n)$:

$$\Delta u \equiv \sum_{i=1}^{n} u_{ii} = 0. \qquad (9.2.5)$$

The symmetries of Eq. (9.2.5) comprise the finite-dimensional Lie algebra spanned by

$$X_i = \frac{\partial}{\partial x^i}, \quad X_{ij} = x^j \frac{\partial}{\partial x^i} - x^i \frac{\partial}{\partial x^j},$$

$$Y_i = (2x^i x^j - |x|^2 \delta^{ij}) \frac{\partial}{\partial x^j} + (2 - n) x^i u \frac{\partial}{\partial u}, \qquad (9.2.6)$$

$$Z_1 = x^i \frac{\partial}{\partial x^i}, \quad Z_2 = u \frac{\partial}{\partial u} \quad (i, j = 1, \ldots, n)$$

and the infinite-dimensional algebra with the generator

$$X_\tau = \tau \frac{\partial}{\partial u},$$

where $\tau = \tau(x)$ is an arbitrary solution of the Laplace equation. In what follows, we let however $\tau(x) = 0$ and use only generators (9.2.6).

Let us find the fundamental solution $\mathcal{E}(x)$. We will apply the invariance principle to Eq. (9.2.2) for the fundamental solution:

$$\Delta \mathcal{E} = \delta(x). \qquad (9.2.7)$$

Namely, we will treat Eq. (9.2.7) as a boundary-value problem with the fixed singular point (the origin) where the δ-function is given. We will first single out from (9.2.6) the operators that leave invariant the singular point $x = 0$. It is easy to see that this property satisfied by all operators (9.2.6) except the translation generators X_i $(i = 1, \ldots, n)$. For the generators of rotation X_{ij} and dilation Z_1 and Z_2 it is obvious, while the invariance test $Y_i(x^k)\big|_{x=0} = 0$ for Y_i is clear from the equation $Y_i(x^k) = 2x^i x^k - |x|^2 \delta^{ik}$.

Now we turn to the invariance of Eq. (9.2.7). First of all, we note that Eq. (9.2.7) admits the operators X_{ij}, since the Laplacian and the δ-function are invariant with respect to rotations. As far as the dilation generators Z_1 and Z_2 are concerned, they are not admitted separately. Therefore, we test for the invariance their linear combination

$$Z = x^i \frac{\partial}{\partial x^i} + ku \frac{\partial}{\partial u} . \tag{9.2.8}$$

We prolong operator (9.2.8) to the second derivatives u_{ii} and extend its action to the δ-function according to (8.4.23). Noting that in this case we have $D_i(\xi^i) = n$ and performing the prolongation we obtain the operator

$$\tilde{Z} = x^i \frac{\partial}{\partial x^i} + ku \frac{\partial}{\partial u} + (k-1)u_i \frac{\partial}{\partial u_i} + (k-2)u_{ij} \frac{\partial}{\partial u_{ij}} - n\delta \frac{\partial}{\partial \delta} . \tag{9.2.9}$$

It follows that

$$\tilde{Z}(\Delta u - \delta) = (k-2)\Delta u + n\delta.$$

Hence, the invariance condition is written

$$\tilde{Z}(\Delta u - \delta)|_{\Delta u = \delta} = (k - 2 + n)\delta = 0$$

and yields $k = 2 - n$. Thus Eq. (9.2.7) admits the following operators:

$$X_{ij} = x^j \frac{\partial}{\partial x^i} - x^i \frac{\partial}{\partial x^j} , \quad Z = x^i \frac{\partial}{\partial x^i} + (2-n)u \frac{\partial}{\partial u} . \tag{9.2.10}$$

Likewise one can verify that the operators Y_i are admitted as well. But we will not need them and will regard them as an *excess symmetry* of the fundamental solution.

In accordance with the invariance principle, we shall look for invariant solutions of Eq. (9.2.7) with respect to generators (9.2.10). The generators X_{ij} have two independent invariants, namely u and $r = \sqrt{(x^1)^2 + \ldots + (x^n)^2}$. We write the operator Z in these invariants:

$$Z = r \frac{\partial}{\partial r} + (2-n)u \frac{\partial}{\partial u} ,$$

and solve the equation $Z(J(r,u)) = 0$ to obtain the following invariant for operators (9.2.10):

$$J = ur^{n-2}.$$

The invariant solution is written in the form $J = C = \text{const.}$, whence

$$u = Cr^{2-n}. \tag{9.2.11}$$

We know from Section 8.3 that function (9.2.11) satisfies Eq. (9.2.7) up to a constant factor. Namely, comparing (9.2.11) with (8.3.8) yields the following value of the constant C :

$$C = \frac{1}{(2-n)\omega_n} ,$$

where ω_n is the surface area of the unit sphere. Hence, the fundamental solution $\mathcal{E}_n$ for Eq. (9.2.5) with $n \geq 3$ has the form (cf. (8.3.12))

$$\mathcal{E}_n = \frac{1}{(2-n)\,\omega_n}\, r^{2-n} = \frac{\Gamma(n/2)}{2(2-n)\sqrt{\pi^n}}\, r^{2-n}. \qquad (9.2.12)$$

We summarize: *The fundamental solution for the Laplace equation (9.2.5) is determined by the invariance condition up to a constant factor. Equation (9.2.7) plays the role of a normalizing condition only.*

Likewise, one can use the invariance principle in the case $n = 2$ and obtain the fundamental solution (cf. (8.3.13))

$$\mathcal{E}_2 = \frac{1}{2\pi}\, \ln r \qquad (9.2.13)$$

for the Laplace equation in two variables (5.2.25),

$$\Delta u \equiv u_{xx} + u_{yy} = 0. \qquad (9.2.14)$$

In the physically important case $n = 3$, formula (9.2.12) gives the fundamental solution (cf. (8.3.13))

$$\mathcal{E}_3 = -\frac{1}{4\pi\, r} \qquad (9.2.15)$$

for the Laplace equation in three variables (1.3.20),

$$\Delta u \equiv u_{xx} + u_{yy} + u_{zz} = 0. \qquad (9.2.16)$$

9.2.4 Application to the heat equation

Consider the heat equation with n spatial variables $x = (x^1, \ldots, x^n)$:

$$u_t - \Delta u = 0, \qquad (9.2.17)$$

where Δ is the n-dimensional Laplacian in x^i (see Eq. (9.2.5)).

The symmetries of Eq. (9.2.17) comprise the finite-dimensional Lie algebra spanned by

$$X_0 = \frac{\partial}{\partial t}, \quad X_i = \frac{\partial}{\partial x^i}, \quad X_{ij} = x^j \frac{\partial}{\partial x^i} - x^i \frac{\partial}{\partial x^j},$$

$$Z_1 = 2t\frac{\partial}{\partial t} + x^i \frac{\partial}{\partial x^i}, \quad Z_2 = u\frac{\partial}{\partial u}, \quad Z_{0i} = 2t\frac{\partial}{\partial x^i} - x^i u \frac{\partial}{\partial u}, \qquad (9.2.18)$$

$$Y = t^2 \frac{\partial}{\partial t} + t x^i \frac{\partial}{\partial x^i} - \frac{1}{4}(2nt + |x|^2) u \frac{\partial}{\partial u}$$

and the infinite-dimensional algebra with the generator

$$X_\tau = \tau \frac{\partial}{\partial u},$$

where $\tau = \tau(t, x)$ is an arbitrary solution of the heat equation. We will let $\tau(t, x) = 0$ and use only generators (9.2.18).

Equation (9.2.2) for the fundamental solution $\mathcal{E}(t, x)$ has the form

$$\mathcal{E}_t - \Delta\mathcal{E} = \delta(t, x) \tag{9.2.19}$$

and is invariant under the group with the following generators:

$$X_{ij} = x^j \frac{\partial}{\partial x^i} - x^i \frac{\partial}{\partial x^j}, \quad Z_{0i} = 2t \frac{\partial}{\partial x^i} - x^i u \frac{\partial}{\partial u},$$

$$Z = 2t \frac{\partial}{\partial t} + x^i \frac{\partial}{\partial x^i} - nu \frac{\partial}{\partial u}. \tag{9.2.20}$$

Operators (9.2.20) are obtained by repeating the procedure used in the case of the Laplace equation. Namely, they are singled out from generators (9.2.18) by demanding the invariance of the point ($t = 0$, $x = 0$) and the invariance of Eq. (9.2.19).

Let us apply the invariance principle. Invariants for X_{ij} are t, r, u. In the space of these invariants the operators Z_{0i} are written in the form

$$Z_{0i} = x^i \left(2 \frac{t}{r} \frac{\partial}{\partial r} - u \frac{\partial}{\partial u} \right).$$

Solving the equations $Z_{0i}(J) = 0$, one obtains two invariants of the rotations and the Galilean transformations, namely t and $p = ue^{r^2/(4t)}$. Rewriting the last operator of (9.2.20) in these invariants:

$$Z = 2t \frac{\partial}{\partial t} - np \frac{\partial}{\partial p}.$$

and solving the equation $Z(J) = 0$, we obtain the following invariant:

$$J = t^{n/2} p \equiv u(\sqrt{t})^n e^{r^2/(4t)}.$$

The invariant solution is given by $J = \text{const.}$, i.e., has the form

$$u = \frac{C}{(\sqrt{t})^n} e^{-\frac{r^2}{4t}}, \quad t > 0.$$

We extend it to $t < 0$ by setting $u = 0$ when $t < 0$. In other words,

$$u = \frac{C\theta(t)}{(\sqrt{t})^n} e^{-\frac{r^2}{4t}}, \tag{9.2.21}$$

where $\theta(t)$ is the Heaviside function (8.1.5) of t. Equation (9.2.19) plays again the role of a normalizing condition. Namely, it is shown in Problem 9.10 and in the next section that $C = (2\sqrt{\pi})^{-n}$. Thus, we arrive at the following fundamental solution for the heat equation with n spatial variables:

$$\mathcal{E}_n = \frac{\theta(t)}{(2\sqrt{\pi t})^n} e^{-\frac{r^2}{4t}}. \tag{9.2.22}$$

9.3 Cauchy's problem for the heat equation

In this section, we introduce the concept of the fundamental solution for the Cauchy problem and calculate it for the heat equation by using the invariance principle. It follows from these calculations that the heat diffusion can be directly derived from the invariance principle.

9.3.1 Fundamental solution for the Cauchy problem

Definition 9.3.1. The distribution $E(t, x)$ is called the *fundamental solution of the Cauchy problem* for the heat equation if it solves the following initial value problem:

$$E_t - \Delta E = 0, \quad E\big|_{t=0} = \delta(x). \tag{9.3.1}$$

Theorem 9.3.1. (See Problem 9.7). Let $E(t, x)$ be the fundamental solution of the Cauchy problem. Then the solution to the Cauchy problem

$$u_t - \Delta u = 0, \quad u\big|_{t=0} = u_0(x) \tag{9.3.2}$$

is given by the convolution of E and the initial function $u_0(x)$:

$$u(t, x) = E * u_0 = \int_{\mathbb{R}^n} u_0(\xi) E(t, x - \xi) d\xi. \tag{9.3.3}$$

Theorem 9.3.2. Let $E(t, x)$ be the fundamental solution of the Cauchy problem and $\theta(t)$ the Heaviside function. Then

$$\mathcal{E} = \theta(t) E(t, x) \tag{9.3.4}$$

is the fundamental solution of the heat equation, i.e., $\mathcal{E}_t - \Delta\mathcal{E} = \delta(t, x)$.

Proof. Indeed, we have

$$\mathcal{E}_t - \Delta\mathcal{E} = \theta'(t) E(t, x) + \theta(t)(E_t - \Delta E).$$

Whence, invoking Eqs. (9.3.1) and using properties (8.2.4), (8.2.2) and (8.2.8) of distributions, we ultimately arrive at Eq. (9.2.19):

$$\mathcal{E}_t - \Delta\mathcal{E} = \delta(t) E(t, x) = \delta(t) E(0, x) = \delta(t) \otimes \delta(x) = \delta(t, x).$$

9.3.2 Derivation of the fundamental solution for the Cauchy problem from the invariance principle

Lemma 9.3.1. The Lie algebra admitted by the initial value problem (9.3.1) is spanned by the following operators from (9.2.18):

$$X_{ij}, \; Z_{0i}, \; Z_1 - nZ_2, \; Y, \quad i, j = 1, ..., n. \tag{9.3.5}$$

Proof. Since the algebra L is admitted by the differential equation (9.2.17), one should consider only the invariance of the initial condition. Here, the initial manifold S is given by $t = 0$. Further, the invariance of the initial data (9.3.1) presupposes, in particular, that the support of $\delta(x)$, i.e., the point $x = 0$, remains unaltered. Thus, Definition 9.2.1 requires the invariance of the equations $t = 0$ and $x = 0$. This requirement removes from the algebra L the translation generators X_i, X_0, and hence reduces (9.2.18) to

$$X_{ij}, \quad Z_{0i}, \quad Z_1, \quad Z_2, \quad Y.$$

The initial condition in (9.3.1) is invariant under the operators X_{ij}, Z_{0i}, and Y. It is not invariant under the two-dimensional algebra spanned by Z_1, Z_2. Therefore, we inspect the invariance test for the linear combination:

$$(Z_1 + kZ_2)\big|_{t=0} = x^i \frac{\partial}{\partial x^i} + ku\frac{\partial}{\partial u}, \quad k = \text{const.}$$

Under this operator, the variable u and the δ-function are subjected to the transformations:

$$\bar{u} \approx u + aku, \quad \bar{\delta} \approx \delta - an\delta.$$

It follows, that $\bar{u} - \bar{\delta} = u - \delta + a(ku + n\delta) + o(a)$, and that

$$(\bar{u} - \bar{\delta})\big|_{u=\delta} = a(k + n)\delta + o(a).$$

Hence, the invariance condition of (9.3.1) has the form $k + n = 0$. Thus, we arrive at operators (9.3.5).

Theorem 9.3.3. The fundamental solution of the Cauchy problem for the heat equation with n spatial variables has the form

$$E_n(t, x) = \frac{1}{\left(2\sqrt{\pi t}\right)^n} e^{-\frac{r^2}{4t}}. \tag{9.3.6}$$

It is the only function $E = u(t, x)$ which satisfies the initial condition (9.3.1) and is invariant under the group of rotations, Galilean transformations and dilations with the infinitesimal generators

$$X_{ij}, \quad Z_{0i}, \quad Z_1 - nZ_2, \quad i, j = 1, ..., n. \tag{9.3.7}$$

Proof. We first notice that the functionally independent invariants of the rotations are t, r, u. Then we write the restriction of the Galilean operators Z_{0i} to functions of these invariants as follows:

$$Z_{0i} = x^i \left(2\frac{t}{r}\frac{\partial}{\partial r} - u\frac{\partial}{\partial u}\right).$$

For these operators, the independent invariants are t and $p = u \exp[r^2/(4t)]$. The last operator of (9.3.7) is written in these variables in the form:

$$Z_1 - nZ_2 = 2t\frac{\partial}{\partial t} - np\frac{\partial}{\partial p}.$$

It has the only independent invariant $J = t^{n/2}p = t^{n/2}u\exp[r^2/(4t)]$. We conclude that the general form of the invariant function $E = u(t,x)$ is obtained from the equation $J = C$, whence

$$u = \frac{C}{(\sqrt{t})^n}e^{-\frac{r^2}{4t}}, \quad t > 0.$$

Letting here $t \to 0$, using the initial condition (9.3.1) and invoking Eq. (8.1.12), we obtain $C = (2\sqrt{\pi})^{-n}$. Hence, we arrive at (9.3.6).

Remark 9.3.1. The fundamental solution (9.2.22) for the heat equation is obtained by substituting (9.3.6) into Eq. (9.3.4).

9.3.3 Solution of the Cauchy problem

Equations (9.2.22), (9.3.6), (9.3.3) and (9.2.3) yield the following results.

Theorem 9.3.4. The fundamental solution $\mathcal{E}$ of the heat equation (9.2.17) and the fundamental solution E of the Cauchy problem for the heat equation have the form (9.2.22)

$$\mathcal{E} = \frac{\theta(t)}{(2\sqrt{\pi t})^n}e^{-\frac{r^2}{4t}}$$

and (9.3.6)

$$E(t,x) = \frac{1}{(2\sqrt{\pi t})^n}e^{-\frac{r^2}{4t}},$$

respectively. The solution to the Cauchy problem (9.3.2),

$$u_t - \Delta u = 0, \quad u\big|_{t=0} = u_0(x),$$

with a continuous initial data $u_0(x)$ is given by the following formula:

$$u(t,x) = \frac{1}{(2\sqrt{\pi t})^n}\int_{\mathbb{R}^n} u_0(\xi)e^{\frac{-|x-\xi|^2}{4t}}d\xi, \quad t > 0. \tag{9.3.8}$$

Consider the Cauchy problem for the non-homogeneous heat equation:

$$u_t - \Delta u = f(t,x), \quad u\big|_{t=0} = u_0(x), \tag{9.3.9}$$

with a twice continuously differentiable function $f(t,x)$ $(t \geq 0)$ and a continuous initial data $u_0(x)$. The solution to problem (9.3.9) is given by

$$u(t,x) = \int_0^t \int_{\mathbb{R}^n} \frac{f(\tau,\xi)}{(2\sqrt{\pi(t-\tau)})^n}e^{\frac{-|x-\xi|^2}{4(t-\tau)}}d\xi d\tau$$

$$+ \frac{1}{(2\sqrt{\pi t})^n}\int_{\mathbb{R}^n} u_0(\xi)e^{\frac{-|x-\xi|^2}{4t}}d\xi, \quad t > 0. \tag{9.3.10}$$

9.4 Wave equation

9.4.1 Preliminaries on differential forms

Let $\phi = \phi(x)$ be a differentiable function, where $x = (x^1, \ldots, x^n)$. The differential of ϕ,

$$d\phi = \sum_{i=1}^{n} \frac{\partial \phi}{\partial x^i} dx^i$$

is written in the compact form

$$d\phi = \phi_i dx^i,$$

where $\phi_i = \partial\phi(x)/\partial x^i$. The following generalization of the differential is useful.

Definition 9.4.1. A *differential 1-form* is the expression

$$\omega = a_i(x) dx^i, \tag{9.4.1}$$

where $a_i(x)$, $i = 1, \ldots, n$, are arbitrary functions.

If, in particular, the coefficients $a_i(x)$ are partial derivatives of a function ϕ, then the 1-form ω is a differential, namely, the differential of ϕ :

$$\omega = d\phi.$$

Let S be a surface in the three-dimensional space (x, y, z). Denote by dS the *element of surface area*, and by $\nu = (\nu^1, \nu^2, \nu^3)$ the unit outward normal to dS, where ν^1, ν^2 and ν^3 are the components of ν along the x, y and z axes, respectively. In the theory of surface integrals, the notions of oriented surfaces and surface elements are of common use. Namely, an *oriented surface element* (known also as an *element of vectorial area*) is defined by

$$\nu dS = (\nu^1 dS, \ \nu^2 dS, \ \nu^3 dS).$$

Its components are the projections of the vector νdS to the coordinate planes (y, z), (z, x) and (x, y), respectively, and are written in the following form (cf. Section 1.3.4):

$$\nu^1 dS = dy \wedge dz, \quad \nu^2 dS = dz \wedge dx, \quad \nu^3 dS = dx \wedge dy. \tag{9.4.2}$$

This notation presumes that we use the orientation on dS in accordance with the usual right-handed system of vectors dx, dy, dz. If one changes the orientation, one exchanges y and z and replaces $\nu^1 dS$ by $-\nu^1 dS$. Hence, $dz \wedge dy = -\nu^1 dS$, and therefore one should take $dy \wedge dx = -dx \wedge dy$, etc. Accordingly, one can associate the exterior product $dy \wedge dx$ with the vector product of the vectors **dy** and **dx** directed along the y and x axes, respectively.

In the case of n independent variables x^i, the generalization of the above construction leads to a formal *exterior multiplication* $\wedge$ obeying the rule

$$dx^i \wedge dx^j = -dx^j \wedge dx^i \quad \text{(in particular,} \quad dx^i \wedge dx^i = 0). \tag{9.4.3}$$

Definition 9.4.2. A *differential 2-form* is the expression

$$\omega = \sum_{i,j=1}^{n} a_{ij}(x)dx^i \wedge dx^j, \qquad (9.4.4)$$

where $a_{ij}(x)$, $i,j = 1,\ldots,n$, are arbitrary functions. In virtue of property (9.4.3) of the exterior product, sum (9.4.4) can be reduced to the form

$$\omega = \sum_{i<j} a_{ij}(x)dx^i \wedge dx^j. \qquad (9.4.5)$$

Definition 9.4.3. A differential p-form (simply, a p-form) is written

$$\omega = \sum_{i_1<\cdots<i_p} a_{i_1\cdots i_p}(x)dx^{i_1} \wedge \cdots \wedge dx^{i_p}, \qquad (9.4.6)$$

where $x = (x^1,\ldots,x^n)$, $dx = (dx^1,\ldots,dx^n)$, and $a_{i_1\cdots i_p}(x)$ are continuously differentiable functions. The summation is extended over all values $i_1,\ldots,i_p = 1,\ldots,n$ such that $i_1 < \cdots < i_p$.

The *exterior differential calculus* concerns manipulations with differential forms and is determined by a formal *exterior multiplication* $\wedge$ obeying law (9.4.3) and by the *exterior differentiation* defined as follows:

$$d\omega = \sum_{i_1<\cdots<i_p} \sum_{j=1}^{n} \frac{\partial a_{i_1\cdots i_p}}{\partial x^j} dx^j \wedge dx^{i_1} \wedge \cdots \wedge dx^{i_p}. \qquad (9.4.7)$$

According to (9.4.7), the differential $d\omega$ of a p-form ω is a $(p+1)$-form. The exterior differentiation and multiplication of forms obey the following rules:

$$d^2\omega \equiv d(d\omega) = 0, \qquad (9.4.8)$$

$$\omega \wedge \eta = (-1)^{pq}\eta \wedge \omega, \qquad (9.4.9)$$

$$d(\omega \wedge \eta) = d\omega \wedge \eta + (-1)^p\omega \wedge d\eta, \qquad (9.4.10)$$

where ω is a p-form and η a q-form. If $p = n$, then any n-form is written $\omega = a(x)dx^1 \wedge \cdots \wedge dx^n$, and its integral is defined by

$$\int \omega = \int a(x)dx^1 \cdots dx^n. \qquad (9.4.11)$$

Definition 9.4.4. A form ω is said to be *closed* if

$$d\omega = 0, \qquad (9.4.12)$$

and *exact* if there exists a $(p-1)$-form η such that

$$\omega = d\eta. \qquad (9.4.13)$$

Equation (9.4.8) shows that any exact form is closed. The following statement known as Poincaré's theorem asserts that the converse is valid as well.

Theorem 9.4.1. A differential form ω is closed if and only if it is locally exact, i.e., Eq. (9.4.13) holds in a neighborhood of a generic point x.

Example 9.4.1. In the notation of exterior differential calculus, Definition 3.2.1 of an exact equation (3.2.3) means that the left-hand side

$$\omega \equiv M(x,y)\mathrm{d}x + N(x,y)\mathrm{d}y$$

of Eq. (3.2.3) is an exact 1-form (cf. (9.4.1) with $n = 2$) i.e., $\omega = \mathrm{d}\Phi$. Invoking (9.4.7) and (9.4.3), the differential of ω is written

$$\mathrm{d}\omega = \left(\frac{\partial N}{\partial x} - \frac{\partial M}{\partial y}\right)\mathrm{d}x \wedge \mathrm{d}y. \tag{9.4.14}$$

Since Theorem 9.4.1 states that the exactness of ω is equivalent to $\mathrm{d}\omega = 0$, Eq. (9.4.14) yields the classical condition (3.2.6) of exactness.

The concept of an integrating factor discussed in Section 3.2.3 applies to differential forms as well. Namely, we say that $\mu(x)$ is an integrating factor for a differential form ω if $\mathrm{d}(\mu\omega) = 0$. According to Section 3.2.3, an integrating factor exists for any 1-form.

The exterior differential calculus allows one to extend three classical integral theorems formulated in Section 1.3.4 to higher dimensions as follows.

Theorem 9.4.2. Let V be a p-dimensional manifold (e.g. V is a p-dimensional domain in an n-dimensional Euclidean space $\mathbb{R}^n$, $p \leq n$) with the boundary ∂V, and let ω be a $(p-1)$-form. Then the following *Stokes' formula* holds:

$$\int_{\partial V} \omega = \int_V \mathrm{d}\omega. \tag{9.4.15}$$

Let us verify that Eq. (9.4.15) encapsulates the Green, Stokes and divergence theorems as particular cases.

Derivation of Green's theorem. In this case, V is a two-dimensional domain in $\mathbb{R}^2$. We consider a 1-form

$$\omega = P(x,y)\mathrm{d}x + Q(x,y)\mathrm{d}y. \tag{9.4.16}$$

The exterior differentiation (9.4.7) yields

$$\mathrm{d}\omega = \frac{\partial P}{\partial x}\mathrm{d}x \wedge \mathrm{d}x + \frac{\partial P}{\partial y}\mathrm{d}y \wedge \mathrm{d}x + \frac{\partial Q}{\partial x}\mathrm{d}x \wedge \mathrm{d}y + \frac{\partial Q}{\partial y}\mathrm{d}y \wedge \mathrm{d}y,$$

whence using property (9.4.3) of the exterior multiplication one obtains

$$\mathrm{d}\omega = \left(\frac{\partial Q}{\partial x} - \frac{\partial P}{\partial y}\right)\mathrm{d}x \wedge \mathrm{d}y. \tag{9.4.17}$$

Substituting (9.4.16) and (9.4.17) into (9.4.15), invoking definition (9.4.11) of integrals of differential forms, one arrives at Green's equation (1.3.16):

$$\int_{\partial V} P(x,y)dx + Q(x,y)dy = \int_V \left(\frac{\partial Q}{\partial x} - \frac{\partial P}{\partial y}\right) dxdy.$$

Derivation of Stokes' theorem. Let V be a two-dimensional domain in $\mathbb{R}^3$ and let

$$\omega = P(x,y,z)dx + Q(x,y,z)dy + R(x,y,z)dz. \qquad (9.4.18)$$

Applying the exterior differentiation (9.4.7) and invoking that

$$dx \wedge dx = dy \wedge dy = dz \wedge dz = 0$$

one obtains

$$d\omega = \frac{\partial P}{\partial y}dy \wedge dx + \frac{\partial P}{\partial z}dz \wedge dx + \frac{\partial Q}{\partial x}dx \wedge dy$$

$$+ \frac{\partial Q}{\partial z}dz \wedge dy + \frac{\partial R}{\partial x}dx \wedge dz + \frac{\partial R}{\partial y}dy \wedge dz,$$

or

$$d\omega = \left(\frac{\partial Q}{\partial x} - \frac{\partial P}{\partial y}\right) dx \wedge dy$$

$$+ \left(\frac{\partial R}{\partial y} - \frac{\partial Q}{\partial z}\right) dy \wedge dz + \left(\frac{\partial P}{\partial z} - \frac{\partial R}{\partial x}\right) dz \wedge dx. \qquad (9.4.19)$$

Substituting (9.4.18) and (9.4.19) into (9.4.15) one arrives at Eq. (1.3.17).

Derivation of the divergence theorem ($\dim V = 3$, $V \subset \mathbb{R}^3$). Rewriting the left-hand side of Eq. (1.3.18) by using Eqs. (9.4.2), we consider the following 2-form

$$\omega = A^1 dy \wedge dz + A^2 dz \wedge dx + A^3 dx \wedge dy \equiv (\boldsymbol{A} \cdot \boldsymbol{\nu})dS. \qquad (9.4.20)$$

Applying differentiation (9.4.7) and invoking that, e.g., $dy \wedge dy \wedge dz = 0$, one obtains

$$d\omega = \frac{\partial A^1}{\partial x}dx \wedge dy \wedge dz + \frac{\partial A^2}{\partial y}dy \wedge dz \wedge dx + \frac{\partial A^3}{\partial z}dz \wedge dx \wedge dy.$$

Using property (9.4.9), one can write

$$d\omega = \operatorname{div} \boldsymbol{A} \, dx \wedge dy \wedge dz. \qquad (9.4.21)$$

Substituting (9.4.20) and (9.4.21) into (9.4.15) one arrives at Eq. (1.3.18).

9.4.2   Auxiliary equations with distributions

Consider a surface in $\mathbb{R}^n$ given by $P(x) = 0$ with a continuously differentiable function $P(x)$. We assume that $\nabla P \neq 0$ on the surface $P(x) = 0$.

Definition 9.4.5. *Leray's form* ([24], Ch. IV, §1) on the surface $P(x) = 0$ is an $(n - 1)$-form ω satisfying the following equation:

$$dP \wedge \omega = dx^1 \wedge \cdots \wedge dx^n.$$

It can be represented in the form (see [24])

$$\omega = (-1)^{i-1} \frac{dx^1 \wedge \cdots \wedge dx^{i-1} \wedge dx^{i+1} \wedge \cdots \wedge dx^n}{P_i} \tag{9.4.22}$$

for any fixed i such that $P_i \equiv \partial P(x)/\partial x^i \neq 0$.

The *Heaviside function* $\theta(P)$ on the surface $P(x) = 0$ is defined by

$$\theta(P) = \begin{cases} 1, & P \geq 0, \\ 0, & P < 0. \end{cases}$$

It can be identified with the distribution

$$(\theta(P), \varphi) = \int_{P \geq 0} \varphi(x) dx. \tag{9.4.23}$$

Dirac's δ-function $\delta(P)$ on the surface $P(x) = 0$ is defined by

$$(\delta(P), \varphi) = \int_{P=0} \varphi \omega, \tag{9.4.24}$$

where ω is Leray's form. These two distributions are related by Eq. (8.2.4):

$$\theta'(P) = \delta(P). \tag{9.4.25}$$

Using the above distributions, we will formulate and solve *auxiliary differential equations*, namely first-order involving distributions. They will be employed for solving the Cauchy problem for wave equations. We start with the simplest equation of this kind. Namely, let us consider the equation

$$x f' = 0 \tag{9.4.26}$$

with one independent variable x. Its only classical solution is $f = $ const. It has, however, more solutions in the space of distributions. Indeed, taking $\alpha(x) = x$ in Eq. (8.2.2), $\alpha(x)\delta = \alpha(0)\delta$, one obtains

$$x\delta(x) = 0. \tag{9.4.27}$$

Invoking that $\theta'(x) = \delta(x)$, where $\theta(x)$ is the Heaviside function with one variable x, we conclude that Eq. (9.4.26) has, in distributions, the solution $f = \theta(x)$ different from $f = $ const. Since Eq. (9.4.26) is linear, the linear combination

$$f = C_1\theta(x) + C_2 \tag{9.4.28}$$

provides a distribution solution involving two arbitrary constants C_1 and C_2. The following theorems generalize equations (9.4.27) and (9.4.26).

Theorem 9.4.3. Dirac's δ-function (9.4.24) satisfies the following equations:

$$P\delta(P) = 0, \tag{9.4.29}$$

$$P\delta^{(m)}(P) + m\delta^{(m-1)}(P) = 0, \quad m = 1, 2, \ldots . \tag{9.4.30}$$

Proof. Using Definition (9.4.24), one has

$$\left(P\delta(P),\ \varphi\right) = \left(\delta(P),\ P\varphi\right) = \int_{P=0} P\varphi\omega = 0,$$

i.e., Eq. (9.4.29). Furthermore, assuming that $\partial P/\partial x^i \neq 0$ (for some i it must be true in view of the condition $\mathrm{grad}P \neq 0$), one obtains from (9.4.29), by differentiating it with respect to x^i, the equation

$$\frac{\partial P}{\partial x^i}\,\delta(P) + P\delta'(P)\frac{\partial P}{\partial x^i} = 0,$$

which upon division by $\partial P/\partial x^i$ yields Eq. (9.4.30) with $m = 1$. The consecutive differentiation leads to Eqs. (9.4.30) with $m = 2, 3, \ldots$ thus completing the proof.

Theorem 9.4.4. The first-order differential equation

$$Pf'(P) + mf(P) = 0 \tag{9.4.31}$$

has the general solution in distributions given by

$$f = C_1\theta(P) + C_2 \qquad \text{for} \quad m = 0, \tag{9.4.32}$$

$$f = C_1\delta^{(m-1)}(P) + C_2P^{-m} \qquad \text{for} \quad m = 1, 2, \ldots . \tag{9.4.33}$$

Proof. Use Theorem 9.4.3 and note that $f = P^{-m}$ is the classical solution to Eq. (9.4.31). For more details, see, e.g. [8].

9.4.3 Symmetries and definition of fundamental solutions for the wave equation

Consider the wave equation with several spatial variables (see Eq. (2.6.18)):

$$u_{tt} - \Delta u = 0, \tag{9.4.34}$$

where Δ is the n-dimensional Laplacian in the variables $x = (x^1, ..., x^n)$.

The symmetries of the wave equation (9.4.34) comprise the finite-dimensional Lie algebra spanned by (cf. generators (7.3.49) of the Lorentz group)

$$X_0 = \frac{\partial}{\partial t}, \quad X_i = \frac{\partial}{\partial x^i}, \quad X_{ij} = x^j \frac{\partial}{\partial x^i} - x^i \frac{\partial}{\partial x^j}, \quad X_{0i} = t \frac{\partial}{\partial x^i} + x^i \frac{\partial}{\partial t},$$

$$Z_1 = t \frac{\partial}{\partial t} + x^i \frac{\partial}{\partial x^i}, \quad Z_2 = u \frac{\partial}{\partial u}, \quad Y_0 = (t^2 + |x|^2) \frac{\partial}{\partial t} + 2t x^i \frac{\partial}{\partial x^i} - (n-1) t u \frac{\partial}{\partial u},$$

$$Y_i = 2t x^i \frac{\partial}{\partial t} + (2x^i x^j + (t^2 - |x|^2) \delta^{ij}) \frac{\partial}{\partial x^j} - (n-1) x^i u \frac{\partial}{\partial u}, \qquad (9.4.35)$$

where $i, j = 1, ..., n$, and the infinite-dimensional algebra with the generators

$$X_\tau = \tau(t, x) \frac{\partial}{\partial u},$$

where $\tau(t, x)$ is an arbitrary solution of the wave equation. We will let $\tau(t, x) = 0$ and use only generators (9.4.35).

Equation (9.2.2) defining the fundamental solution $\mathcal{E}(t, x)$ for the wave equation has the form

$$\mathcal{E}_{tt} - \Delta \mathcal{E} = \delta(t, x). \qquad (9.4.36)$$

Let us give the definition of the fundamental solution of the Cauchy problem for the wave equation similar to that for the heat equation. Note that the general Cauchy problem with arbitrary initial conditions:

$$u_{tt} - \Delta u = 0, \quad t > 0,$$

$$u\big|_{t=0} = u_0(x), \quad u_t\big|_{t=0} = u_1(x) \qquad (9.4.37)$$

can be reduced to the following particular Cauchy problem:

$$u_{tt} - \Delta u = 0, \quad u\big|_{t=0} = 0, \quad u_t\big|_{t=0} = h(x). \qquad (9.4.38)$$

Indeed, one can readily verify the following statement.

Lemma 9.4.1. Let $v(t, x)$ and $w(t, x)$ be the solutions to the particular Cauchy problem (9.4.38) with $h(x) = u_0(x)$ and $h(x) = u_1(x)$, respectively. Then the function

$$u(t, x) = w(t, x) + \frac{\partial v(t, x)}{\partial t} \qquad (9.4.39)$$

solves the general Cauchy problem (9.4.37).

Definition 9.4.6. The distribution $E(t, x)$ is called the fundamental solution of the Cauchy problem for the wave equation if it solves the following particular Cauchy problem:

$$E_{tt} - \Delta E = 0 \ (t > 0), \quad E\big|_{t=0} = 0, \quad E_t\big|_{t=0} = \delta(x), \qquad (9.4.40)$$

where

$$E\big|_{t=0} \equiv \lim_{t \to +0} E(t, x), \quad E_t\big|_{t=0} \equiv \lim_{t \to +0} \frac{\partial E(t, x)}{\partial t}.$$

Theorem 9.4.5. Let $E(t, x)$ be the fundamental solution of the Cauchy problem. Then the solution to the Cauchy problem (9.4.38) is given by the convolution of E and the initial function $h(x)$:

$$u(t, x) = E * h(x).$$

Remark 9.4.1. Let $E(t, x)$ be the fundamental solution of the Cauchy problem. Then $\mathcal{E} = \theta(t)E(t, x)$ is the fundamental solution of the wave equation, i.e., $\mathcal{E}_{tt} - \Delta\mathcal{E} = \delta(t, x)$.

9.4.4 Derivation of the fundamental solution

Proceeding as in Section 9.3.2, Lemma 9.3.1, one arrives at the following.

Lemma 9.4.2. The Lie algebra admitted by Eqs. (9.4.40) for the fundamental solution is spanned by the following operators from (9.4.35):

$$X_{ij}, \ X_{0i}, \ Z_1 + (1 - n)Z_2, \ Y_0, \ Y_i, \quad i, j = 1, ..., n. \tag{9.4.41}$$

We will derive here the fundamental solution of the Cauchy problem for the wave equations with odd n. The fundamental solution for the wave equation with even n can be obtained by means of a simple method known as Hadamard's *method of descent* presented in [12] (see also [4], [14]).

Theorem 9.4.6. The fundamental solution of the Cauchy problem for the wave equation (9.4.34) with an odd number of spatial variables has the form:

$$E_1 = \frac{1}{2}\theta(t^2 - x^2) \qquad \text{for} \quad n = 1, \tag{9.4.42}$$

$$E_3 = \frac{1}{2\pi}\delta\left(t^2 - r^2\right) \qquad \text{for} \quad n = 3, \tag{9.4.43}$$

$$E_n = \frac{1}{2\sqrt{\pi}^{n-1}}\delta^{\left(\frac{n-3}{2}\right)}\left(t^2 - r^2\right) \qquad \text{for} \quad n > 3. \tag{9.4.44}$$

It is determined uniquely by the invariance principle. Namely, $E(t, x)$ given by (9.4.42)—(9.4.44) is the only distribution which solves the particular Cauchy problem (9.4.40) and is invariant under the group of rotations, Lorentz transformations and dilations with the infinitesimal generators

$$X_{ij}, \quad X_{0i}, \quad Z_1 + (1 - n)Z_2, \quad i, j = 1, ..., n. \tag{9.4.45}$$

Proof. The generators X_{ij}, X_{0i} of the rotations and Lorentz transformations have two independent invariants, u and $p = t^2 - r^2$, where $r = |x|$. We restrict the dilation generator from (9.4.45) to these invariants:

$$Z_1 + (1 - n)Z_2 = 2p\frac{\partial}{\partial p} + (1 - n)u\frac{\partial}{\partial u} \tag{9.4.46}$$

and look for invariant distributions of the form $u = f(p)$. The invariance test under operator (9.4.46) yields the ordinary differential equation

$$2pf'(p) + (n-1)f(p) = 0.$$

We set $n = 2m + 1$ and rewrite it in form (9.4.31):

$$pf'(p) + mf(p) = 0, \quad m = 0, 1, 2, \ldots.$$

Its general solution is given by (9.4.32), $f(p) = C_1\theta(p) + C_2$, when $m = 0$ and by (9.4.33), $f(p) = C_1\delta^{(m-1)}(p) + C_2 p^{-m}$, when $m \neq 0$. Thus,

$$u = C_1\theta(p) + C_2 \qquad (n = 1),$$

$$u = C_1\delta^{\left(\frac{n-3}{2}\right)}(p) + C_2 p^{\frac{1-n}{2}} \qquad (n \geq 3).$$

The initial conditions in (9.4.40), together with the known equations

$$\lim_{t \to 0} \delta^{\left(\frac{n-3}{2}\right)}(p) = 0, \quad \lim_{t \to 0} \theta(p) = 0,$$

lead to

$$C_1 = \frac{1}{2\sqrt{\pi}^{\,n-1}}, \quad C_2 = 0.$$

Hence, we have obtained the fundamental solutions (9.4.42)~(9.4.44). Note that using Eq. (8.4.8) one can write (9.4.43) in the form

$$E_3 = \frac{1}{4\pi r}\left[\delta(t-r) - \delta(t+r)\right].$$

9.4.5 Solution of the Cauchy problem

The solution to the Cauchy problem for the one-dimensional wave equation is given by (5.4.11). Using the fundamental solution, one can obtain the solution for the wave equation with $x = (x^1, \ldots, x^n)$. We formulate the result for $n = 3$ and $n = 2$.

Theorem 9.4.7. The solution to the Cauchy problem

$$u_{tt} - k^2\Delta u = f(t,x), \quad u\big|_{t=0} = u_0(x), \quad u_t\big|_{t=0} = u_1(x) \qquad (9.4.47)$$

is given by

$$u(t,x) = \frac{1}{4\pi k^2}\left[\frac{1}{t}\int_{|\xi-x|=kt} u_1(\xi)\mathrm{d}S + \frac{\partial}{\partial t}\left(\frac{1}{t}\int_{|\xi-x|=kt} u_0(\xi)\mathrm{d}S\right)\right.$$

$$\left. + \int_{|\xi-x|<kt} f\left(t - \frac{|\xi-x|}{k}, \xi\right)\frac{\mathrm{d}\xi}{|\xi-x|}\right], \quad \text{when} \quad n = 3, \qquad (9.4.48)$$

$$u(t,x) = \frac{1}{2\pi k} \left[\int\limits_{|\xi-x|<kt} \frac{u_1(\xi)d\xi}{\sqrt{k^2t^2 - |\xi - x|^2}} + \frac{\partial}{\partial t} \int\limits_{|\xi-x|<kt} \frac{u_0(\xi)d\xi}{\sqrt{k^2t^2 - |\xi - x|^2}} \right.$$

$$\left. + \int\limits_0^t \int\limits_{|\xi-x|<k(t-\tau)} \frac{f(\tau,\xi)d\xi d\tau}{\sqrt{k^2(t-\tau)^2 - |\xi - x|^2}} \right], \quad \text{when } n = 2. \quad (9.4.49)$$

9.5 Equations with variable coefficients

The invariance principle furnishes an effective method for deriving fundamental solutions for linear equations with variable coefficients. As an example, We present here the fundamental solution for the Black-Scholes equation obtained by using the invariance principle. One can find in [20], Chapter 3, the application of the invariance principle to hyperbolic equations with variable coefficients, namely to the wave equations in curved space-times with non-trivial conformal group.

The fundamental solution of the Cauchy problem for the Black-Scholes equation (2.4.15) is a distribution $E(t, x; t_0, x_0)$ satisfying the following initial value problem:

$$E_t + \frac{1}{2}A^2x^2E_{xx} + BxE_x - CE = 0 \ (t < t_0), \quad E\big|_{t \to t_0} = \delta(x - x_0). \quad (9.5.1)$$

Using the invariance principle, the following fundamental solution has been obtained (see [15], English or Swedish edition, and the references therein)

$$E(x,t;x_0,t_0) = \frac{1}{Ax_0\sqrt{2\pi(t_0 - t)}} \exp\left[-\frac{(\ln x - \ln x_0)^2}{2A^2(t_0 - t)} - \left(\frac{K^2}{2A^2} + C\right)(t_0 - t) \right.$$

$$\left. -\frac{K}{A^2}(\ln x - \ln x_0) \right], \quad K = B - \frac{A^2}{2}. \quad (9.5.2)$$

Problems to Chapter 9

9.1. Obtain the fundamental solution (9.2.14) of the Laplace equation in two variables, $u_{xx} + u_{yy} = 0$, by means of the invariance principle.

9.2. Prove property (9.4.8) for the 1-form (9.4.16) with two variables.

9.3. Prove property (9.4.8) for the 1-form (9.4.18) with three variables.

9.4. Prove property (9.4.8) for an arbitrary p-form (9.4.6).

9.5. Prove property (9.4.9).

9.6. Prove Eq. (9.4.25).

9.7. Prove Theorem 9.3.1, i.e., verify that if $E(t, x)$ is the fundamental solution of the Cauchy problem for the heat equation, then the convolution

$$u(t, x) = E * u_0 = \int_{\mathbb{R}^n} u_0(y) E(t, x - y) dy$$

solves the Cauchy problem

$$u_t - \Delta u = 0, \quad u\big|_{t=0} = u_0(x).$$

9.8. Prove Theorem 9.4.5, i.e., verify that if $E(t, x)$ is the fundamental solution of the Cauchy problem for the wave equation, then the convolution $u(t, x) = E * h$ solves the Cauchy problem (9.4.38):

$$u_{tt} - \Delta u = 0, \quad u\big|_{t=0} = 0, \quad u_t\big|_{t=0} = h(x).$$

9.9. Derive solution (9.4.48) of the Cauchy problem for the three-dimensional wave equation.

9.10. Derive solution (9.4.49) of the Cauchy problem for the two-dimensional wave equation by applying the method of descent to solution (9.4.48) of the three-dimensional wave equation.

9.11. Prove that function (9.2.21),

$$u = \frac{C\theta(t)}{(\sqrt{t})^n} e^{-\frac{r^2}{4t}},$$

satisfies Eq. (9.2.19) precisely in the case $C = (2\sqrt{\pi})^{-n}$.

Answers

Chapter 1

1.2. (i) $\operatorname{arcsinh} x = \ln(x + \sqrt{x^2 + 1})$,   (ii) $\operatorname{arctanh} x = \dfrac{1}{2} \ln \dfrac{1+x}{1-x}$ $(|x| < 1)$,

(iii) $\operatorname{arccosh} x = \ln(x \pm \sqrt{x^2 - 1})$ $(x \geq 1)$.

1.4. $x_1 = -1$, $x_2 = 2 + i$, $x_3 = 2 - i$.

1.5. We have

$$\int_{-\pi}^{\pi} \sin(kx) \sin(mx) \, dx = \int_{-\pi}^{\pi} \cos(kx) \cos(mx) \, dx = 0, \quad m \neq k,$$

$$\int_{-\pi}^{\pi} \cos(kx) \sin(mx) \, dx = 0, \quad m, k = 0, 1, 2, \ldots,$$

$$\int_{-\pi}^{\pi} \sin^2(kx) \, dx = \int_{-\pi}^{\pi} \cos^2(kx) \, dx = \pi, \quad k = 1, 2, \ldots.$$

1.8. Functions (ii) are functionally independent whereas functions (i) and (iii) are functionally dependent.

1.9. $(\sinh x)' = \cosh x$, $(\cosh x)' = \sinh x$, $(\tanh x)' = 1/\cosh^2 x$.

1.10. Functions (i), (iv) and (v) are linearly independent whereas functions (ii) and (iii) are linearly dependent.

1.11. $i^i = e^{-\pi/2}$.

1.12. $\nabla \cdot \boldsymbol{x} = 3$, $\nabla \times \boldsymbol{x} = 0$.

1.13. We have

(i) $\nabla \times (\nabla \phi) = 0$,   (ii) $\nabla \cdot (\nabla \times \boldsymbol{a}) = 0$,

(iii) $\nabla \cdot (\boldsymbol{a} \times \boldsymbol{x}) = \boldsymbol{x} \cdot (\nabla \times \boldsymbol{a})$,

(iv) $\nabla \times (\nabla \times \boldsymbol{a}) = \nabla(\nabla \cdot \boldsymbol{a}) - \nabla^2 \boldsymbol{a}$,

(v) $\nabla \cdot (\phi \nabla \psi - \psi \nabla \phi) = \phi \Delta \psi - \psi \Delta \phi$.

1.14. We have

$$\int_V (\phi \Delta \psi - \psi \Delta \phi) \, dx \, dy \, dz = \int_{\partial V} (\phi \nabla \psi - \psi \nabla \phi) \cdot \boldsymbol{\nu} dS.$$

1.15. We have

$$\bar{y}' = \frac{\psi_x + y'\psi_y}{\varphi_x + y'\varphi_y}, \quad \bar{y}'' = \frac{\begin{vmatrix} \varphi_x + y'\varphi_y & \varphi_{xx} + 2y'\varphi_{xy} + y'^2\varphi_{yy} + y''\varphi_y \\ \psi_x + y'\psi_y & \psi_{xx} + 2y'\psi_{xy} + y'^2\psi_{yy} + y''\psi_y \end{vmatrix}}{(\varphi_x + y'\varphi_y)^3}.$$

1.16. $\bar{y}' = -e^{-x}y^{-2}y', \quad \bar{y}'' = (yy' + 2y'^2 - yy'')/(y^3 e^{2x}).$

1.17. $\bar{y}''' = (3y''^2 - y'y''')/y'^5.$

1.18. The three solutions of the equation $w^3 + 1 = 0$ are

$$w_1 = \frac{1}{2}(1 + i\sqrt{3}), \quad w_2 = -1, \quad w_3 = \frac{1}{2}(1 - i\sqrt{3}).$$

1.21. The spherically symmetric solution is

$$\phi(r) = \frac{C_1}{r} + C_2.$$

1.25. The equation for circles

$$y''' - 3\frac{y'y''^2}{1 + y'^2} = 0, \quad y = y(x),$$

and the equation for hyperbolas

$$z''' - \frac{3}{2}\frac{z''^2}{z'} = 0, \quad z = z(t),$$

are connected by the complex transformation $x = z + it$, $y = t + iz$.

1.29. $\Gamma(-1/2) = -2\sqrt{\pi}.$

1.30. $\displaystyle\int_0^\infty e^{-s^2}ds = \sqrt{\pi}/2.$

Chapter 2

2.1. The corresponding Euler-Lagrange equations are:

  (i) $u_{tt} - \Delta u = f(t, x, y, z)$, see Eq. (2.6.16) with $k = 1$,

  (ii) $2u_{tx} + u_x u_{xx} - u_{yy} = 0$, see Eq. (2.3.37),

  (iii) $u_{tt} + \mu u_{xxxx} = f$, see Eq. (2.1.4),

  (iv) $u_{tt} + u_{xxxx} + 2u_{xxyy} + u_{yyyy} = f(t, x, y)$, cf. Eq. (2.6.28).

2.6. $P = \alpha C e^{\alpha t}/(\beta C e^{\alpha t} - 1).$

2.7. Equations (2.3.30) and equation 9 from (1.3.3) yield: $D_t(\text{div }\boldsymbol{E}) = \text{div }\boldsymbol{E}_t$ $= c\,\text{div}\,(\nabla \times \boldsymbol{H}) = 0, D_t(\text{div }\boldsymbol{H}) = \text{div }\boldsymbol{H}_t = -c\,\text{div}\,(\nabla \times \boldsymbol{E}) = 0.$

Chapter 3

3.1. (iv) $y = Ce^x - (x^2 + 2x + 2)$, (v) $y = e^{-C_1 x} \left[C_2 - \int (x + x^2) e^{C_1 x} dx \right]$.

3.2. (vi) $y = C_1 x - \ln \left| C_2 - \int (x + x^2) e^{C_1 x} dx \right|$.

3.3. (iii) The general solution of the equation $y''' + y = 0$ is

$$y = K_1 e^{-x} + e^{x/2} \left[K_2 \cos \left(\frac{\sqrt{3}}{2} x \right) + K_3 \sin \left(\frac{\sqrt{3}}{2} x \right) \right], \quad K_i = \text{const.}, i = 1, 2, 3.$$

3.4. The equation $y' + y^2 = C x^{-2}$ is invariant under the dilation $\bar{x} = ax$, $\bar{y} = a^{-1} y$ and is the only homogeneous equation of the form $y' + y^2 = C x^s$.

3.11. The solutions are:

(i) $y = -\cos x \ln \left| \dfrac{1 + \sin x}{\cos x} \right| + C_1 \cos x + C_2 \sin x$.

Hint : Use the integral $\displaystyle \int \frac{dx}{\cos x} = \ln \left| \frac{1 + \sin x}{\cos x} \right|$.

(ii) $y = \cos x \ln |\cos x| + x \sin x + C_1 \sin x + C_2 \cos x$.

(iii) $y = \sin x \ln |\sin x| - x \cos x + C_1 \sin x + C_2 \cos x$.

3.13. The general second-order equations $f(x, y, y', y'') = 0$ reducible to the form $g(x, y, y') y''' = 0$ through differentiation, has the form

$$(ax^2 + bx + c + ky) y'' - \frac{k}{2} y'^2 - (2ax + b) y' + 2ay = 0,$$

where $a, b, c, k = \text{const.}$ Upon differentiation, this equation becomes

$$(ax^2 + bx + c + ky) y''' = 0$$

and hence reduces to the linear equation $y''' = 0$.

3.14. The standard substitution $t = \ln x$ reduces the equation in question to $\dfrac{d^2 y}{dt^2} + \dfrac{dy}{dt} + 4y = 0$. The characteristic equation $\lambda^2 + \lambda + 4 = 0$ has the roots $\lambda_{1,2} = \dfrac{-1 \pm i\sqrt{15}}{2}$. Hence, the fundamental system of solutions:

$$y_1 = e^{-\frac{t}{2}} \sin \left(\frac{\sqrt{15}}{2} t \right), \quad y_2 = e^{-\frac{t}{2}} \cos \left(\frac{\sqrt{15}}{2} t \right).$$

Thus, the general solution to the equation in question is

$$y = \frac{1}{\sqrt{x}} \left[C_1 \sin \left(\frac{\sqrt{15}}{2} \ln x \right) + C_2 \cos \left(\frac{\sqrt{15}}{2} \ln x \right) \right].$$

3.15. Multiplying the equation in question by x^2, we obtain Euler's equation $x^2 y'' - 3xy' + 3y = 0$ and integrate it by the standard change of the independent variable, $t = \ln x$. Simple calculations give the solution $y = C_1 x + C_2 x^3$.

Chapter 4

4.9. The system is not complete. Cf. Example 4.5.1.

4.13. $u = \phi(z - xy)$.

Chapter 5

5.9. The equation $u_{tt} - x^2 u_{xx} = 0$ is mapped to the telegraph equation $v_{\xi\eta} + v = 0$ by the transformation

$$\xi = \frac{1}{4}(t + \ln x), \quad \eta = \frac{1}{4}(t - \ln x), \quad u(t, x) = \sqrt{x}\, v(\xi, \eta).$$

5.13. Since the initial condition involves only a trigonometric function, $\sin x$, the solution can be obtained by taking only one term in (5.5.13). Invoking that $l = 2\pi$, we let $u(t, x) = [a \cos(kt/2) + b \sin(kt/2)] \sin(kx/2)$. Differentiating in t and using the condition $u_t(0, x) = 0$, we obtain $b = 0$. Hence, $u(t, x) = a \cos(kt/2) \sin(kx/2)$. The initial condition $u|_{t=0} = \sin x$ is written $a \sin(kx/2) = \sin x$, whence $a = 1$, $k = 2$. Thus, the solution is $u(t, x) = \cos t \sin x$.

5.14. $u(t, x) = \sin t \sin x$.

5.15. Formula (5.5.26) is written

$$c_k = \frac{1}{\sqrt{\pi}} \int_0^{2\pi} \sin x \sin \frac{k\,x}{2}\,\mathrm{d}x.$$

Hence, $c_1 = c_3 = c_4 = \cdots = 0$, and

$$c_2 = \frac{1}{\sqrt{\pi}} \int_0^{2\pi} \sin^2 x\mathrm{d}x = \frac{1}{\sqrt{\pi}}\,\pi = \sqrt{\pi}.$$

Thus, series (5.5.26) for the solution has only one term:

$$u(t, x) = \frac{1}{\sqrt{\pi}}\, c_2\, e^{-(2\pi/2\pi)^2\, t} \sin x = e^{-t} \sin x.$$

Alternative solution. Seek the solution in the form $u = T(t)X(x)$ and satisfy the boundary and initial conditions. Substitution in the heat equation yields the first-order ODE $T' + \lambda T = 0$ for $T(t)$ and the second-order ODE $X'' + \lambda X = 0$ for $X(x)$. The boundary conditions are written $X(0) = X(2\pi) = 0$. The boundary value problem $X'' + \lambda X = 0, X(0) = X(2\pi) = 0$ yields $\lambda = (k/2)^2$, $X(x) = C_1 \sin(kx/2)$, $k = 1, 2, \ldots$. Now, we solve the equation $T' + (k/2)^2 T = 0$ and obtain $T(t) = C_2 e^{-(k/2)^2\, t}$. Finally, we have

$$u = C e^{-(k/2)^2\, t} \sin(kx/2),$$

where $C = C_1 C_2 = \text{const}$. The initial condition yields $C \sin \frac{kx}{2} = \sin x$, whence $k = 2, C = 1$, and we arrive at the solution $u(t, x) = e^{-t} \sin x$.

5.16. The problem is ill posed since the consistency conditions (5.5.21) are not satisfied.

Chapter 6

6.1. No.

6.2. The most general invariant equations are:

(i) $y' = \Phi\left(\frac{y}{x}\right)$, $\quad y'' = \frac{1}{x} F\left(\frac{y}{x}, y'\right)$;

(ii) $y' + P(x)y = 0$, $\quad y'' = yF\left(x, \frac{y'}{y}\right)$;

(iii) $y' = \frac{1}{x} \phi(y)$, $\quad y'' = \frac{1}{x^2} F(y, xy')$.

6.3. The third-order equation

$$\mu^2 \mu''' = \nu(2\mu\mu'' - \mu'^2), \quad \nu = \text{const.} \neq 0$$

for $\mu = \mu(x)$ can be reduced to the first-order equation

$$pss' = s(p - s) + \nu(2s - p)$$

for $s = s(p)$ by setting first $\mu' = p(\mu)$ and then $\tau = \ln\mu$, $dp/d\tau = s(p)$.

6.4. Equation (P6.2) (i), $y'' + (y'/x) - e^y = 0$, has two symmetries

$$X_1 = x\ln x \frac{\partial}{\partial x} - 2(1 + \ln x)\frac{\partial}{\partial y}, \quad X_2 = x\frac{\partial}{\partial x} - 2\frac{\partial}{\partial y}.$$

Equation (P6.2) (ii), $y'' - (y'/x) + e^y = 0$, has one symmetry

$$X = x\frac{\partial}{\partial x} - 2\frac{\partial}{\partial y}.$$

6.5. Equations (P6.3) (i), (ii), (iii) and (vi) are linearizable, while (iv) and (v) are not linearizable.

6.6. (i) The equations $X_2(t) = 1$, $X_2(u) = u$ yield $t = \ln x$, $u = y/\sqrt{x}$.

(ii) Equation (6.5.6) is written in the canonical variables t, u in the form

$$u'' = \frac{u'}{u^2} - \frac{1}{2u} + \frac{u}{4}.$$

(iii) The substitution $u' = p(u)$ yields $u'' = pp'$, and hence reduces the above equation to the first-order equation

$$pp' = \frac{p}{u^2} - \frac{1}{2u} + \frac{u}{4}.$$

6.7. In the case of arbitrary parameters k and ω, the solution is

$$y = A\sqrt{1 + \omega^2 x^2} \cos(C - p \arctan(\omega x)),$$

where $p = \sqrt{1 + (k/\omega^2)}$, and A and C are constants. In particular, when $k = 3\omega^2$ the solution can be written, invoking the elementary formulae $\arctan s = \arccos(1/\sqrt{1 + s^2}) = \arcsin(s/\sqrt{1 + s^2})$, in the form

$$y = A\left(\frac{1 - \omega^2 x^2}{\sqrt{1 + \omega^2 x^2}} \cos C + \frac{2\omega x}{\sqrt{1 + \omega^2 x^2}} \sin C\right).$$

6.10. The first-order equations admitting the operator

$$X = \sqrt{2}\,x\frac{\partial}{\partial x} + y\frac{\partial}{\partial y}$$

have the form

$$\frac{dy}{dx} = \frac{y}{x} F\left(\frac{y^{\sqrt{2}}}{x}\right).$$

The solution is given implicitly by quadrature

$$\int \frac{du}{\Phi(u) - u} = t + C.$$

6.11. The double invariant first-order equation has the form

$$y' = C\frac{y}{x}, \quad C = \text{const}.$$

6.12. The double homogeneous second-order equation has the form

$$y'' = \frac{y}{x^2} H\left(\frac{xy'}{y}\right).$$

6.14. Looking for polynomial solutions $y = A_0 + A_1 x + A_2 x^2 + \cdots + A_n x^n$ of the equation $y'' = xy' - 4y$, one can verify that such a solution exists for $n = 4$ and is given by $y = x^4 - 6x^2 + 3$. Applying one of methods of Section 6.5.5, one obtains (cf. formula (6.5.29) with $z(x) = x^4 - 6x^2 + 3$) the following solution to the equation in question:

$$y = (x^4 - 6x^2 + 3)\left[C_1 \int \frac{e^{x^2/2}}{(x^4 - 6x^2 + 3)^2}\,dx + C_2\right].$$

6.16. The algebra L_2 belongs to type II. Therefore, we solve equations $X_1(t) = 0$, $X_1(u) = 1$; $X_2(t) = 0$, $X_2(u) = t$ and obtain the canonical variables $t = y/x$, $u = -1/x$. Since the variable t involves the dependent variable y, it can be a new independent variable only if one excludes the singular solutions of equation in question along which t is identically constant. These singular solutions are the straight lines, $y = Kx$, $K = $ const. In the variables t, u our equation is written $u'' = 2$, whence $u = t^2 + C_1 t + C_2$. Substituting the expressions for t and u, we have $y^2 + C_1 xy + C_2 x^2 + x = 0$. Solving this equation with respect to y and setting $A = -C_1/2$, $B = A^2 - C_2$, we obtain the general solution $y = Kx$, $y = Ax \pm \sqrt{Bx^2 - x}$.

Since the equation in question admits an L_2 of type II, it follows from Table 6.5.2 in Section 6.5.4 that it is linearizable (indeed, it is mapped to the equation $u'' = 2$), and hence admits an eight-dimensional Lie algebra.

6.17. (i) $y = -\dfrac{x}{2} + \sqrt{\dfrac{5}{4} x^2 - x}$.

6.18. The general solution of Eq. (6.6.53) has the form (see (6.6.56)):

$$y = \frac{1}{K_1 x + K_2 x^2 + K_3 x^3}, \qquad K_i = \text{const.}$$

6.19. It follows from Eqs. (6.6.82), (6.6.83) and Problem 3.3(iii) that the general solution to Eq. (6.6.74) is given implicitly by

$$x = K_1 e^{-y} + e^{y/2} \left[K_2 \cos\left(\frac{\sqrt{3}}{2} y\right) + K_3 \sin\left(\frac{\sqrt{3}}{2} y\right) \right], \qquad K_i = \text{const.}$$

Chapter 7

7.2. The infinitesimal symmetries of the linear heat equation comprise the infinite-dimensional algebra with the generator

$$X_\tau = \tau \frac{\partial}{\partial u},$$

where $\tau = \tau(t, x)$ is an arbitrary solution of the heat equation, and:
(i) the 6-dimensional Lie algebra spanned by

$$X_1 = \frac{\partial}{\partial t}, \quad X_2 = \frac{\partial}{\partial x}, \quad X_3 = 2t\frac{\partial}{\partial t} + x\frac{\partial}{\partial x}, \quad X_4 = u\frac{\partial}{\partial u},$$

$$X_5 = 2t\frac{\partial}{\partial x} - xu\frac{\partial}{\partial u}, \quad X_6 = t^2\frac{\partial}{\partial t} + tx\frac{\partial}{\partial x} - \frac{1}{4}(2t + x^2)u\frac{\partial}{\partial u}$$

for the one-dimensional heat equation $u_t - u_{xx} = 0$,
(ii) the 9-dimensional Lie algebra spanned by

$$X_1 = \frac{\partial}{\partial t}, \quad X_2 = \frac{\partial}{\partial x}, \quad X_3 = \frac{\partial}{\partial y}, \quad X_4 = 2t\frac{\partial}{\partial t} + x\frac{\partial}{\partial x} + y\frac{\partial}{\partial y}, \quad X_5 = u\frac{\partial}{\partial u},$$

$$X_6 = y\frac{\partial}{\partial x} - x\frac{\partial}{\partial y}, \quad X_7 = 2t\frac{\partial}{\partial x} - xu\frac{\partial}{\partial u}, \quad X_8 = 2t\frac{\partial}{\partial y} - yu\frac{\partial}{\partial u},$$

$$X_9 = t^2\frac{\partial}{\partial t} + tx\frac{\partial}{\partial x} + ty\frac{\partial}{\partial y} - \frac{1}{4}(4t + x^2 + y^2)u\frac{\partial}{\partial u}$$

for the two-dimensional heat equation $u_t = u_{xx} + u_{yy}$,
  (iii) the 13-dimensional Lie algebra spanned by

$$X_1 = \frac{\partial}{\partial t}, \quad X_2 = \frac{\partial}{\partial x}, \quad X_3 = \frac{\partial}{\partial y}, \quad X_4 = \frac{\partial}{\partial z}, \quad X_5 = 2t\frac{\partial}{\partial t} + x\frac{\partial}{\partial x} + y\frac{\partial}{\partial y} + z\frac{\partial}{\partial z},$$

$$X_6 = u\frac{\partial}{\partial u}, \quad X_7 = y\frac{\partial}{\partial x} - x\frac{\partial}{\partial y}, \quad X_8 = z\frac{\partial}{\partial y} - y\frac{\partial}{\partial z}, \quad X_9 = x\frac{\partial}{\partial z} - z\frac{\partial}{\partial x},$$

$$X_{10} = 2t\frac{\partial}{\partial x} - xu\frac{\partial}{\partial u}, \quad X_{11} = 2t\frac{\partial}{\partial y} - yu\frac{\partial}{\partial u}, \quad X_{12} = 2t\frac{\partial}{\partial z} - zu\frac{\partial}{\partial u},$$

$$X_{13} = t^2\frac{\partial}{\partial t} + tx\frac{\partial}{\partial x} + ty\frac{\partial}{\partial y} + tz\frac{\partial}{\partial z} - \frac{1}{4}(6t + x^2 + y^2 + z^2)u\frac{\partial}{\partial u}$$

for three-dimensional heat equation $u_t = u_{xx} + u_{yy} + u_{zz}$.

7.3. The operator $X = \frac{\partial}{\partial t} + ku\frac{\partial}{\partial u}$ has two independent invariants: x and $v = ue^{-kt}$. Hence, we look for the invariant solutions in the form $v = \psi(x)$, whence $u = \psi(x)\,e^{kt}$. Substituting in the heat equation, we get $\psi'' - k\psi = 0$. If $k < 0$, we let $k = -\alpha^2$ and obtain $\psi = C_1\cos(\alpha x) + C_2\sin(\alpha x)$. Then the invariant solution is

$$u = \left[C_1\cos(\alpha x) + C_2\sin(\alpha x)\right]e^{-\alpha^2 t}.$$

If $k > 0$, we let $k = \beta^2$ and obtain $\psi = C_1 e^{\beta t} + C_2 e^{-\beta t}$. Then the invariant solution is

$$u = \left(C_1 e^{\beta t} + C_2 e^{-\beta t}\right)e^{\beta^2 t}.$$

7.4. (i) Proceeding as in Problem 7.3, one can show that the invariance under $X = \frac{\partial}{\partial t} + ku\frac{\partial}{\partial u}$ yields $u = \psi(x, y)\,e^{kt}$. Substituting into the heat equation $u_t = u_{xx} + u_{yy}$, we get $\psi_{xx} + \psi_{yy} = k\psi$.
  (ii) The two-dimensional Lie algebra spanned by X, Y has the invariants $r = \sqrt{x^2 + y^2}$ and $v = ue^{-kt}$. Hence, we look for the invariant solutions in the form $v = \phi(r)$, whence $u = \phi(r)\,e^{kt}$. Substituting into the heat equation and multiplying by r, we obtain the equation $r\phi'' + \phi' - kr\phi = 0$. Let us assume that $k < 0$. Then setting $k = -\alpha^2$ and $\tilde{r} = \alpha r$, we arrive at the Bessel equation (see Section 3.3.5) for the Bessel function $J_0(\tilde{r})$ of order zero: $\tilde{r}\phi'' + \phi' + \tilde{r}\phi = 0$, where the "prime" denotes the differentiation with respect to $\tilde{r}$. Hence, $\phi = J_0(\tilde{r})$ and the invariant solution is given by

$$u = J_0(\alpha r)\,e^{-\alpha^2 t}.$$

7.8. We have: $T_5 = 2tE - m\boldsymbol{x}\cdot\boldsymbol{v}, \quad T_6 = 2t^2 E + m\boldsymbol{x}\cdot(\boldsymbol{x} - 2t\boldsymbol{v})$, where $E = \frac{m}{2}|\boldsymbol{v}|^2 + kr^{-2}$ is the energy.

7.10. $s = n$.

7.11. One of simple solution is

$$A_1 = \frac{x}{r}\left(0,\ \frac{z}{y^2 + z^2},\ -\frac{y}{y^2 + z^2}\right).$$

A more symmetric solution is

$$A = \frac{1}{3r}\left(\frac{zy(z^2 - y^2)}{(r^2 - z^2)(r^2 - y^2)},\ \frac{xz(x^2 - z^2)}{(r^2 - x^2)(r^2 - z^2)},\ \frac{yx(y^2 - x^2)}{(r^2 - y^2)(r^2 - x^2)}\right).$$

7.14. The first prolongation of $\widehat{X}_1$ is obtained by adding to the original cperator $\widehat{X}_1$ the following terms:

$$-\left(x^2\dot{v}^2 + (v^2)^2 + x^3\dot{v}^3 + (v^3)^2\right)\frac{\partial}{\partial v^1}$$

$$+\left(2x^1\dot{v}^2 + v^1v^2 - x^2\dot{v}^1\right)\frac{\partial}{\partial v^2} + \left(2x^1\dot{v}^3 + v^1v^3 - x^3\dot{v}^1\right)\frac{\partial}{\partial v^3}.$$

7.15. After introducing the variable $y = \ln|x|$, the Black-Scholes equation becomes $u_t + \frac{1}{2}A^2 u_{yy} + Ku_y - Cu = 0$, where $K = B - (A^2/2)$ (cf. (9.5.2)).

7.16. The first equation, $\xi_y = 0$, yields that $\xi = \xi(x)$. Then the fourth equation, $\xi_{xx} = 0$, becomes $\xi''(x) = 0$, whence $\xi = K_1 + K_2 x$, where $K_1, K_2 =$ const. Likewise, we obtain from the third equation, $\eta_x = 0$, that $\eta = \eta(y)$. Finally, substituting the expressions for ξ and η into the second equation, $3(\eta_y + \eta) - 2\xi_x = 0$, we obtain the first-order linear non-homogeneous ordinary differential equation $\eta' + \eta = \frac{2}{3}K_2$, whence $\eta = \frac{2}{3}K_2 + K_3 e^{-y}$. Denoting $K_3 = C_1$, $K_1 = C_2$, $3C_3 = K_2$, we have the following general solution of the over-determined system in question: $\xi = C_2 + 3C_3 x$, $\eta = 2C_3 + C_1 e^{-y}$.

Chapter 8

8.1. We have

    (i) $(\delta'(x), \varphi(x)) = (\delta(x), -\varphi'(x)) = -\varphi'(0),$

    (iii) $(\delta'(x - x_0), \varphi(x)) = (\delta(x - x_0), -\varphi'(x)) = -\varphi'(x_0).$

8.2. We have (see Section 8.1.3)

    (i) $\displaystyle\lim_{\varepsilon \to 0} \frac{2\varepsilon x}{\pi(x^2 + \varepsilon^2)^2} = -\delta'(x),$ (ii) $\displaystyle\lim_{\nu \to \infty} \int_{-\nu}^{\nu} i\xi e^{i\xi x} d\xi = 2\pi\delta'(x),$

    (iii) $\displaystyle\lim_{\nu \to \infty} \int_{-\nu}^{\nu} \xi^2 e^{i\xi x} d\xi = -2\pi\delta''(x).$

Chapter 9

9.11. We have for function (9.2.21),

$$u_t - \Delta u = \theta'(t)\,\frac{C}{(\sqrt{t})^n}\,e^{-\frac{r^2}{4t}} + \theta(t)\left[\left(\frac{C}{(\sqrt{t})^n}\,e^{-\frac{r^2}{4t}}\right)_t - \Delta\left(\frac{C}{(\sqrt{t})^n}\,e^{-\frac{r^2}{4t}}\right)\right].$$

Invoking that $\theta'(t) = \delta(t)$, $F(t,x)\delta(t) = F(0,x)\delta(t)$ and that

$$\left(\frac{C}{(\sqrt{t})^n}\,e^{-\frac{r^2}{4t}}\right)_t - \Delta\left(\frac{C}{(\sqrt{t})^n}\,e^{-\frac{r^2}{4t}}\right) = 0, \quad t > 0,$$

we obtain, using Eqs. (8.1.12) and (8.2.8),

$$u_t - \Delta u = \delta(t)\,\lim_{t\to+0}\frac{C}{(\sqrt{t})^n}\,e^{-\frac{r^2}{4t}} = C(2\sqrt{\pi t})^n\,\delta(t)\delta(x) = C(2\sqrt{\pi t})^n\,\delta(t,x).$$

Hence, Eq. (9.2.19) requires that $C(2\sqrt{\pi t})^n = 1$.

Bibliography

[1] AMES, W.F. *Nonlinear Partial Differential Equations in Engineering*, Vol. I. Academic Press, New York, 1965.

[2] BLUMAN, G.W., AND KUMEI, S. *Symmetries and Differential Equations*. Springer, New York, 1989.

[3] CANTWELL, B.J. *Introduction to Symmetry Analysis*. Cambridge University Press, Cambridge, 2002.

[4] COURANT, R., AND HILBERT, D. *Methods of Mathematical Physics. Vol. II: Partial Differential Equations, By R. Courant.* Interscience Publishers, John Wiley & Sons, New York, 1962. Wiley Classics Edition Published in 1989.

[5] COURANT, R., AND HILBERT, D. *Methods of Mathematical Physics*, Vol. I. Interscience Publishers, John Wiley & Sons, New York, 1989.

[6] DUFF, G.F.D. *Partial Differential Equations*. University of Toronto Press, Toronto, 1956.

[7] EULER, L. *Integral Calculus*. Vol. III, Part 1, Chapter II, 1769/1770.

[8] GEL'FAND, I.M., AND SHILOV, G.E. *Generalized functions*, Vol. 1. Fizmatgiz, Moscow, 1959. English translation by E. Saletan, Academic Press, New York., 1964.

[9] GOURSAT, E. *Differential equations*. Ginn and Co, Boston, 1917. English translation by E.R. Hedrick and O. Dunkel of Goursat's classical *Cours d'analyse mathématique*, Vol. 2, Part II.

[10] GOURSAT, E. *Cours d'analyse Mathématique*, Tome 1, 5th ed. Gauthier-Villars, Paris, 1956. English transl. *A Course in Mathematical Analysis*.

[11] GREENBERG, M. *Advanced Engineering Mathematics*, 2nd ed. Prentice Hall, New Jersey, 1998.

[12] HADAMARD, J. *Lectures on Cauchy's Problem in Linear Partial Differential Equations.* Yale University Press, New Haven, 1923. Reprinted in Dover Publications, New York, 1952. See also revised French edition: *Le problème de Cauchy,* Paris, 1932.

[13] IBRAGIMOV, N.H. Invariant variational problems and conservation laws. *Teoreticheskaya i Matematicheskaya Fizika 1, No. 3* (1969), 350–359. English transl., Theor. Math. Phys., 1, No. 3, (1969), 267–276. Reprinted in: N.H. Ibragimov, *Selected Works,* Vol. I, ALGA Publications, Karlskrona. 2006, Paper 8.

[14] IBRAGIMOV, N.H. *Transformation Groups Applied to Mathematical Physics.* Nauka, Moscow, 1983. English transl., Riedel, Dordrecht, 1985.

[15] IBRAGIMOV, N.H. *Primer of Group Analysis.* Znanie, No. 8, Moscow, 1989. (Russian). Revised edition in English: *Introduction to Modern Group Analysis,* Tau, Ufa, 2000. Available also in Swedish: *Modern Gruppanalys: En Inledning Till Lies Lösningsmetoder av Ickelinjära Differentialekvationer,* Studentlitteratur, Lund, 2002.

[16] IBRAGIMOV, N.H. *Essay in Group Analysis.* Znanie, No. 7, Moscow, 1991. (Russian).

[17] IBRAGIMOV, N.H. Group analysis of ordinary differential equations and the invariance principle in mathematical physics (for the 150th anniversary of Sophus Lie). *Uspekhi Mat. Nauk 47, No. 4* (1992), 83–144. English transl., Russian Math. Surveys, 47:2 (1992), 89-156. Reprinted in: N.H. Ibragimov, *Selected Works,* Vol. I, ALGA Publications, Karlskrona. 2006, Paper 21.

[18] IBRAGIMOV, N.H., Ed. *CRC Handbook of Lie Group Analysis of Differential equations.* Vol. 1: *Symmetries, Exact Solutions and Conservation laws.* CRC Press Inc., Boca Raton, 1994.

[19] IBRAGIMOV, N.H., Ed. *CRC Handbook of Lie Group Analysis of Differential Equations.* Vol. 2: *Applications in Engineering and Physical sciences.* CRC Press Inc., Boca Raton, 1995.

[20] IBRAGIMOV, N.H., Ed. *CRC Handbook of Lie Group Analysis of Differential Equations.* Vol. 3: *New trends in Theoretical Developments and Computational Methods.* CRC Press Inc., Boca Raton, 1996.

[21] IBRAGIMOV, N.H. *Elementary Lie Group Analysis and Ordinary Differential equations.* John Wiley & Sons, Chichester, 1999.

[22] IBRAGIMOV, N.H. Integrating factors, adjoint equations and Lagrangians. *Journal of Mathematical Analysis and Applications 318, No. 2* (2006), 742–757.

[23] LAPLACE, P.S. Recherches sur le calcul intégral aux différences partielles. *Mémoires de l'Académie Royale des Sciences de Paris 23, No. 24* (1773), 341–402. Reprinted in: P. S. Laplace, *Oevres complètes*, t. IX, Gauthier-Villars, Paris, 1893, pp. 5–68; English Translation, New York, 1966.

[24] LERAY, J. *Hyperbolic Differential Equations*. Lecture Notes, Institute for Advanced Study, Princeton, 1953. Available in a book form in Russian translation by N.H. Ibragimov, Nauka, Moscow, 1984.

[25] LIE, S. Klassifikation und Integration von gewönlichen Differentialgleichungen zwischen x, y, die eine Gruppe von Transformationen gestatten. III. *Archiv for Matematik og Naturvidenskab (*Abbr. *Arch. for Math.) 8, Heft 4* (1883), 371–458. Reprinted in Lie's Ges. Abhandl., vol. 5, 1924, paper XIV, pp. 362–427.

[26] LIE, S. *Vorlesungen über Differentialgleichungen mit Bekannten Infinitesimalen Transformationen*. (Bearbeited und herausgegeben von Dr. G. Scheffers), B. G. Teubner, Leipzig, 1891.

[27] MURRAY, J.D. *Mathematical Biology. I: An Introduction*, 3rd ed. Springer, New York, 2002.

[28] MURRAY, J.D. *Mathematical Biology. II: Spatial Models and Biomedical Applications*, 3rd ed. Springer, New York, 2003.

[29] NEWTON, I. *Mathematical Principles of Natural Philosophy*. Benjamin Motte, Middle-Temple-Gate, in Fleet Street, 1929. Translated into English by Andrew Motte, to which are added, *The laws of Moon's motion, according to gravity* by John Machin. 1st ed., 1687; 2nd ed., 1713; 3rd ed., 1726.

[30] NOETHER, E. Invariante Variationsprobleme. *Königliche Gesellschaft der Wissenschaften zu Göttingen, Nachrichten. Mathematisch-Physikalische Klasse Heft 2* (1918), 235–257. English transl., *Transport Theory and Statistical Physics*, Vol. 1, No. 3, 1971, 186–207.

[31] OLVER, P.J. *Applications of Lie groups to Differential Equations*. Springer, New York, 1986. 2nd ed., 1993.

[32] OVSYANNIKOV, L.V. *Group Analysis of Differential Equations*. Nauka, Moscow, 1978. English transl., ed. W.F. Ames, Academic Press, New York, 1982. See also L. V. Ovsyannikov, *Group Properties of Differential equations*, Siberian Branch, USSR Academy of Sciences, Novosibirsk, 1962.

[33] PETROVSKY, I.G. *Lectures on Partial Differential Equations*, 3 ed. Fizmatgiz, Moscow, 1961. English transl., Interscience, New York, 1964. Republished by Dover, 1991. Translated from Russian by A. Shenitzer.

[34] SCHWARTZ, L. *Métodes Mathématiques de la Physique*. Hermann, Paris, 1961. English transl., Mathematics for the physical sciences, Addison-Wesley, Reading, 1966.

[35] SIMMONS, G.F. *Differential Equations with Applications and Historical notes*, 2nd ed. McGraw-Hill, New York, 1991.

[36] SMIRNOV, V.I. *A Course of Higher Mathematics*, Vol. IV. Pergamon Press, New York, 1964. Translated from Russian by D.E. Brown, edited by I.N. Sneddon.

[37] SOBOLEV, S.L. *Partial Differential Equations of Mathematical Physics*. Dover, New York, 1989. Translated from Russian by E.R. Dawson, edited by T.A.A. Broadbent.

[38] SOMMERFELD, A. *Partial Differential Equations in Physics*. Academic Press, New York, 1964. English translation by E.G. Straus of A. Sommerfeld's *Lectures on Theoretical Physics, vol. VI.*

[39] TIKHONOV, A.N., AND SAMARSKII, A.A. *Equations of Mathematical Physics*, 2nd ed. Dover, New York, 1990. Translated from Russian.

[40] WHITTAKER, E., AND WATSON, G. *A Course of Modern Analysis*, 4th ed. Cambridge University Press, Cambridge, 1927.

Index